PRACTICAL STATISTICS FOR STUDENTS

An Introductory Text

Louis Cohen and Michael Holliday

Loughborough University

P·C·P

Paul Chapman
Publishing Ltd

*'That must be wonderful; I don't
understand it at all.'*

Molière

 Paul Chapman Publishing Ltd
A SAGE Publications Company
6 Bonhill Street
London EC2A 4PU

SAGE Publications Inc.
2455 Teller Road
Thousand Oaks, California 91320

SAGE Publications India Pvt Ltd
32, M-Block Market
Greater Kailash-I
New Delhi 110 048

British Cataloguing in Publication Data

Cohen, Louis
  Practical statistics for students
  1. Statistics 2. Mathematical statistics
  I. Title II. Holliday, Michael III. Statistics for social scientists
  519.5'024375

  ISBN 1 85396 329 1

Typeset by Globe Graphics London
Printed and bound in Great Britain by Athenaeum Press, Gateshead, Tyne & Wear

B C D E F 3 2 1 0 9 8

CONTENTS

LIST OF TABLES

Practical Statistics for Students

LIST OF FIGURES

PREFACE

Few students look forward with relish to a course in statistics. The word itself evokes feelings of apprehension and the thought of working with mathematical formulae causes many to anticipate difficulties that need never arise.

Why the concern?

Many fear that statistics will be too difficult for them to grasp because their knowledge of mathematics is rudimentary and much of what they once knew is now forgotten. We can reassure readers that the level of arithmetical skills and the knowledge of algebra necessary to follow this text are within the reach of those who have pursued courses of study to G.C.S.E. 'O' level standard. That said, statistics is a branch of mathematics and the statistical methods and formulae that we introduce are mathematically derived. However, it is neither our intention to present mathematical explanations of statistical techniques nor at the other extreme to provide a cookbook of statistical recipes.

The purpose of this text is to help students understand statistical methods so that they can tackle their research problems successfully. By working through the book carefully and systematically students should gain a thorough grasp of a range of statistical tests and applications widely used in contemporary society. The text teaches basic statistical methods and their application to those who may lack a strong background in mathematics. In doing so, mathematical derivations of formulae are completely omitted and apart from a minimal number of essential 'shorthand' mathematical symbols, familiar examples drawn from actual data are presented in a non-mathematical form throughout.

In planning the scope and content of the text we have in mind the needs of students who are required to undertake small scale empirical research as part fulfilment of the requirements for the award of academic and vocational qualifications. The research needs of these students are often quite specific; first to find an appropriate design for their particular research topic and second, to apply appropriate statistical tests in the analysis of their data.

The first sections of the text teach basic statistics and prepare students to tackle research problems with confidence and intelligence. The latter half of the book ('Research Designs') provides students with concrete examples which illustrate many commonly used research designs. A standard format is employed throughout. It consists of:

1   a tabular presentation of research design

2   a concrete example of the research design using actual data

3   that example is then worked through using an appropriate statistical test and illustrating each stage of the computation in full.

1

CHAPTER 1
INTRODUCTION

1.1 What do we mean by statistics?

From earliest times men and women have collected and made use of statistics. They used them six thousand years ago in Upper Egypt, assessing the size of the harvest in order to fix the price of corn until the next flooding of the Nile valley. They used them in Judaea and Samaria two thousand years ago, taking head counts of the Jewish tribes in order to exact taxation for Rome. And they used them daily three hundred or so years ago for the grim purpose of assessing the number of dead as the dreaded plague spread across London and the Home Counties. Statistics has had a venerable and varied history!

Today, statistics is a familiar and accepted aspect of our modern world. We have statistics in the shape of sports results, long range weather forecasts, stockmarket trends, consumer reports, cost of living indices and life insurance premiums. It is impossible to imagine life without some form of statistical information being readily at hand.

The word *statistics* is used in two senses. It refers to collections of quantitative information and methods of handling these sort of data. Domesday Book, that mammoth stocktaking exercise of William the Conqueror, is an example of this first sense in which the word is used. Statistics also refers to the drawing of inferences about large groups on the basis of observations made on smaller ones. The calculation of swings in the nation's party political preferences before an election is based upon information gathered in carefully selected interviews and illustrates the second sense in which the word statistics is used.

Statistics then, is to do with ways of organizing, summarizing and describing quantifiable data, and methods of drawing inferences and generalizing upon them.

1.2 Why is statistics necessary?

Apart from the intrusion of statistics upon so many aspects of our daily lives there are two reasons why some knowledge of statistics is an important part of the professional competence of every researcher. First, statistical literacy is necessary if we are to read and evaluate research critically and intelligently. Take for example a social worker required to make recommendations in respect of the behaviour problems of preschool children from disadvantaged homes. What will she make of a research note such as, 'differences between boys and girls on the behaviour screening schedule resulted in a chi square value of 9.45, d.f. = 1, $p > .01$'? And how will a clinical psychologist respond to the findings of an experiment in which 'ANOVA gave an F value of 14.36 with 2 and 33 degrees of freedom

($p >.001$)'? Will both social worker and psychologist be forced to ignore such information because they lack the statistical understanding to appreciate its meaning and its value?

A second reason why statistical literacy is important is that if individuals are to undertake research on their own account, then a grasp of statistical methods is essential. With such understanding our hypothetical social worker could, for example, use the behaviour screening schedule in a replication study and in light of her own findings and those of the reported research, make her recommendations with greater knowledge and certainty. Similarly, the clinical psychologist's grasp of the significance of the experimental findings could help him choose a course of treatment and be cautiously optimistic that his patients would be likely to benefit as a result.

Whether actively involved in research or not, membership of a profession carries with it an obligation to keep abreast of developments in specific areas of interest and expertise. Studying research reports and evaluating new techniques and approaches generally require familiarity with statistical principles and methods.

1.3 The purpose of the text

Two objectives of *Practical Statistics For Students* follow directly from our discussion in 1.2 above.

First, the text aims to give students sufficient grounding in statistical principles and methods to enable them to read and understand research reports. Second, the text aims to present students with a variety of statistical techniques relevant to their own research and to help them select statistical tests appropriate to the solution of their problems.

1.4 The limitations of statistics

Statistical techniques can assist researchers to describe data, design experiments, and test hunches about relationships among the things or events in which they are interested. At best however, such techniques can only partly inform their professional judgements and decisions. Statistics helps researchers accept or reject hunches about data within recognized degrees of confidence—no more, no less. Statistics is simply a tool that can be used to answer problems that are amenable to quantification. It hardly needs mentioning that there are innumerable situations that do not lend themselves to quantification or to statistical analysis.

CHAPTER 2
MEASUREMENT CONCEPTS

2.1 Sources of data

It's often the case that various human attributes and accomplishments are somewhat arbitrarily 'parcelled-up' as the particuar provinces of various disciplines. Psychology, for example, deals with intelligence, learning, and personality; Social Psychology with values, attitudes, and beliefs; Sociology with the family and socioeconomic circumstances; and so on. We have no wish to promote such arbitrary divisions and are therefore concerned to illustrate statistical techniques with data from a wide variety of soures in order to make our examples both interesting and relevant to our readers.

2.2 Populations, parameters, samples, and statistics

We are often required to assemble and evaluate information regarding some common characteristic(s) of a *population*. It must be stressed at this stage that in the widest sense a population is more than just a collection of people. It could refer to a large collection of objects or events which vary in respect of some characteristic(s). All the employees of a large chocolate factory for example, who serve as subjects for a medical student's research into the ratio of height to weight are a population. All the Friday afternoon absentees in a shipyard are a population. A population need not be very large although the procedures set out in this text generally assume the existence of extremely large or infinite populations.

Characteristics of a population which differ from person to person or object to object are called *variables*. Height, weight, age, intelligence, anxiety differences, reading ability, fitness, to name but a few, are examples of human variables and to these variables we can assign numbers or values.

Once numbers or values are assigned to the population characteristics we can measure them. The measures which describe population characteristics are called *parameters*.

It is not always practical to obtain measures from a total population due to factors such as expense, time, accessibility, etc., and so the researcher has to collect his information from a smaller group or 'sub set' of the population, assuming that the information gained is representative of the whole population under study. This smaller group or 'sub set' is known as a *sample*.

Suppose the population of employees of our hypothetical chocolate factory is 2000. This is a very large number of subjects for even the most ambitious student of obesity to handle on his own. He would probably settle for some part of that population, choosing a representative sample, of,

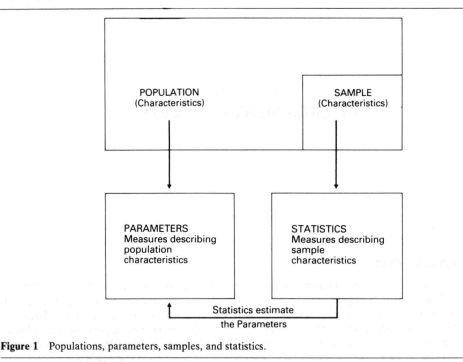

Figure 1   Populations, parameters, samples, and statistics.

say, 200 subjects on the basis of their age, sex, access to chocolates and so on. (The various methods of selecting representative samples from their populations are discussed in Sections 10.1 to 10.7.)

The measures taken from a sample describing the sample characteristics are called *statistics* and it is from these statistics that the population parameters are estimated. The height to weight ratios of the chocolate factory sample are used to predict the chocolate factory population height to weight parameter.

The relationship between populations, parameters, samples, and statistics is represented in Figure 1. A simple way of remembering this relationship is that statistic is to sample as parameter is to population.

2.3   Descriptive and inferential statistics

Descriptive statistics are used to organize, summarize, and describe measures of a sample. No predictions or inferences are made regarding the population parameters. For example, from our sample of chocolate factory employees we can work out the average height to weight ratio, or the most common height to weight ratio, to name but two descriptive measures. At this descriptive level no attempt is made to predict population parameters.

Inferential or inductive statistics, on the other hand, are used to infer or predict population parameters from sample measures. This is done by a process of inductive reasoning based on the mathematical theory of probability. From our chocolate factory sample it would be possible, using appropriate mathematical techniques, to estimate or predict from the average obesity score what the population average obesity score would probably be. However, no one can predict exactly the population measures from sample measures, only a 'probable' measure. How closely this probable measure would be to the exact measure would depend upon the sample selection procedures and the statistical techniques used.

2.4   Parametric and non-parametric statistics

It stands to reason that in order to predict from sample data, the population measures have to a lesser or greater extent to be predictable. The mathematician must know or assume certain factors regarding the population measures in order to have a base from which to derive his statistical method. Statistical methods of inference which depend upon knowing or assuming certain population factors are called *parametric* methods and are the oldest and most often used.

In practice however it is not always possible to be able to predict or make assumptions about the population measures. More recently methods have been evolved which are not based upon the stringent assumptions of population measures. These methods are called *non-parametric* methods.

It would be inappropriate at this point in the text when foundations are being laid, to enlarge on the complexities of the two basic methods until we have discussed in detail the properties of population and sample measures. The advantages and disadvantages of parametric and non-parametric techniques are discussed in detail in Section 10.18.

CHAPTER 3
CLASSIFYING DATA

3.1 Scales of measurement

Almost every page of this book is concerned with classifying, categorizing and quantifying individuals, objects, and events in order to examine relationships between them. Whether the objective is to record the verbal behaviour of adolescents, the height of newly born infants or the mortality rates of the various social classes, underlying the specific classification scheme is the fact that the variables themselves are *measurable*. Variables that are measurable are capable of being placed at some point along a continuum against which numerals may be assigned according to certain rules. Measurement transforms organized and classified data into familiar amenable things called *numbers*.

Generally, four levels of measurement or four ways of assigning numerals can be distinguished. They are referred to as *nominal, ordinal, interval,* and *ratio* scales. Each level has its own rules and restrictions; each level moreover is hierarchical in that each higher level or scale incorporates the properties of the lower.

3.2 The nominal scale

The most elementary scale in measurement is one which does no more than identify the categories into which individuals, objects, or events may be classified. Those categories have to be mutually exclusive of course. That is to say, one cannot place an individual into more than one category. When numbers are assigned to such categories, as they often are in research, they have no numerical meaning; rather they are simply convenient ways of labelling or coding the information. For example, in coding up a fieldwork project, the researcher may assign the label '1' to adult males, '2' to adult females, '3' to male children, and '4' to female children. Such labels have no quantitative meaning. They cannot be added, subtracted, multiplied or divided; their function is purely to identify categories that are different. Notice one other feature of a nominal scale illustrated in our '1', '2', '3', '4' example. A nominal scale should also be *complete,* that is, it should include all possible classifications of a particular type.

3.3 The ordinal scale

The ordinal scale of measurement incorporates the classifying and labelling function of the nominal scale, but in addition it brings to it a sense of order. Ordinal numbers are used to indicate rank order and nothing more. The ordinal scale is used to arrange individuals or objects in a series ranging from the highest to the lowest according to the particular characteristic being measured. The ordinal numbers assigned to such a series do not indicate absolute quantities, nor can it be assumed that the intervals between the numbers are equal. For example, in a group of children rated by an observer on the degree of their co-operativeness and ranged from highest to lowest according to that attribute, it cannot be assumed that the difference in the degree of co-operativeness between subjects one and two is the same as that obtaining between subjects nine and ten; nor can it be taken that subject one possesses ten times the quantity of co-operativeness of subject ten.

3.4 The interval scale

As the term 'interval' implies, in addition to rank-ordering the data, the interval scale allows us to state *precisely how far apart* are the individuals, the objects or the events that form the focus of our inquiry. Interval scales (or, as they are sometimes called, *equal-interval* scales) permit certain mathematical procedures previously untenable at nominal and ordinal levels of measurements.

Because we can legitimately conclude that the difference between the scores of, say, the eighth and ninth individuals is equal to the difference between the scores of the second and third, it follows that the intervals can be added or subtracted.

In our example of individual performance scores above we cannot, of course, assume that the score of the third individual is three times the score of the ninth individual, for one limitation of the interval scale is that there is no absolute zero point.

In most tests however, it is not necessary to define absolute zero achievement, since the object of the testing procedure is to compare individuals or groups one with another or to compare their performance with a conventionally accepted level such as 40% = pass and 70% = distinction.

3.5 The ratio scale

The highest level of measurement, and the one which subsumes interval, ordinal, and nominal levels, is the ratio scale. A ratio scale includes an absolute zero, it provides equal intervals, it gives a rank ordering, and it can be used for simple labelling purposes. Because there is an absolute zero, all of the arithmetical processes of addition, subtraction, multiplication and division are possible. Whereas few, if any, educational and psychological tests can assume ratio level measurement, it is in the physical sciences that zero ratio scales are meaningfully utilized. Using weight as an illustration of a ratio scale we can say that no mass at all is a zero measure and that 1000 grams is 400 grams heavier than 600 grams and twice as heavy as 500.

3.6 Discrete and continuous variables

The task of the student researcher is to collect, organize, and analyse data from variables. As mentioned earlier these variables are capable of being assigned numbers or values. The nature of these numbers or values will depend upon whether the variables are classified as *discrete* or *continuous.*

A discrete variable is a variable which can take on numerals or values that are specific, distinct points on a scale. For example *sex* is a discrete variable. A person is either male or female and cannot be assigned any value between the two. The number of players in a football team is a discrete variable. It is possible to have only 1, 2, 3, 4, 5, 6, 7, 8, 9, 10, or 11 players, but not $7\frac{1}{2}$.

A continuous variable is a variable that, theoretically at least, can take on any value between two points on a scale. It can be measured with differing degrees of exactness depending on the measuring instrument. For example, weight is a continuous variable. It can take on any number of possible values between 0 and infinity. Height, time, distance jumped, percentage body fat are a few other examples.

It must be made clear however, that this distinction between discrete and continuous variables is often only a theoretical one and in practice, because of the lack of suitable measuring instruments, many continuous variables have to be assigned whole number values. Intelligence Quotient theoretically is a continuous variable, but in practice the tests used to measure it give whole numbers or discrete scores.

3.7   Limits of numbers

Because of certain problems in measuring on continuous scales, most variables are given discrete values. In order to interpret such values, account is taken of the mathematical *true limits* of a number. In general, true limits of a number are said to extend one half a unit below and one half a unit above that number and any score within those limits is rounded off to the number.

For example, if after a performance test it was found that ten students scored 90, does this really mean that they are all equal in performance ability? The likelihood of them all being exactly alike in ability is extremely small. It is more likely that some scored just below 90 and some scored just above 90, but the measuring instrument was not sophisticated enough to detect the differences.

To overcome this problem the number 90 is said to have limits ranging from 89.5 to 90.5 and any scores between those limits are given to score 90.

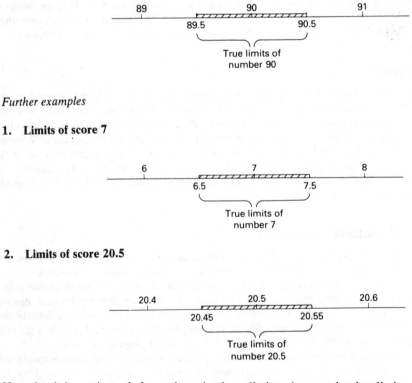

Further examples

1.   Limits of score 7

2.   Limits of score 20.5

Note that it is not just *whole* numbers that have limits. Any number has limits.

This 'true limit' concept is particularly useful when classifying scores into groups or intervals. For example, if we wanted to find out from the test just mentioned how many people scored between 80 and 90 (Interval A) and 90 and 100 (Interval B), in which interval do we include those people who scored 90? The problem is usually solved by considering the true limits of the intervals. Interval A really extends from 79.5 to 89.5 and Interval B from 89.5 to 99.5.

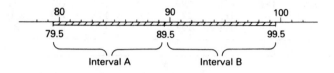

Therefore all those scoring 90 would be included in Interval B.

3.8 The frequency table

The problem that confronts the researcher once he has collected his results is what to do with them. His first priority must be to organize and summarize the data in a form that allows further interpretation and analysis. This is done by constructing a frequency table or frequency distribution which arranges scores into groups or classes.

Suppose a survey was conducted in a particular town to investigate the marital status of women aged between 25 and 35. The data could be organized, summarized and presented by constructing a frequency table as in Table 1.

Table 1 Frequency distribution: marital status of females aged between 25–35. Classified as single, married, widowed or divorced

	Column 1	Column 2	Column 3
	Marital status	**Frequency (f)**	**Percentage (p)**
	Single	389	10.9
	Married	3068	85.9
	Widowed	14	0.4
	Divorced	100	2.8
	Total	3571	100%

Table 1 shows how the four categories, single, married, widowed, and divorced have been separated (Column 1). The number of women have been entered in Column 2, next to the appropriate category. The numbers of observations in particular categories are called frequencies (f). These frequencies have been converted to percentages in Column 3.

In the example above the data are discrete. Since many data are collected from continuous rather than discrete variables, raw scores have to be dealt with. Raw scores can also be organized and summarized by means of a frequency table.

Suppose the following scores were obtained after measuring the heights of 50 factory workers.

Table 2 Height scores of 50 factory workers (measured in cms)

162	188	173	168	174	183	167	186	177	187
170	174	164	174	159	177	172	163	180	196
171	156	184	179	190	181	166	181	182	176
169	172	174	162	175	192	178	177	200	191
188	168	165	179	193	175	160	180	187	176

The data in Table 2 can be summarized and grouped as shown in the frequency table below.

Table 3 Frequency distribution of the height scores of 50 factory workers

Step intervals (True limits)	Step intervals (Limits rounded off)	Tallies	Frequency (f)
197.5–200.5	198–201	1	1
194.5–197.5	195–198	1	1
191.5–194.5	192–195	11	2
188.5–191.5	189–192	11	2
185.5–188.5	186–189	11111	5
182.5–185.5	183–186	11	2
179.5–182.5	180–183	11111	5
176.5–179.5	177–180	111111	6
173.5–176.5	174–177	11111111	8
170.5–173.5	171–174	1111	4
167.5–170.5	168–171	1111	4
164.5–167.5	165–168	111	3
161.5–164.5	162–165	1111	4
158.5–161.5	159–162	11	2
155.5–158.5	156–159	1	1
			$N = 50$

3.9 Steps in the construction of Table 3

1 Determine the range of scores: the highest score minus the lowest score plus 1. (One is added to take into account the true limits of the numbers.)

$$\text{Range} = \text{Highest score (200)} - \text{Lowest score (156)} + 1$$
$$= (200 - 156) + 1$$
$$= 45$$

2 Decide on how many categories or step intervals are required. Normally the number of intervals used is not less than 10 and not more than 20. In this case we have selected 15 as a convenient number of step intervals.

3   Divide the range by the number of step intervals giving the actual size of each step interval.

$$\text{Step Interval size} = \frac{45}{15} = 3$$

If the step interval size does not work out to be a whole number it is advisable to round it off, easing the arithmetical load. For example, if we had chosen 12 step intervals, the step interval size would have worked out to be $\frac{45}{12} = 3.7$ and this would round off to 4.

4   Construct the step interval column starting with the lower limit of the lowest score (155.5).Add the step interval size (3) to this lower limit. The range of the lowest step interval becomes 155.5–158.5. The lower limit of the next interval becomes 158.5 to which 3 is added to give the interval range 158.5–161.5. This procedure is repeated, moving up the table until the step interval column includes an interval into which the highest score (200) can be placed, namely, 197.5–200.5.

5   Insert, in the column provided, a tally for each individual score in the raw data table. For example, for the score 162, a tally or mark is inserted to show it falls into the step interval 161.5–164.5.

6   Total up the tallies within each step interval and place in the frequency column in line with the appropriate interval.

7   Total the frequency column (N). This serves as a useful check that all the data have been included in the table.

CHAPTER 4
PRESENTING DATA

4.1   Introduction

One drawback in presenting data in the form of a frequency table is that the information contained there does not become immediately apparent unless the table is studied in detail. To simplify the interpretation of the information, the data are often processed further and transformed into a visual presentation which can be more readily comprehended.

The most common methods of presenting data are based upon graphical techniques. In this section we describe five methods suitable for presenting social science data.

4.2   Bar graph

Portraying information by means of a bar graph is particularly useful when dealing with data gathered from discrete variables that are measured on a nominal scale such as the data on marital status shown in Table 1.

A bar graph uses rectangles (i.e. bars) to represent discrete categories of data, the length of the bar being proportional to the number of frequencies within that category. Using data from Table 1 the bar graph shown in Figure 2 can be constructed.

The categories are placed on the horizontal base line (the x axis or abscissa) with each category being assigned a bar. The width of the bar is arbitrary but it is recommended that all bars should be the same width. The vertical line (the y axis or ordinate) is marked off to scale, indicating the observed frequencies in each category.

4.3   Histogram

The histogram is similar to the bar graph, the only difference in presentation being that the bars are joined together.

Joining the bars together makes the histogram a suitable method for presenting data gathered from continuous variables measured on interval or ratio scales.

The height score data set out in Table 3 can be represented by the histogram in Figure 3.

Each bar represents a step interval as defined in Table 3, the exact limits of the intervals producing a continuous scale along the base line of the graph. The width of each bar is the size of

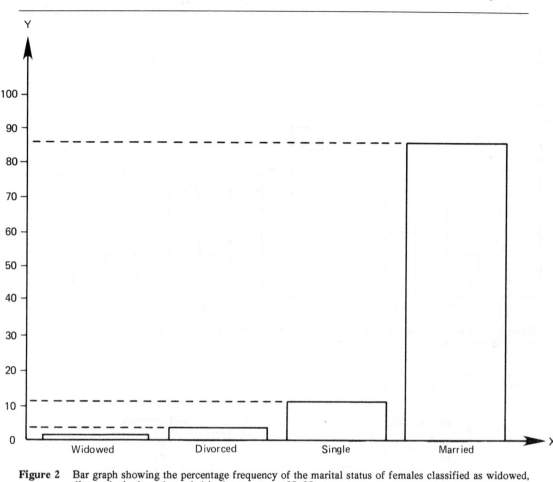

Figure 2 Bar graph showing the percentage frequency of the marital status of females classified as widowed, divorced, single and married in the age group 25–35 years.

each step interval (3) and the height of each bar is proportional to the frequencies within that interval. The histogram gives an immediate picture of how the scores are distributed.

4.4 Frequency polygon

Another commonly used graphical technique is the frequency polygon. This form of presentation dispenses with the use of bars and employs single points joined together by a series of straight lines. The reader will recognize the similarity of this approach to basic graphing techniques in mathematics.

The single point in the frequency polygon is placed at the midpoint of the step interval at a height proportional to the frequencies within an interval.

Prior to constructing a frequency polygon, the midpoints of each interval are calculated. This involves adding a further column to the frequency table.

Using the data in our previous example, Table 4 shows the frequency scores with the interval midpoints included (using true limits only). The midpoints are calculated by adding half the step

15

Figure 3   Histogram showing the frequency distribution of the height scores of 50 factory workers.

Table 4   Frequency distribution of height scores (interval midpoints)

Step intervals	Tallies	*f*	Interval midpoints
197.5–200.5	1	1	199
194.5–197.5	1	1	196
191.5–194.5	11	2	193
188.5–191.5	11	2	190
185.5–188.5	11111	5	187
182.5–185.5	11	2	184
179.5–182.5	11111	5	181
176.5–179.5	111111	6	178
173.5–176.5	11111111	8	175
170.5–173.5	1111	4	172
167.5–170.5	1111	4	169
164.5–167.5	111	3	166
161.5–164.5	1111	4	163
158.5–161.5	11	2	160
155.5–158.5	1	1	157

interval size (1.5) to the lower limit of each interval. The interval midpoints are plotted against frequencies as in Figure 4. Notice that the end points of the polygon have been positioned on the axis at the midpoints of the intervals on either side of the two extreme intervals. The area under the frequency polygon is equal to the area under a histogram constructed from the same data.

16

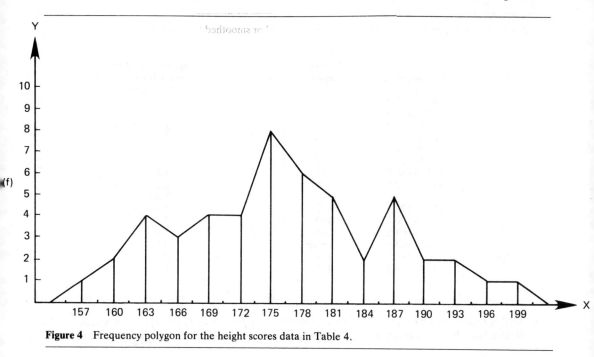

Figure 4 Frequency polygon for the height scores data in Table 4.

4.5 Smoothed frequency polygon

Occasionally the researcher wishes to know what sort of frequency distribution he would have obtained had the data been collected from a larger sample or a whole population. He can get some indication of this by smoothing the original polygon. In theory, the larger the number of results that one processes, the nearer the distribution approximates a curve. Care must be taken however, as smoothing can only give an impression of what the distribution might have been, *not* what it actually is.

Smoothing a frequency polygon requires the calculation of average frequencies for each interval. The *average frequency* for any interval is calculated by adding the frequency of that interval to the frequencies of the two adjacent intervals, and dividing the total by 3.

For example, using the data in Table 5, the average frequency for the interval 158.5–161.5 is the frequency of that interval (2), added to the frequency of the interval 155.5–158.5 (1) and the frequency of the interval 161.5–164.5 (4), divided by 3.

$$\frac{2+1+4}{3}=2.3$$

This is repeated for each of the step intervals. Where an interval has only one adjacent interval, then the missing interval is given a frequency of 0.

For example, for the interval 155.5–158.5, the average frequency is

$$\frac{0+1+2}{3}=1$$

Table 5 shows the height score frequency distribution with a column added for average (sometimes called *smoothed*) frequencies. Frequency tallies have been omitted. The smoothed frequencies are

17

Table 5 Frequency distribution of height scores (averaged or smoothed frequencies)

Step intervals	Interval midpoints	f	Average or smoothed frequencies
197.5–200.5	199	1	0.6
194.5–197.5	196	1	1.3
191.5–194.5	193	2	1.6
188.5–191.5	190	2	3.0
185.5–188.5	187	5	3.0
182.5–185.5	184	2	4.0
179.5–182.5	181	5	4.3
176.5–179.5	178	6	6.3
173.5–176.5	175	8	6.0
170.5–173.5	172	4	5.3
167.5–170.5	169	4	3.6
164.5–167.5	166	3	3.6
161.5–164.5	163	4	3.0
158.5–161.5	160	2	2.3
155.5–158.5	157	1	1.0

plotted for each interval and superimposed over the original frequency polygon, as shown in Figure 5. Notice how the original frequency polygon is jagged and irregular as compared with the smoothed one.

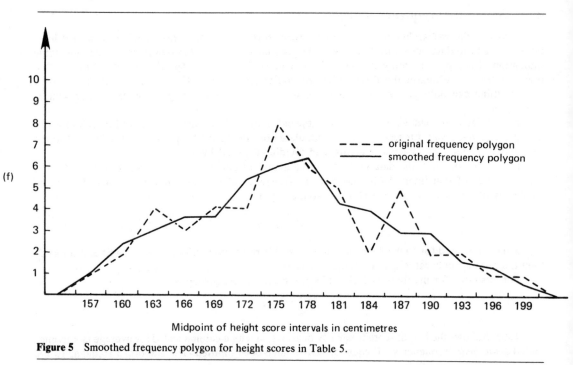

Midpoint of height score intervals in centimetres

Figure 5 Smoothed frequency polygon for height scores in Table 5.

4.6 Cumulative frequency graph or ogive

A fourth method of presenting data in a graphical form is the cumulative frequency graph or ogive. This method is particularly useful when the researcher wishes to know how many scores lie below either an individual or a step interval score. The utility of this technique will become clear later in the text when we discuss measures of relative position.

We begin to construct the ogive by adding the scores in our frequency table *serially*. Scores that are added serially are accumulated progressively starting with the bottom interval of the distribution. The cumulative frequency for any given interval is the frequency of that interval added to the total frequencies below that interval. For example, using our height data once again, the cumulative frequency for interval 164.5–167.5 is

$$3 + 4 + 2 + 1 = 1 0$$

Table 6 shows the original frequency distribution with a cumulative frequency column included. The cumulative frequency graph is plotted with the scores or step interval along the horizontal axis and the *cumulative frequencies* along the vertical axis (see Figure 6).

Table 6   Cumulative frequency distribution of height scores

Step intervals	Tallies	*f*	Cumulative frequencies (cf)
197.5–200.5	1	1	50
194.5–197.5	1	1	49
191.5–194.5	11	2	48
188.5–191.5	11	2	46
185.5–188.5	11111	5	44
182.5–185.5	11	2	39
179.5–182.5	11111	5	37
176.5–179.5	111111	6	32
173.5–176.5	11111111	8	26
170.5–173.5	1111	4	18
167.5–170.5	1111	4	14
164.5–167.5	111	3	10
161.5–164.5	1111	4	7
158.5–161.5	11	2	3
155.5–158.5	1	1	1

The difference between this graph and the normal frequency polygon is that it is not the actual frequencies that are plotted but cumulative frequencies. Moreover, the cumulative frequencies must be plotted at the upper limit of each step interval as they include the frequencies in that interval. For example, the cumulative frequency for the interval 170.5–173.5, (18), is plotted at point 173.5 on the vertical axis.

Cumulative frequency graphs can be smoothed in the same way that frequency polygons can be smoothed.

4.7   The circle or pie graph

The last of the five methods of presenting data is best suited to simple comparisons of data to do with discrete variables. It is based on proportioning a circle to equivalent percentage proportions of

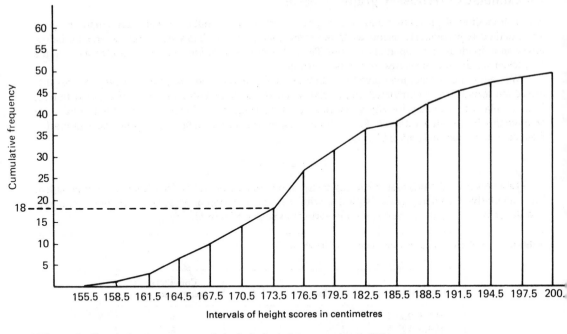

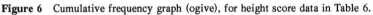

Figure 6 Cumulative frequency graph (ogive), for height score data in Table 6.

the frequency distribution. For example, from the data in Table 1, we can determine the angles of the circle that represent married, single, divorced, and widowed females as follows:

Since 360° represents 100%, then

$$married = \frac{85.9}{100} \times 360° = 309°,$$

$$single = \frac{10.9}{100} \times 360° = 39°,$$

$$divorced = \frac{2.8}{100} \times 360° = 10°, \text{ and}$$

$$widowed = \frac{0.4}{100} \times 360° = 2°.$$

The choice of method used in presenting data in a graphical form must rest upon the nature of the initial data and the amount of detailed visual information that is required. Remember, pictorial methods of presentation add nothing to the data that isn't already there to begin with! Their task is simply to display it more effectively.

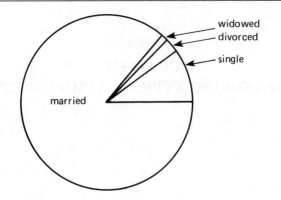

Figure 7   Circle or pie graph of marital status of females.

CHAPTER 5
MEASURING TYPICAL ACHIEVEMENT

5.1   Introduction

We said in 1.1 above that one meaning of statistics is to do with describing quantitative data. Suppose those data are scores on a personality measure. Describing such a set of scores involves giving information about two aspects of their composition.

Personality scores, like most other measures of human dispositions and abilities, tend to cluster together about the centre. One description of our data, then, is to do with the *central tendency* of the scores. The three descriptive statistics that we will consider, the mean, the median and the mode, provide information about how a distribution of scores is centred or grouped together.

A second aspect of the composition of our personality data is to do with the *variability* of the scores, that is, the way in which they are spread out from the centre of the distribution. We deal with variability in Sections 6.2 to 6.9.

5.2   Calculating the mean from ungrouped data

The mean (M) is the most familiar and useful measure used to describe the central tendency or average of a distribution of scores for any group of individuals, objects or events. The mean is computed by dividing the sum of the scores by the total number of scores.
The formula is given as:

$$M = \frac{\Sigma X}{N}$$

where X is each score, N is the total number of scores and Σ means 'the sum of'.

Example The scores of 8 students on an achievement motivation scale are recorded below:

$$16$$
$$11$$
$$13$$
$$20$$
$$15$$
$$18$$
$$10$$
$$17$$

$$\overline{\sum X = 120}$$

$$M = \frac{\sum X}{N} = \frac{120}{8} = 15$$

5.3 Calculating the mean from grouped data

In the example above we use the *ungrouped* scores of 8 students as our data. Suppose, however, that we are to compute the mean when we have scores that have already been *grouped*. The calculation of the mean from grouped data is slightly different.

Example The scores of 20 psychiatric patients on a curiosity measure have been arranged in *groups of five*, ranging from a group scoring 5–9 who exhibit little interest in what is going on around them, to a group scoring 30–34 who manifest a high degree of curiosity in their environment.

The formula for computing the mean from grouped data is given as:

$$M = \frac{\sum fX}{N}$$

where X is the midpoint of a class interval, f is the number of cases in that interval, N is the total number of scores and $\sum$ means 'the sum of'.

Table 7 Computing the mean from grouped data

Curiosity measure (exact limits)	*X* Midpoint of the interval	*f*	*fX*
29.5–34.5	32	2	64
24.5–29.5	27	2	54
19.5–24.5	22	5	110
14.5–19.5	17	6	102
9.5–14.5	12	4	48
4.5– 9.5	7	1	7
		$N = 20$	$385 = \sum fX$

$$M = \frac{\sum fX}{N} = \frac{385}{20} = 19.25$$

5.4 A short method of calculating the mean from grouped data

There is a shorter way of calculating the mean from grouped data which saves time and labour in computation, particularly when one has to deal with a large number of cases. It involves making a guess at identifying the interval in which the mean probably falls. Look at the frequency distribution (column f) in Table 7 above. The mean is probably contained either in interval 19.5–24.5 or in interval 14.5–19.5. Let's assume that it occurs in interval 14.5–19.5. Our *guessed* or *assumed* mean (A.M.) then becomes the midpoint of that interval, that is to say, A.M. = 17.

Table 8 Computing the mean from grouped data (short method)

Curiosity measure (exact limits)	X Midpoint of the interval	f	d	fd	
29.5–34.5	32	2	+3	+6	
24.5–29.5	27	2	+2	+4	
19.5–24.5	22	5	+1	+5	(+15)
14.5–19.5 ← A.M.	17	6	0		
9.5–14.5	12	4	−1	−4	
4.5– 9.5	7	1	−2	−2	(−6)
		$N = 20$		$\sum fd = +9$	

1   In Table 8 in the column headed d, we place the value 0 next to the interval containing our assumed mean (A.M. = 17). The first interval above the one containing the assumed mean is assigned the value +1, the one above is given the value +2, and the third interval above is given the value +3. Below the interval containing the assumed mean, the first interval is assigned the value −1, the second, −2. These values represent the deviations (d) of the midpoints of the various intervals from the assumed mean in units of class intervals.

2   The deviations in column d are multiplied by their respective frequencies (f) to give the products recorded in column fd. These are summed algebraically, the 'positives' first, then the 'negatives'. The total ($\sum fd$) is recorded at the bottom of column fd.

3   By dividing $\sum fd$ by N, the total number of scores in the frequency column (f) we obtain C^i a correction factor in class interval units:

$$C^i = \frac{\sum fd}{N} = \frac{9}{20} = 0.45$$

4   By multiplying C^i by our interval ($i = 5$) we obtain the correction factor C which can then be applied to our assumed mean (A.M.) in order to transform it into the true mean (M).

$$C = C^i i = .45 \times 5 = 2.25$$

5   The assumed mean plus the correction factor gives the true mean

$$\text{A.M.} + C = M$$

$$17 + 2.25 = 19.25$$

5.5 The median

Another useful measure of central tendency is the *median* or *middle score*.

The median is simply that point on a scale of measurement above which there are exactly half the scores and below which there are the other half of the scores.

In the distribution 1, 3, 5, 7, 9, 11, 13, the middle score is 7. Thus the score of seven indicates the median point at which there are three scores above and three scores below. We obtain the median by arranging the scores in ascending order from the smallest to the greatest score and selecting that point above and below which there are an equal number of scores.

The median point in a distribution of scores, the total of which comes to an odd number is the middle score in that distribution providing that the middle score has a frequency of one.

$$2, \quad 3, \quad 5, \quad 7, \quad 9, \quad 10, \quad \underset{\uparrow}{12}, \quad 13, \quad 14, \quad 16, \quad 18, \quad 20, \quad 21$$

median point

Where the middle score in a distribution has a frequency greater than one, it is necessary to understand the meaning of the *interval of a score* in order to calculate the median point.

The *interval of a score* defines the *exact limits* of that score. The interval ranges from 0.50 units below to 0.50 units above the score. Thus the score six includes all values within the limits of 5.50 to 6.50. The exact midpoint of the interval having lower and upper limits ranging from 5.50 to 6.50 is 6.

In the distribution below, the score of five has a frequency of 3:

$$2, \quad 3, \quad 3, \quad 4, \quad 5, \quad 5, \quad 5, \quad 7, \quad 9, \quad 10$$

The exact limits, that is, the interval range of 5, includes all values from 4.50 to 5.50. We assume that the scores 5, 5, 5 are equally spread throughout the interval of 4.50 to 5.50. Thus, each 5 occupies $\frac{1}{3}$ (0.33) of that interval as illustrated below:

$$\boxed{\begin{array}{c|c|c} 5 & 5 & 5 \end{array}}$$

4.50 4.83 5.16 5.50

The median point of the distribution 2, 3, 3, 4, 5, 5, 5, 7, 9, 10 is to be found between the fifth and the sixth scores. Below the interval 4.50 to 5.50 there are four scores (2, 3, 3, 4). The fifth score (5) is thus $\frac{1}{3}$ of the distance into the interval 4.50 to 5.50, that is 4.50 + 0.33 = 4.83. The median is shown in the diagram below:

$$\langle .5 \ldots 5 \ldots 5. \rangle$$

2   3   3   4   4.50 4.83 5.16 5.50⌐   9   10

↑

Median

Look at the distribution below:

$$3, \quad 4, \quad 4, \quad 5, \quad 5, \quad 5, \quad 5, \quad 6, \quad 7$$

↑

There is an odd number of scores in the distribution, the median being located at the point indicated by the arrow, that is, within the group of four 5s. Again, we assume that the scores 5, 5, 5, 5 are equally spread throughout the interval 4.50 to 5.50. Thus each 5 occupies $\frac{1}{4}$ (0.25) of the interval as illustrated below:

| 5 | 5 | 5 | 5 |

4.50   4.75   5.00   5.25   5.50

The median point of the distribution 3, 4, 4, 5, 5, 5, 5, 6, 7 is to be found midway between 4.75 and 5.00, that is, at a point 4.75 + $\frac{1}{2}$ of (0.25) = 4.875 (4.88).

⟨.5 . . . 5 . . . 5 . . . 5 .⟩

```
  :    :    :    :    :    :    :    :    :    :
 ___  ___  ___  ____ ____ ____ ____ ____ ___  ___
  3    4    4   4.50 4.75 5.00 5.25 5.50  6    7
```

↑

Median (4.88)

Look at the distribution below:

53,   54,   55,   59,   62,   67,   70,   71

The median is located between the fourth and the fifth scores, that is, at a point between the upper exact limit of 59 (59.50) and the lower exact limit of 62 (61.50). The diagram below shows that the median is located at 60.50.

```
  :     :     :     :     :     :     :
 ___  _____  ___  _____  ___  _____  ___
 59   59.50  60   60.50  61   61.50  62
```

↑

Median (60.50)

5.6   Calculating the median

Finding the median score in the frequency distribution below involves five steps.

Table 9   Incidence of aggressive acts in a study of behaviour problems in three-year-old children

Class interval	Frequency
48–50	2
45–47	3
42–44	4
39–41	6
36–38	8
33–35	8
30–32	7
27–29	6
24–26	3
21–23	2
18–20	1

$N = 50$

1 Divide the total number of scores by two.

$$(50 \div 2 = 25)$$

2 Start at the low end of the frequency distribution and sum the scores in each interval until the interval containing the median (i.e. the 25th score) is reached.

$$(1 + 2 + 3 + 6 + 7 = 19)$$

3 Subtract the sum obtained in step two above from the number necessary to reach the median.

$$(25 - 19 = 6)$$

4 Now calculate the proportion of the median interval that must be added to its lower limit in order to reach the median score. This is done by dividing the number obtained in step 3 above by the number of scores in the median interval and then multiplying by the size of the class interval.

$$\tfrac{6}{8} \times 3 = \tfrac{18}{8} = 2.25$$

5 Finally, add the number obtained in step 4 above to the exact lower limit of the median interval.

$$(32.5 + 2.25 = 34.75)$$

$$\text{Median} = 34.75$$

5.7 Summary

The formula below summarizes the calculation of the median:

$$\text{Median} = L + \left(\frac{\frac{N}{2} - S}{f} \right) \times i$$

where:

L = the exact lower limit of the median interval

N = the total number of scores

S = the sum of the scores in the intervals below L

f = the number of scores in the median interval

i = the size of the class interval.

5.8 The mode

The mode is yet another measure of central tendency. In its most common usage this measure is sometimes referred to as the *crude mode,* that is, the most frequently occurring score in a distribution. The crude mode is rarely applied to ungrouped data. The reason is obvious enough. Any particular score might occur haphazardly at any point in a series of observations and be quite unrepresentative of the data as a whole.

With grouped data one cannot identify the score that occurs most frequently, because that score is lost within a specific interval of the grouped data. One can, however, identify the *modal class,* that is the interval that contains more scores than any other. When the mean and the median have

27

also been determined, the *computed mode* can be obtained from the formula:

$$\text{Computed mode} = 3 \times \text{Median} - 2 \times \text{Mean}$$

Alternatively, when a histogram has been constructed, an approximation of the mode can be made by dividing the modal class geometrically as in Figure 8.

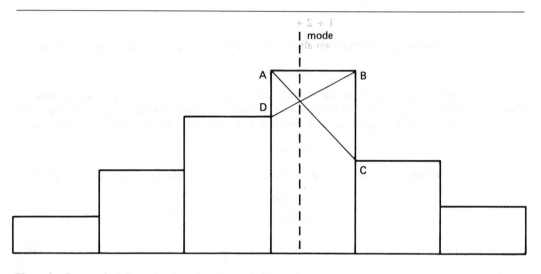

Figure 8   Geometrical determination of the mode.

5.9   Choosing a measure of central tendency

It's sometimes a problem for the researcher to decide which of the three measures of central tendency to use—the *mean,* the *median,* or the *mode*—as being most appropriate to his particular problem. The following advice may be of help.

The mean is doubtless the most commonly used measure of central tendency. It is the only one of the three measures which uses all of the information available in a set of data, that is to say, it reflects the value of each score in a distribution. It has the decided advantage of being capable of combination with the means of other groups measured on the same variable. For example, from the average unemployment levels in twenty-two Priority Areas we can compute the overall mean unemployment rate in those designated regions. Because neither the median nor the mode is arithmetically-based, this useful application is not possible. The precisely defined mathematical value of the mean allows other advanced statistical techniques to be based on it too.

There are occasions, however, when taking into account the value of every score in a distribution can give a distorted picture of the data. Suppose, for example, that the annual number of marriage ceremonies performed in five Scottish border communities is as follows: Dornock (40), East Riggs (60), Canonbie (60), Annan (80), Gretna Green (810). Without the very atypical score of 810, the mean score of the group is 60 and the median, likewise, is 60. The effect of introducing the score of 810 is to pull the mean in the direction of that extreme value. The mean now becomes 210, a value that is unrepresentative of connubial proclivity in the five communities. The median remains 60, providing a more realistic description of the distribution than the mean.

With these observations in mind:

5.10   Use the mean

1   When the scores in a distribution are more or less symmetrically grouped about a central point.

2   When the research problem requires a measure of central tendency that will also form the basis of other statistics such as measures of variability (see Chapter 6) or measures of association (see Chapter 8).

3   When the research problem requires the combination of the mean with the means of other groups measured on the same variable.

4   Recalling our discussion of the relationship between a *sample* and a *population* (see Section 2.2), use the mean to measure the central tendency in a sample of observations when one needs to estimate the value of a corresponding mean of the population from which the sample is taken.

5.11   Use the median

1   When the research problem calls for knowledge of the exact midpoint of a distribution.

2   When extreme scores distort the mean as in our hypothetical example of annual marriage ceremonies. The mean reflects extreme values, the median does not.

3   When dealing with 'oddly-shaped' distributions, for example, those in which a high proportion of extremely high scores occur as well as a low proportion of extremely low ones.

5.12   Use the mode

1   When all that is required is a quick and appropriate way of determining central tendency.

2   When in referring to what is 'average', the word is used in the sense of the 'typical' or the 'most usual'. For example, in talking about the average take-home pay of the coalface worker, it is the modal wage that is being alluded to rather than an exact arithmetic average.

Finally, referring to our earlier discussion of scales of measurement in Section 3.1, the *mode* would be the appropriate statistic to use as a measure of the 'most fashionable' or 'most popular' when data are collected using a nominal scale. The *median* would generally be associated with ordinal level data. The mean would be used with interval level or ratio level data providing that the distribution of scores approximates a normal curve. It is to a description of the normal curve that we now turn.

5.13   The normal curve

At the beginning of most introductory textbooks concerned with research the reader is confronted with a lovely bell-shaped curve known as the *normal probability curve* (see Figure 9).

To be strictly correct, one refers to normal probability *curves* rather than the *normal curve*. Three such curves are shown in Figure 10. They differ from each other in their respective 'spreads', a point that we enlarge upon shortly.

Very often, coin-tossing is the way in which the student is introduced to the idea of probability and to some of the important properties of the normal curve.

Imagine that we have ten pennies all perfectly symmetrical so that none of them is more likely to fall 'heads' than 'tails' when tossed.

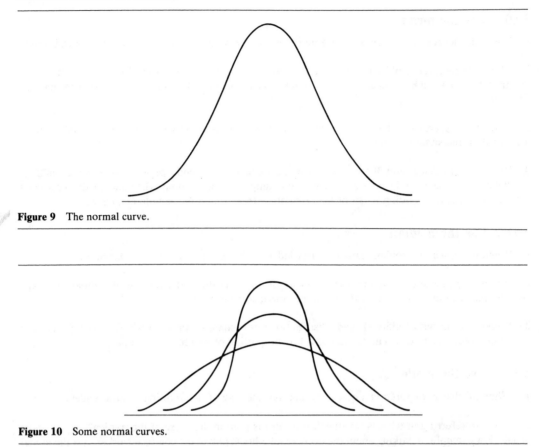

Figure 9   The normal curve.

Figure 10   Some normal curves.

When one of those pennies is tossed we know that the chances of tossing 'heads' is one in two; equally, the chances of tossing 'tails' is one in two. With two pennies tossed simultaneously four times, there are four possible results. These are shown below:

Distribution of chances of falls when two coins are tossed
4 times

Chances in 4 throws	Two heads (2H)	One heads and one tails (1H)(1T)	Two tails (2T)
	1	2	1

Set out below are the chances (or the probabilities) of the distributions of falls when 3, 4, 5, 6, 7, 8, 9, and 10 coins are simultaneously tossed.

Distribution of chances of falls
3 coins

Chances in 8 throws	3H	2H 1T	1H 2T	3T
	1	3	3	1

4 coins

Chances in 16 throws	4H	3H 1T	2H 2T	1H 3T	4T
	1	4	6	4	1

5 coins

Chances in 32 throws	5H	4H 1T	3H 2T	2H 3T	1H 4T	5T
	1	5	10	10	5	1

6 coins

Chances in 64 throws	6H	5H 1T	4H 2T	3H 3T	2H 4T	1H 5T	6T
	1	6	15	20	15	6	1

7 coins

Chances in 128 throws	7H	6H 1T	5H 2T	4H 3T	3H 4T	2H 5T	1H 6T	7T
	1	7	21	35	35	21	7	1

8 coins

Chances in 256 throws	8H	7H 1T	6H 2T	5H 3T	4H 4T	3H 5T	2H 6T	1H 7T	8T
	1	8	28	56	70	56	28	8	1

9 coins

Chances in 512 throws	9H	8H 1T	7H 2T	6H 3T	5H 4T	4H 5T	3H 6T	2H 7T	1H 8T	9T
	1	9	36	84	126	126	84	36	9	1

10 coins

Chances in 1,024 throws	10H	9H 1T	8H 2T	7H 3T	6H 4T	5H 5T	4H 6T	3H 7T	2H 8T	1H 9T	10T
	1	10	45	120	210	252	210	120	45	10	1

Look at the distribution of chances of falls for ten pennies. The chances of throwing five heads and five tails are 252 out of 1,024, that is, approximately one in four chances. The chances of throwing ten heads or ten tails are only one in over a thousand!

But what possible relevance have the distributions of tossed pennies to everyday matters?

5.14 A practical application of the normal probability curve

For illustrative purposes only, the histogram below represents the numbers of pairs of men's shoes of varying sizes carried by a shoe shop. The average size in men's shoes is size eight. Our hypothetical shop owner (albeit intuitively) has correctly applied some awareness of the normal probability curve to his stock request order. The chances of customers with size three or size thirteen feet visiting his premises are sufficiently low as to deter him from keeping all but a few outsize shoes!

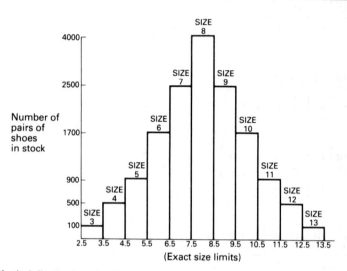

Figure 11   Hypothetical distribution of men's shoe sizes in stock.

Histograms drawn to represent such diverse properties as height, weight, intelligence quotients, and achievement test scores would be similar in shape to the one illustrating the distribution of men's shoe sizes.

Suppose we were to construct such a histogram in respect of children's weight and that we were to use very fine gradations in weight (fractions of grams) in plotting the distribution of weight for 10,000 children. The effect of using very small weight categories over a very large population of children would be to transform the small step-like sides of the histogram into a smooth continuous curve: the normal curve of distribution.

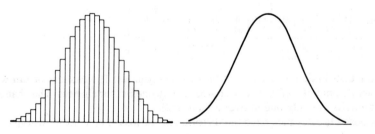

5.15   Some mathematical characteristics of the normal probability curve

On the normal curve of distribution there are two points, called *points of inflection* where the curve changes direction from convex to concave. If a perpendicular line is dropped from these points of inflection to the base line (see diagram below) it is possible to measure off one *unit of distance* from either side of the central axis of the normal curve marked by the dotted line *XY*.

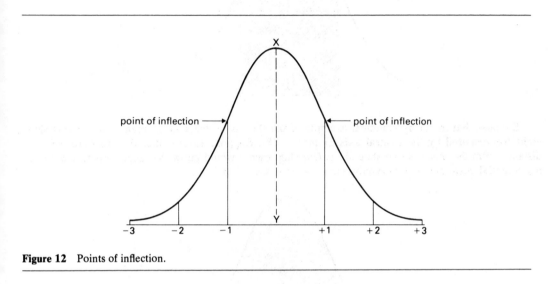

Figure 12   Points of inflection.

This unit of distance can be used as a *standard* by which to divide the base line into equal segments.

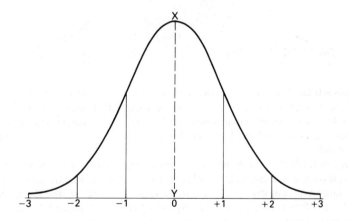

Each segment is *one standard unit of distance* or *one standard deviation* from the central axis *XY*.
If we let the total area under the normal curve equal 100%, then one of the mathematical properties of the normal curve is that the area bounded by one standard deviation on either side of the central axis (or mean) is approximately 68.26% of the total area.

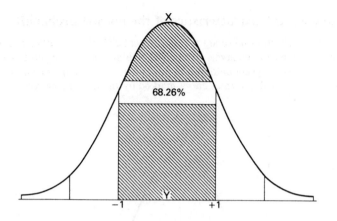

Suppose that in our hypothetical example of the shoe shop, the average man's shoe size is size eight (represented by the central axis or mean, *XY*). Suppose further that the standard unit of distance from the mean is two shoe sizes. Our shop owner would know that approximately 68% of his potential customers would require shoes of sizes six to ten.

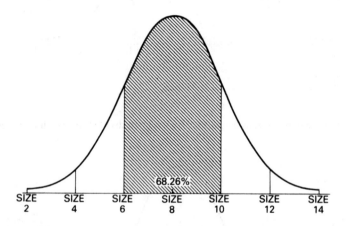

Given the hypothetical distribution of male foot sizes we can estimate from the diagram below the probability of the shoe-shop owner encountering customers with outsize feet!

99.74% of men have shoe sizes between two and fourteen. The chances of a customer requesting size two or size fourteen shoes is about 1 in 300. The shop owner can complete his stock order with this knowledge in mind.

So far, we have described characteristics of normal probability curves where we have shown perfectly symmetrical balances between both sides of distributions as illustrated in Figures 9 and 10. In such curves, the three averages discussed earlier, the *mean,* the *median,* and the *mode,* all coincide.

Many data collected however are not normally distributed. In Figures 14 and 15 we show two non-normal distributions commonly encountered by the researcher. We indicate the typical positions of the mean, median, and mode in these distributions.

Figure 14 shows a *negatively-skewed distribution,* that is, *skewed to the left.* Such a distribution might be found on an Introductory Psychology quiz in which the majority of students in the group do well and only a few fail to grasp particular concepts.

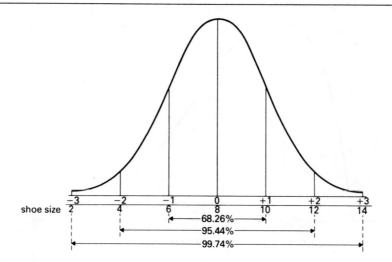

Figure 13 Distribution of shoe sizes.

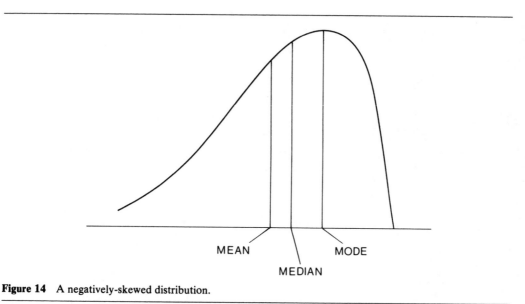

MEAN

MEDIAN

MODE

Figure 14 A negatively-skewed distribution.

Figure 15 shows a *positively-skewed distribution,* that is, *skewed to the right.* Such a distribution might be found on a very difficult end of term examination where only a small number of students demonstrate their thorough grasp of the coursework while most of their fellows perform poorly.

Notice how, in Figures 14 and 15 the mean tends to be 'pulled' away from the mode in the direction of extreme values.

35

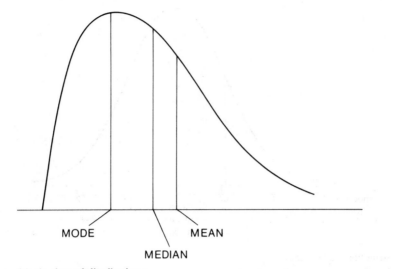

Figure 15   A positively-skewed distribution.

CHAPTER 6
MEASURING VARIATIONS IN ACHIEVEMENT

6.1   Introduction

The measures of central tendency outlined in Sections 5.1 to 5.11 describe the typical achievement of a group. They tell us nothing however, of the way in which individuals within the group differ from one another in their achievement. To know how individuals vary we need measures of variability.

Variability refers to the extent to which individuals or scores differ from each other. As we shall see, certain measures of central tendency and of variability, when used together, enable us to compute precisely how different the achievements of two or more groups are.

6.2   The range

The range is the simplest measure of variability. It tells us about the interval between the highest and the lowest scores in a distribution. The range shows how wide the distribution is over which our scores are spread. It is computed by taking the lowest from the highest score and adding 1. The addition of 1 allows for the exact limits of the lowest and highest scores in the range. In the distribution 2, 4, 5, 7, 7, 8, 9, the range is $9 - 2 + 1 = 8$. Because the range takes account only of the two most extreme scores, it is limited in its usefulness since it gives no information on how the scores are distributed.

Look at the following distributions. While each has a range of 50, the distribution of the scores across the interval varies considerably.

3, 9, 16, 28, 35, 43, 52, Range = 50

11, 20, 21, 22, 23, 25, 60, Range = 50

Use the range when scores are too scattered to warrant the calculation of a more precise measure of variability, or when all that is required is knowledge of the overall spread of the scores.

6.3   Average deviation (A.D.)

In discussing the normal curve in Section 5.14 we introduced the idea of units of distance which are measured off on either side of the central axis or mean of the bell-shaped curve. A unit of distance or a *deviation* from the mean helps describe the spread of the curve. That is to say, it provides an index of the amount of dispersion of the scores of a group about the mean. A deviation is simply the distance any score is from the mean of its distribution. Scores that are higher than the mean deviate positively, scores that are lower than the mean deviate negatively. If d = the deviation of a score, then the formula for calculating the deviation of any score (X) from the mean of its distribution (M) is given as:

$$d = X - M$$

Because scores are spread positively and negatively around the mean of any distribution, the true average deviation will always be zero. A *'true' average deviation* tells us nothing about the distribution. Therefore, to calculate the *average deviation* (note the difference!) we proceed as follows:

1   Ignore the signs attached to deviation (d) of each score. (The symbol $|d|$ means the *absolute value* of d, that is, without signs.)

2   Add up or sum (Σ) the d's.

3   Divide by N – the number of scores in the distribution.

The formula for the average deviation (A.D.) is given as:

$$A.D. = \frac{\Sigma |d|}{N}$$

The distribution 2, 3, 5, 6, 7, 8, 8, 9, has a mean of 6. The scores and their deviations are set out in Table 10.

Table 10   Calculating the average deviation

X	d
9	3
8	2
8	2
7	1
6	0
5	−1
3	−3
2	−4

$$A.D. = \frac{\Sigma |d|}{N} = \frac{16}{8} = 2$$

Notice how the *average deviation* improves upon the *range* as a measure of variability in that it takes account of all the scores in a distribution.

6.4   The standard deviation (S.D.)

Although the average deviation (A.D.) is rarely used in modern statistics, it has been included here as a useful introduction to the concept of deviation. Moreover, it leads directly into a consideration of the most commonly used measure of variability, the *standard deviation* (S.D.).

How does the average deviation (A.D.) differ from the standard deviation (S.D.)? Recall that in discussing the normal curve in Section 5.14 we identified not a 'unit of distance' but a *'standard unit of distance'*, which we equated with *'standard deviation'*. Furthermore, we showed how that standard deviation was derived from the shape of the normal curve, to be precise, from the points of inflection, that is, where the curve changed direction from convex to concave. A mathematical explanation of how a standard deviation (as contrasted with an average deviation) is derived from the characteristics of the normal curve need not concern us here. At a more practical level, the immediate difference between the average deviation and the standard deviation is that in calculating A.D. we ignore all signs attached to d's and sum their absolute values. In computing S.D. the difficulty of signs is bypassed. We simply square each of the deviations as the following examples show.

The following symbols will be used:

S.D. = the standard deviation

X = the raw score

M = the mean

d = the deviation score obtained by subtracting the mean from each raw score

d^2 = the deviation score squared

N = the number of cases

f = the frequency of cases

i = the class interval

C^i = the correlation factor in terms of intervals

6.5   Calculating the standard deviation from ungrouped data

The formula for the standard deviation from ungrouped data* is given as:

$$\text{S.D.} = \sqrt{\frac{\Sigma d^2}{N-1}}$$

Example   The raw scores of 11 boys on an arithmetic test are arranged in descending order in Table 11.

1   First calculate M, the mean score, by dividing ΣX, the total of the raw scores, by N, the number of cases.

$$M = \frac{\Sigma X}{N} = \frac{110}{11} = 10.0$$

2   Calculate d, the deviation score, by subtracting M, the mean score, from each raw score X. Enter the d score in the column next to the appropriate X score.

3   Square each deviation score and enter in the third column. Note that in squaring minus quantities we obtain plus products. Sum the squared deviations.

---

* Why $N - 1$ in the formula? One of the most easily understood explanations of the division by $N - 1$ is given in M. Hamburg (1974) *Basic Statistics: A Modern Approach,* New York: Harcourt, Brace, Jovanovich, page 64. In brief, when N is large, the subtraction of 1 is neither here nor there. With smaller N's however, $N - 1$ gives us a less-biased estimate of what the variance is in the population from which the particular sample is drawn.

Table 11 Calculating the standard deviation

X	d	d^2
16	6	36
14	4	16
12	2	4
11	1	1
10	0	0
10	0	0
9	-1	1
9	-1	1
8	-2	4
6	-4	16
5	-5	25
$\sum X = 110$	$\sum d = 0.0$	$\sum d^2 = 104$

4 Obtain S.D., the standard deviation, by extracting the square root of the sum of the square deviations divided by $N-1$

$$S.D. = \sqrt{\frac{\sum d^2}{N-1}} \quad \text{In our example, S.D.} = \sqrt{\frac{104}{10}} = 3.2$$

A short-cut* calculation of the standard deviation

A close approximation to the S.D. as computed from the formula for ungrouped data is given by:

$$S.D. = \frac{\text{Sum of high sixth of scores} - \text{sum of low sixth of scores}}{\text{half the number of subjects involved}}$$

Suppose for example that the following job satisfaction scores have been obtained from 20 psychiatric nurses and that we wish to compute the S.D. using the short-cut method. First we arrange the scores from highest to lowest as in Table 12.

Table 12 Computation of the S.D. of job satisfaction scores using the short-cut method

			Job satisfaction scores
45 ⎫	17		Number of subjects $= 20$
43 ⎬ high	16		One-sixth of scores $= \frac{20}{6} = 3\frac{1}{3}$
40 ⎭ sixth	14		
31	13		
30	10		$\sum$ of high sixth of the scores $= 45+43+40+(\frac{1}{3}$ of $31) = 138.3$
26	9 ⎫		
25	7 ⎬ low		
24	6 ⎭ sixth		
22	3		$\sum$ of low sixth of the scores $= 3+6+7+(\frac{1}{3}$ of $9) = 19$
19			
18			

P. Diederich (1960) *Short-cut Statistics for Teacher-Made Tests.* Princeton: Educational Testing Service (Evaluation and Advisory Service Series No 5).

Substituting into the formula:

$$S.D. = \frac{138.3 - 19}{10} = \frac{119.3}{10} = 11.93$$

Compare this value with that obtained by the more usual calculation of the standard deviation from ungrouped data.

Table 13 Computation of the standard deviation of the job satisfaction scores of 20 psychiatric nurses

X	d	d²
45	24.1	580.81
43	22.1	488.41
40	19.1	364.81
31	10.1	102.01
30	9.1	82.81
26	5.1	26.01
25	4.1	16.81
24	3.1	9.61
22	1.1	1.21
19	−1.9	3.61
18	−2.9	8.41
17	−3.9	15.21
16	−4.9	24.01
14	−6.9	47.61
13	−7.9	62.41
10	−10.9	118.81
9	−11.9	141.61
7	−13.9	193.21
6	−14.9	222.01
3	−17.9	320.41
$\Sigma X = 418$	$\Sigma d = 0.0$	$\Sigma d^2 = 2829.8$

$$M = \frac{\Sigma X}{N} = \frac{418}{20} = 20.9$$

$$S.D. = \sqrt{\frac{\Sigma d^2}{N-1}} = \sqrt{\frac{2829.8}{19}} = 12.20$$

The S.D. computed by the short-cut method varies slightly from that derived from the full formula, illustrating the fact that there is some loss of accuracy in most short-cut methods as the sample becomes smaller and less normally distributed.

6.6   Calculating the standard deviation from grouped data

Alternative formulae for calculating the standard deviation from grouped data are now given. The reader will see that the second of these formulae closely parallels the short method of computing the mean from grouped data outlined in Section 5.4 above.

Formula 1

$$S.D. = \sqrt{\frac{\Sigma fd^2}{N-1}}$$

The same data as in Table 11 are used.

Table 14 Calculating the standard deviation from grouped data (Formula 1)

X	f	fX	d	d^2	fd^2
16	1	16	6	36	36
14	1	14	4	16	16
12	1	12	2	4	4
11	1	11	1	1	1
10	2	20	0	0	0
9	2	18	−1	1	2
8	1	8	−2	4	4
6	1	6	−4	16	16
5	1	5	−5	25	25
	$N = 11$	$\sum fX = 110$			$\sum fd^2 = 104$
		$M = 10$			

$$\text{S.D.} = \sqrt{\frac{\sum fd^2}{N-1}} \quad \text{In our example, S.D.} = \sqrt{\frac{104}{10}} = 3.2$$

Formula 2

$$\text{S.D.} = i\sqrt{\frac{\sum fd^2}{N-1} - C'^2}$$

The same data as in Table 11 are used.

Table 15 Calculating the standard deviation from grouped data (Formula 2)

Score intervals	f	d	fd	fd^2
14–16	2	+2	+4	8
11–13	2	+1	+2(+6)	2
A.M. → 8–10	5	0		
5–7	2	−1	−2(−2)	2
	$N = 11$		$\sum fd = 4$	$\sum fd^2 = 12$

1   In Table 15, in the column headed d, we have placed the value 0 next to the interval containing our assumed mean (A.M.) and have assigned deviation values +2, +1, −1, as in the computation of the mean for grouped data in Section 5.4.

2   We obtain C', a correction factor in class interval units.

$$C^i = \frac{\sum fd}{N} = \frac{4}{11} = 0.36$$

This is squared to give $C'^2 = 0.13$

3   We obtain Σfd^2 by summing the values in column fd^2

$$\Sigma fd^2 = 12$$

$$\text{S.D.} = i \sqrt{\frac{\Sigma fd^2}{N-1} - C^{i2}} = 3 \sqrt{\frac{12}{10} - 0.13} = 3\sqrt{1.07} = 3.1$$

6.7   Variance

Another measure of variability closely related to the standard deviation is the *variance*. Like the S.D., the variance is related to the size of the difference between each score and the mean of a distribution. The variance is simply the standard deviation squared.

$$\text{Variance} = \text{S.D.}^2 = \frac{\Sigma d^2}{N-1}$$

It's more common to use the S.D. than the variance in describing the variability of a distribution. Why bother computing the variance in that case? The variance, as we shall see, is used in many of the more advanced statistical analyses that we describe in the second part of the text.

6.8   Coefficient of variation (V)

Recall that in our discussion of levels of measurement in Section 3.5 we described the ratio scale as the highest level of measurement and went on to illustrate a ratio scale by reference to weight.
   Ratio scales are useful when an investigator is interested in the variability of a sample on one characteristic as compared with another. Take, for example, the question, 'Does a sample of student apprentices vary as much in respect of their numeracy levels as their literacy levels?' We cannot, of course compare numeracy and literacy levels directly. But we can compare the relative variability of the group over its distribution on numeracy and literacy. The *coefficient of variation (V)*, sometimes known as the *coefficient of relative variation,* is the statistic we employ.
   In computing V, we make use of the mean and the standard deviation of a distribution. In our hypothetical example, since we are dealing with a sample, it is M_s, the mean of a sample that we use.
   The formula for the coefficient of variation is given by

$$V = \frac{100 \text{ S.D.}}{M_s} \%$$

Example   In a sample of student apprentices the mean score on a numeracy test is 15 with a standard deviation of 2 and the mean score on a literacy test is 50 with a standard deviation of 15. We wish to find out on which characteristic the group is more variable.

$$V_{numeracy} = \frac{100 \times 2}{15} \% = 13.3\%$$

$$V_{literacy} = \frac{100 \times 15}{50} \% = 30.0\%$$

The sample of student apprentices is more than twice as variable in *literacy* as in *numeracy.*

6.9   The quartile deviation (Q)

Whenever the median is chosen as an appropriate measure of central tendency, then the quartile deviation is an appropriate measure of variability. Let's see why.

We need first to introduce the *percentile*. Percentiles are used to describe the position of a score in a distribution. Percentiles divide up a distribution into 100 parts. The 10th, the 20th, and the 30th percentiles are those scale values below which 10%, 20%, and 30% of the cases in a distribution fall.

The 50th percentile, written as P_{50}, indicates that this particular scale value has 50% of the total distribution above and below it. P_{50} is, of course, the median.

Quartiles divide up a distribution into 4 parts. Thus, the first quartile (Q_1) is the same as P_{25}, and the third quartile (Q_3) is the same as P_{75}.

The *quartile deviation* or Q is defined as one half the scale distance between the 75th and 25th percentiles in a frequency distribution.

The formula for Q is given as

$$Q = \frac{Q_3 - Q_1}{2}$$

It will be seen that the quartile deviation is simply a restricted version of the range; it tells us about the variability around the middle of a distribution of scores, that is, about the median.

Calculating Q

Example The scores of 40 students on a short-form dogmatism scale are set out in the frequency distribution in Table 16.

Table 16 Calculating the quartile deviation

Class interval	f	
85–89	1	count downwards for Q_3
80–84	1	
75–79	1	
70–74	2	
65–69	3	8 cases
60–64	4	
55–59	6	
50–54	7	
45–49	5	
40–44	3	10 cases
35–39	3	
30–34	2	
25–29	1	
20–24	1	count upwards for Q_1

$N = 40$

To calculate the quartile deviation we must first compute Q_1 and Q_3.

1   $Q_1 = P_{25}$. That is, 25% of the total number of cases (40) or *10 cases*. We begin to count upwards from the bottom of the frequency distribution. It can be seen that below the lower limit of the interval 45–49 (i.e. at 44.5) we have included exactly 10 cases. We have no need therefore to interpolate.

$$Q_1 = 44.5$$

2 $Q_3 = P_{75}$. We need not count upwards however. It's simpler to count downwards from the top of the frequency distribution. It can be seen that by the lower limit of the interval 65–69 (i.e. at 64.5) we have included 8 cases. We need to include 2 more. In the interval below the one containing the 8 cases (i.e. 60–64) there are 4 cases. We need to go down a further $\frac{2}{4}$ of the way through this interval. Arithmetically this is,

$$Q_3 = 64.5 - \tfrac{2}{4}(5)$$

$$Q_3 = 64.5 - 2.5$$

$$Q_3 = 62.0$$

3 To compute Q we proceed as follows:

$$Q = \frac{Q_3 - Q_1}{2} = \frac{62.0 - 44.5}{2} = \frac{17.5}{2} = 8.75\,(8.8)$$

6.10 The usefulness of Q

The usefulness of Q as a measure of variability lies in the fact that one quartile deviation on either side of the median in a normal distribution contains 50% of the cases. In our example in Table 16 (near normally distributed) the median is 53.1 (see Section 5.6 for the calculation of the median from grouped data). Thus 50% of the cases lie between 53.1–8.8 and 53.1+8.8, that is between 44.3 and 61.9.

CHAPTER 7
MEASURING RELATIVE ACHIEVEMENT

7.1   Introduction

Often we are concerned with interpreting and comparing individual scores. For example, we may wish to know whether a student's score is good or bad, whether he does better on Test A than on Test B, and how much better or worse his score is when compared with others in his group.

In order to interpret a particular test score correctly, we need to have a basis for comparison. This can be achieved by measuring the *relative standing* of the score in relation to the total distribution of scores for the group. Various methods are available for measuring the relative standing of a score. Two of the more commonly used are discussed below.

7.2   Percentiles

A frequently used measure of relative standing is the *percentile rank (p)*. The percentile rank indicates the percentage of scores in a distribution that lie below any particular score. That specific score is known as the *percentile point,* designated (P).

For example, if a student scores 45 on a performance test and his percentile rank is calculated to be 70, then 70% of the total distribution of scores would lie below the score of 45. We can express the score 45 as P_{70}, that is, 45 corresponds with the percentile rank 70. Recall that in discussing the quartile deviation we noted that P_{50}, the 50th percentile, is that scale value which has 50% of the total distribution above and below it; in other words, P_{50} is the median.

The advantages of using percentile ranks as measures of relative standing are obvious. Not only can we compare an individual with other members of his group on a particular test, but an individual's performance on different tests can readily be evaluated—a percentile rank of 55 on Test A for example, being clearly superior to a percentile rank 50 on Test B.

There are two approaches to calculating percentiles. The first involves calculating the percentile points which correspond to particular percentile ranks. For example, what are the exact scores (the percentile points) needed to achieve, say, the percentile ranks of 30 and 60?

The second approach involves calculating the percentile rank for a particular score.

7.3 Method 1: Calculating percentile points

The method employed in calculating percentile points is a modified form of that used in calculating the median (see page 28).

Example   Using the frequency distribution of the incidence of aggressive acts from page 26, calculate the scores corresponding to percentile ranks 10, 20, 30, 40, 50, 60, 70, 80, and 90.

Table 17   Calculating percentile points

Step intervals (rounded off)	Step intervals (exact limits)	f	cf
48–50	47.5–50.5	2	50
45–47	44.5–47.5	3	48
42–44	41.5–44.5	4	45
39–41	38.5–41.5	6	41
36–38	35.5–38.5	8	35
33–35	32.5–35.5	8	27
30–32	29.5–32.5	7	19
27–29	26.5–29.5	6	12
24–26	23.5–26.5	3	6
21–23	20.5–23.5	2	3
18–20	17.5–20.5	1	1

For percentile rank 10:

1   Construct a cumulative frequency column *(cf)*, by serially adding the frequencies.

2   Multiply the desired percentile rank (10) by the total number of scores (50) and divide by 100.

$$\frac{p \times N}{100} = \frac{10 \times 50}{100} = 5$$

N.B. 10% of 50 = 5 or 5th score.

3   Moving up the cumulative frequency column note the step interval in which the 5th score lies (23.5–26.5)

$$\text{Lower limit (L)} = 23.5$$

4   Note the number of scores in the step interval in which the 5th score lies (f)

$$f = 3$$

5   Note the number of scores below that particular step interval (3).

6   Substitute into the formula:

$$P_p = L + \left(\frac{\frac{pN}{100} - S}{f} \times i \right)$$

where

$P\hat{p}$ = Percentile point for given percentile rank p

p = Percentile rank desired

L = Lower limit of interval in which the $P_{\hat{p}}$ lies (23.5)

N = Total number of scores (50)

S = Sum of all the scores below L (3)

f = Number of scores within the interval containing P_p (3)

i = Step interval size (3)

$$P_{10} = 23.5 + \left(\frac{\frac{10 \times 50}{100} - 3}{3} \times 3 \right)$$

$$= 23.5 + 2 = 25.5$$

$$P_{10} = 25.5 \text{ aggressive acts}$$

It follows that 10% of the scores lie below score 25.5

7   Repeat procedure for P_{20}, P_{30}, P_{40}, P_{50}, P_{60}, P_{70}, P_{80}, and P_{90}.

Table 18   Summary table: Percentile points

Percentile rank	Calculation	Percentile point (P_p) (aggressive acts)
10	$23.5 + \left(\frac{\frac{10 \times 50}{100} - 3}{3} \times 3 \right)$	$P_{10} = 25.5$
20	$26.5 + \left(\frac{10 - 6}{6} \times 3 \right)$	$P_{20} = 28.5$
30	$29.5 + \left(\frac{15 - 12}{7} \times 3 \right)$	$P_{30} = 30.8$
40	$32.5 + \left(\frac{20 - 19}{8} \times 3 \right)$	$P_{40} = 32.87$
50	$32.5 + \left(\frac{25 - 19}{8} \times 3 \right)$	$P_{50} = 34.75$
60	$35.5 + \left(\frac{30 - 27}{8} \times 3 \right)$	$P_{60} = 36.62$
70	$38.5 + \left(\frac{35 - 35}{6} \times 3 \right)$	$P_{70} = 38.5$
80	$38.5 + \left(\frac{40 - 35}{6} \times 3 \right)$	$P_{80} = 41.0$
90	$44.5 + \left(\frac{45 - 45}{3} \times 3 \right)$	$P_{90} = 44.5$

7.4 Method 2: Calculating percentile ranks for individual scores

The procedure for computing percentile ranks for individual scores is basically the reverse of that outlined above.

Example Calculate the percentile rank (p) for score 34 (P_p) in the distribution of aggressive acts.

We are required to calculate p when P_p = 34. The formula we employ is as follows:

$$p = \frac{\left(\frac{f}{i}\right) \times (P_p - L) + S}{N} \times 100$$

The symbols have the same meaning as given on page 47, substituting from Table 17.

$$p = \frac{\frac{8}{3} \times (34 - 32.5) + 19}{50} \times 100$$

$$= 46 \text{ (rounded off)}$$

A score of 34 has a percentile rank of 46. That is to say, 46% of the scores lie below score 34.

We can see from the preceding sections that a percentile rank provides us with an indication of relative position within a group. At times however this can be misleading. Take, for example, the Percentile rank summary table (Table 18). In that table we note that to achieve percentile rank 10, an individual needs to obtain a percentile point score of 25.5 aggressive acts. To achieve percentile rank 20, he needs a percentile point score of 28.5 aggressive acts. In other words, a difference of 3 aggressive acts moves the child up 10 ranks. Now consider ranks 40 and 50 in the same table. A difference of 1.88 acts of aggression (34.75–32.87) is sufficient to move a child the 10 ranks from 40 to 50.

Consider the following hypothetical results of two examinations taken by each of four Sociology students.

Table 19 Hypothetical examination results for four sociology students

	Social theory Marks out of 100	Demography Marks out of 100
Student A	95	95
Student B	25	93
Student C	20	92
Student D	19	19

If the percentile ranks were calculated they would be the same on both examinations for each of the four students. The distribution of raw scores within each examination however is quite different. In Social Theory, Student A scores far better than all others, whereas in Demography, Students A, B, and C all score very much alike and obtain far higher marks than Student D.

The lesson is plain. Percentiles can provide us with a measure of rank only, *not* a measure of difference between scores.

If a more accurate and meaningful picture of relative achievement is required, we must take into account not only the ranking of scores but also the differences between them, that is, their variability. We can do this by using measures of relative achievement known as standard scores.

7.5   Standard scores or Z scores

A standard score, or Z score, tells us where any particular score lies in relation to the mean score of its distribution. Not only does a standard score indicate whether a particular score lies above or below the mean; it shows how far above or below the mean that score is located.

Before showing how to calculate standard scores it is important that the reader has a thorough grasp of the rationale behind them.

Let's reconsider the mathematical properties of a normal curve that we touched upon in Section 5.14.

Recall that one of the properties of a normal curve is that a score can be placed above or below the mean of a distribution in terms of a standard unit of distance and that the percentage of scores above or below that score can be estimated.

The hypothetical example we provided in our discussion of the normal curve was to do with shoe sizes. We showed how shoe sizes could be described in terms of standard deviations from the mean.

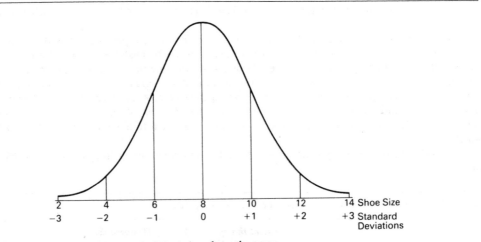

Figure 16   Shoe sizes shown as standard deviations from the mean.

In Figure 16 we see that the average shoe size is 8 and that shoe size 10 is one standard deviation, that is, two shoe sizes, above the mean. Size 12 is two standard deviations or four shoe sizes above the mean.

It follows that shoe size 9 is 0.5 standard deviations or 1 shoe size above the mean.

Our example shows that if we position scores in standard deviation units above or below the mean of a normal distribution, then not only do we get a measure of rank, but also a standard measure of the distance between the scores. The difference between a score one standard deviation above the mean and a score two standard deviations above the mean is the same as the difference between a score three standard deviations above the mean and one at two standard deviations above the mean.

The distance a score lies above or below the mean of a distribution, measured in standard deviations units, is called its *standard score* or Z score.

If we convert actual raw scores into standard scores we obtain an accurate picture of that score's relative position in a distribution.

To convert a raw score into a standard score we need first to calculate the mean and the standard deviation of the distribution. Both these procedures are outlined in Sections 5.2 and 6.4. The mean and standard deviation are then substituted into the formula:

$$Z = \frac{X - M}{S.D.}$$

where

$$Z = \text{standard score}$$

$$X = \text{raw score}$$

$$M = \text{mean}$$

$$S.D. = \text{standard deviation.}$$

7.6 Example 1

The mean score on a test is 50 and the standard deviation is 10. What is the standard score for John who scores 65? (Assume a normal distribution.)

$$Z = \frac{X - M}{S.D.}$$

$$Z = \frac{65 - 50}{10} = +1.5$$

John's score is 1.5 standard deviations above the mean. It can be represented diagrammatically as:

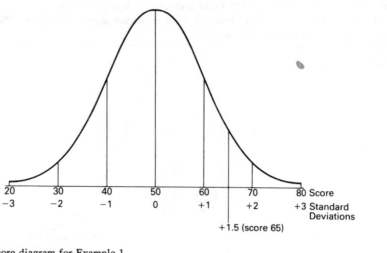

Figure 17 Standard score diagram for Example 1.

Scores can, of course, have negative Z values. All those scores below the mean have negative Z values; all those above the mean have positive Z values.

7.7 Example 2

As part of an apprenticeship selection assessment, school leavers are required to take three tests consisting of Test 1 (numeracy), Test 2 (literacy), and Test 3 (general knowledge).

Given the following results for candidate K, and assuming normal distribution of results, on which of the three tests does that individual do best?

Table 20 Numeracy, literacy, and general knowledge scores

	Raw score	Mean	Standard deviation
Test 1	87	75	12
Test 2	16	13	2
Test 3	31	34	10

1 Calculate Z for numeracy test

$$Z = \frac{X - M}{\text{S.D.}} = \frac{87 - 75}{12} = +1$$

2 Calculate Z for literacy test

$$Z = \frac{X - M}{\text{S.D.}} = \frac{16 - 13}{2} = +1.5$$

3 Calculate Z for general knowledge test

$$Z = \frac{X - M}{\text{S.D.}} = \frac{31 - 34}{10} = -0.3$$

Candidate K achieves better results on the literacy test than on the numeracy or the general knowledge tests in as much as his Z score shows that he is further above average for Test 2. Candidate K's results can be represented diagrammatically:

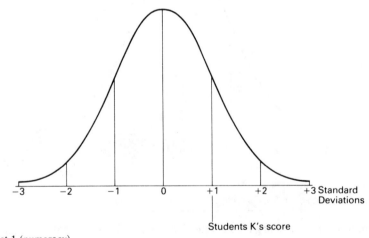

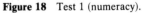

Students K's score

Figure 18 Test 1 (numeracy).

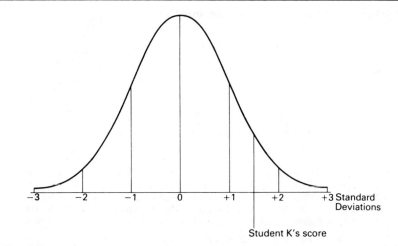

Student K's score

Figure 19   Test 2 (literacy).

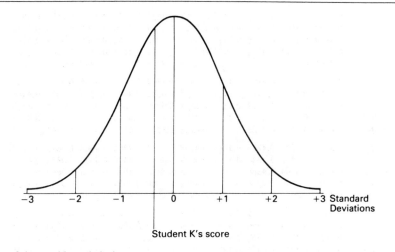

Student K's score

Figure 20   Test 3 (general knowledge).

The examples above show how standard scores can be used to identify a score's position in relation to its distance from the mean. This is not the only information we can obtain about the relative standing of a score. If the distribution is normal, we can also estimate how many scores lie above or below our particular score.

From the properties of the normal curve we know that if the area under it is equal to 100%, then within one standard deviation on either side of the mean, 68.26% of scores lie, within two standard deviations lie 95.44% of scores, and within three standard deviations on either side of the mean lie 99.74% of scores (see Section 5.14).

Table 21   Percentage of scores under the normal curve from 0 to Z

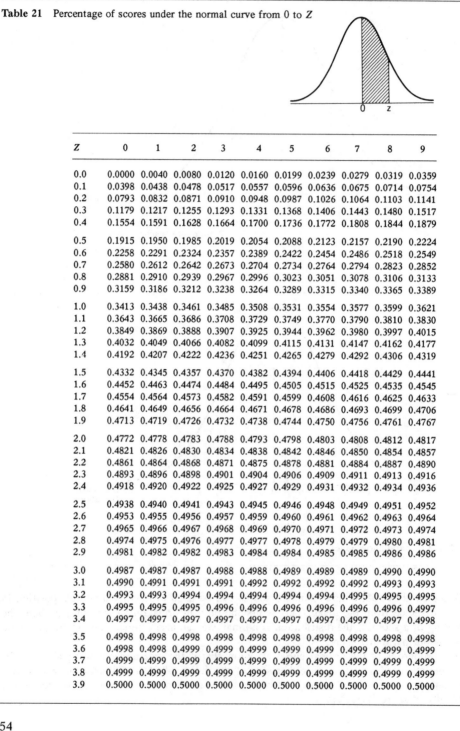

Z	0	1	2	3	4	5	6	7	8	9
0.0	0.0000	0.0040	0.0080	0.0120	0.0160	0.0199	0.0239	0.0279	0.0319	0.0359
0.1	0.0398	0.0438	0.0478	0.0517	0.0557	0.0596	0.0636	0.0675	0.0714	0.0754
0.2	0.0793	0.0832	0.0871	0.0910	0.0948	0.0987	0.1026	0.1064	0.1103	0.1141
0.3	0.1179	0.1217	0.1255	0.1293	0.1331	0.1368	0.1406	0.1443	0.1480	0.1517
0.4	0.1554	0.1591	0.1628	0.1664	0.1700	0.1736	0.1772	0.1808	0.1844	0.1879
0.5	0.1915	0.1950	0.1985	0.2019	0.2054	0.2088	0.2123	0.2157	0.2190	0.2224
0.6	0.2258	0.2291	0.2324	0.2357	0.2389	0.2422	0.2454	0.2486	0.2518	0.2549
0.7	0.2580	0.2612	0.2642	0.2673	0.2704	0.2734	0.2764	0.2794	0.2823	0.2852
0.8	0.2881	0.2910	0.2939	0.2967	0.2996	0.3023	0.3051	0.3078	0.3106	0.3133
0.9	0.3159	0.3186	0.3212	0.3238	0.3264	0.3289	0.3315	0.3340	0.3365	0.3389
1.0	0.3413	0.3438	0.3461	0.3485	0.3508	0.3531	0.3554	0.3577	0.3599	0.3621
1.1	0.3643	0.3665	0.3686	0.3708	0.3729	0.3749	0.3770	0.3790	0.3810	0.3830
1.2	0.3849	0.3869	0.3888	0.3907	0.3925	0.3944	0.3962	0.3980	0.3997	0.4015
1.3	0.4032	0.4049	0.4066	0.4082	0.4099	0.4115	0.4131	0.4147	0.4162	0.4177
1.4	0.4192	0.4207	0.4222	0.4236	0.4251	0.4265	0.4279	0.4292	0.4306	0.4319
1.5	0.4332	0.4345	0.4357	0.4370	0.4382	0.4394	0.4406	0.4418	0.4429	0.4441
1.6	0.4452	0.4463	0.4474	0.4484	0.4495	0.4505	0.4515	0.4525	0.4535	0.4545
1.7	0.4554	0.4564	0.4573	0.4582	0.4591	0.4599	0.4608	0.4616	0.4625	0.4633
1.8	0.4641	0.4649	0.4656	0.4664	0.4671	0.4678	0.4686	0.4693	0.4699	0.4706
1.9	0.4713	0.4719	0.4726	0.4732	0.4738	0.4744	0.4750	0.4756	0.4761	0.4767
2.0	0.4772	0.4778	0.4783	0.4788	0.4793	0.4798	0.4803	0.4808	0.4812	0.4817
2.1	0.4821	0.4826	0.4830	0.4834	0.4838	0.4842	0.4846	0.4850	0.4854	0.4857
2.2	0.4861	0.4864	0.4868	0.4871	0.4875	0.4878	0.4881	0.4884	0.4887	0.4890
2.3	0.4893	0.4896	0.4898	0.4901	0.4904	0.4906	0.4909	0.4911	0.4913	0.4916
2.4	0.4918	0.4920	0.4922	0.4925	0.4927	0.4929	0.4931	0.4932	0.4934	0.4936
2.5	0.4938	0.4940	0.4941	0.4943	0.4945	0.4946	0.4948	0.4949	0.4951	0.4952
2.6	0.4953	0.4955	0.4956	0.4957	0.4959	0.4960	0.4961	0.4962	0.4963	0.4964
2.7	0.4965	0.4966	0.4967	0.4968	0.4969	0.4970	0.4971	0.4972	0.4973	0.4974
2.8	0.4974	0.4975	0.4976	0.4977	0.4977	0.4978	0.4979	0.4979	0.4980	0.4981
2.9	0.4981	0.4982	0.4982	0.4983	0.4984	0.4984	0.4985	0.4985	0.4986	0.4986
3.0	0.4987	0.4987	0.4987	0.4988	0.4988	0.4989	0.4989	0.4989	0.4990	0.4990
3.1	0.4990	0.4991	0.4991	0.4991	0.4992	0.4992	0.4992	0.4992	0.4993	0.4993
3.2	0.4993	0.4993	0.4994	0.4994	0.4994	0.4994	0.4994	0.4995	0.4995	0.4995
3.3	0.4995	0.4995	0.4995	0.4996	0.4996	0.4996	0.4996	0.4996	0.4996	0.4997
3.4	0.4997	0.4997	0.4997	0.4997	0.4997	0.4997	0.4997	0.4997	0.4997	0.4998
3.5	0.4998	0.4998	0.4998	0.4998	0.4998	0.4998	0.4998	0.4998	0.4998	0.4998
3.6	0.4998	0.4998	0.4999	0.4999	0.4999	0.4999	0.4999	0.4999	0.4999	0.4999
3.7	0.4999	0.4999	0.4999	0.4999	0.4999	0.4999	0.4999	0.4999	0.4999	0.4999
3.8	0.4999	0.4999	0.4999	0.4999	0.4999	0.4999	0.4999	0.4999	0.4999	0.4999
3.9	0.5000	0.5000	0.5000	0.5000	0.5000	0.5000	0.5000	0.5000	0.5000	0.5000

When the distances of successive points from the mean of a normal distribution are known and are measured in standard deviation units it is possible to estimate the percentage of scores that lie between the mean and these various points. This information is usually given in the form of a table such as the one set out on the previous page.

Once we have calculated a Z score we can estimate how many scores lie between the mean and our particular score. We do this by reference to Table 21.

Take, for example, a Z score of 0.8. By reference to Table 21 we see that 28.81% of the scores lie between a Z of 0.8 and the mean. We know also that 78.81% of all scores lie below our score. That is, 50.00% + 28.81 % = 78.81 % as shown in the figure below.

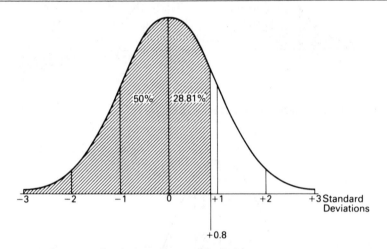

Figure 21   Percentage of scores under the normal curve (Example).

7.8   More examples

1   Calculate the percentage of individuals who score below 32 when the mean of the distribution is 48 and the standard deviation is 10.

$$Z = \frac{32-48}{10} = -1.6$$

From Table 21 we see that 44.52% score between the mean and −1.6 S.D.'s (that is, score 32). It follows that 50.00%−44.52% = 5.48% of the scores lie below score 32. We diagram the calculation in Figure 22 below.

2   If the mean of a distribution is 96 and the standard deviation is 28, what percentage of individuals scored above 110?

$$Z = \frac{110-96}{28} = \frac{14}{28} = +0.5$$

From Table 21 we see that 19.15% score between the mean and +0.5 S.D.'s (that is, score 110). It follows that 100% − (50.00% + 19.15%) = 30.85% of the scores lie above score 110. We diagram the calculation in Figure 23 below.

55

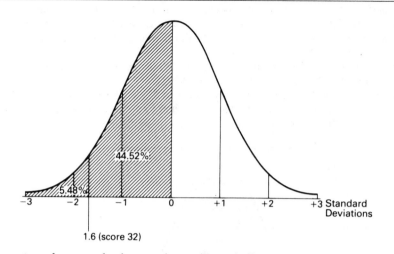

1.6 (score 32)

Figure 22 Percentage of scores under the normal curve (Example 1).

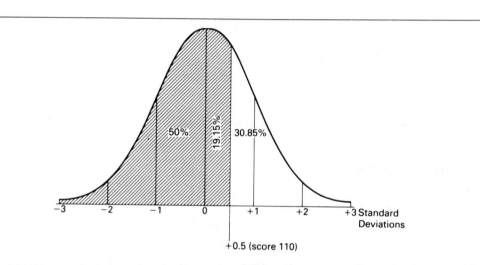

+0.5 (score 110)

Figure 23 Percentage of scores under the normal curve (Example 2).

3 If the mean of a distribution is 45 and the standard deviation is 5, what percentage of individuals score between 40 and 60?

$$Z \text{ for score } 40$$

$$Z = \frac{40-45}{5} = -1$$

From Table 21 we see that 34.13% score between the mean and −1.0 S.D.'s (that is, score 40). We diagram this below.

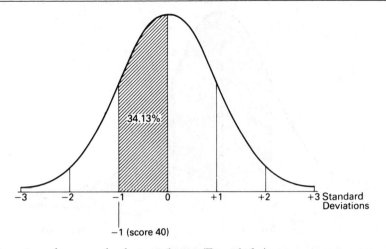

Figure 24 Percentage of scores under the normal curve (Example 3.a).

$$Z \text{ for score } 60$$

$$Z = \frac{60-45}{5} = +3$$

From Table 21 we see that 49.87% score between the mean and +3.0 S.D.'s (that is, score 60). We diagram this below.

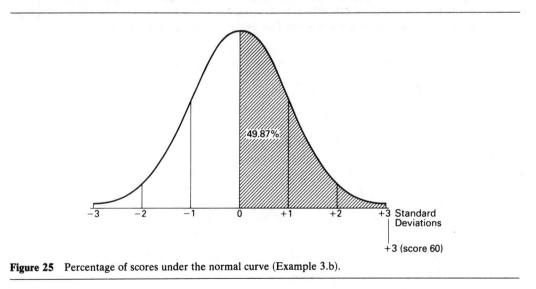

Figure 25 Percentage of scores under the normal curve (Example 3.b).

It follows that 34.13% + 49.87%, that is, 84.00% of the scores lie between scores 40 and 60.

57

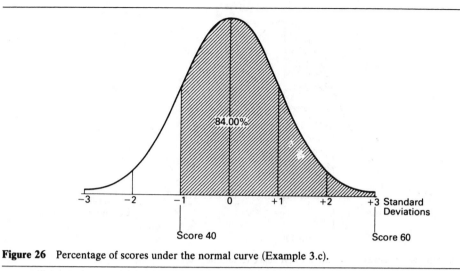

Figure 26   Percentage of scores under the normal curve (Example 3.c).

7.9   Sigma, Hull and *T* scales

One problem in using standard scores as measures of relative position is that we have to work in standard deviation units which may be decimalized, and have positive or negative values.

This can be confusing and may lead to errors. One way round the problem is to transform the *Z* scores so as to arrive at a simple measure of relative position defined in points from 0 to 100.

To make this transformation we have to consider the normal distribution curve, not only in terms of standard deviation units above or below the mean, but as a series of points from 0 to 100 with a mean score set at 50 points. Such a transformation is called a *sigma scale*.

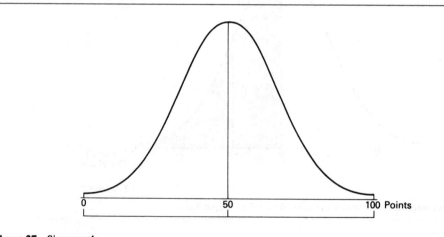

Figure 27   Sigma scale.

7.10 Sigma scale

In the construction of a sigma scale the distribution curve is divided into 100 equal parts along its horizontal axis, commencing with the 0 at 3 standard deviations below the mean and finishing with the 100 at 3 standard deviations above the mean.

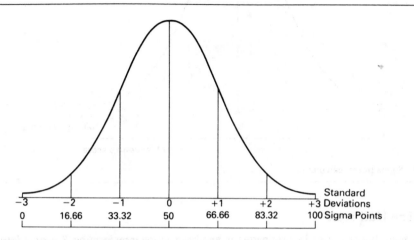

Figure 28 Sigma points and standard deviations.

To calculate the sigma points for a specific test score we use the formula:

$$\text{Sigma points} = 16.66Z + 50$$

where

$$Z = \frac{X - M}{\text{S.D.}}$$

$$X = \text{Raw score}$$

$$M = \text{Mean}$$

$$\text{S.D.} = \text{Standard Deviation}$$

Example Using the data from Example 1 (page 51) we can calculate the points scored by John who gained 65 marks in a test when the mean of the distribution was 50 and the standard deviation 10.

$$\text{Sigma points} = 16.66Z + 50 = 16.66\left(\frac{65 - 50}{10}\right) + 50$$

$$= 24.99 + 50$$

$$= 74.99$$

John scores 74.99 sigma points for the test.

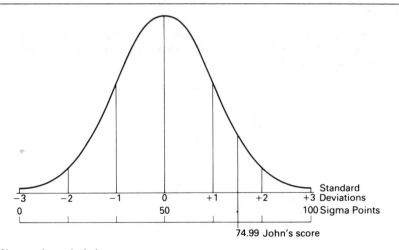

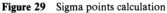

Figure 29 Sigma points calculation

7.11   The Hull scale

The Hull scale (named after its originator) is another way of transforming Z scores into a simpler measure of relative position defined in points from 0 to 100.

In the Hull scale we again divide the distribution curve into 100 equal parts, but this time the starting point (0) is positioned 3.5 standard deviations below the mean and the finishing point (100) 3.5 standard deviations above the mean.

To calculate Hull points for a particular test score we use the formula:

$$\text{Hull points} = 14.28Z + 50$$

where Z has its usual meaning.

Example   Once again using John's mark of 65 in a test when the mean is 50 and the standard deviation is 10 (see page 51), we can calculate how many Hull points he obtains.

$$\text{Hull points} = 14.28Z + 50 = 14.28\left(\frac{65-50}{10}\right) + 50$$

$$= 21.42 + 50$$

$$= 71.42$$

John scores 71.42 Hull points on the test.

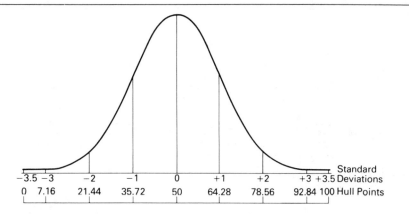

Figure 30 Hull points and standard deviations.

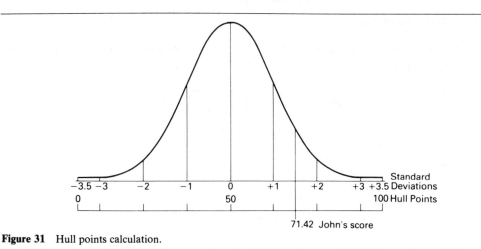

Figure 31 Hull points calculation.

7.12 *T*-scale

Yet another method of scaling which divides the distribution curve into 100 equal parts obtains an even greater spread of raw scores than in the Hull scale. The starting point (0) is placed 5 standard deviations below the mean and the finishing point (100), 5 standard deviations above the mean.

From our previous discussion we know that 99.74% of the scores in a normal distribution lie between ±3 standard deviations from the mean.

Looking at the diagram below, we can see that most raw scores will be given *T*-points between 20 and 80. It might appear that the outside ranges (0–20) and (80–100) are redundant. However the usefulness of the *T*-scale is demonstrated when there are extreme scores in a distribution, that is, scores that are more than 3 standard deviations from the mean.

In a sigma scale no points are made available with the range 0–100 for scores that lie outside ±3 standard deviations from the mean. Using a *T*-scale however, when a score is more than 3 standard deviations away from the mean a points score within the range 0–100 can still be allocated.

61

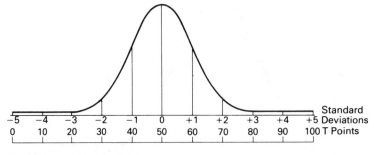

Figure 32 *T*-points and standard deviations.

To calculate *T*-points we use the formula:

$$T\text{-points} = 10Z + 50$$

where Z has its usual meaning.

Example Referring again to John's test mark of 65 when the distribution mean is 50 and standard deviation 10 we can estimate his *T*-points as follows:

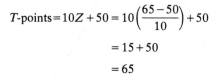

$$T\text{-points} = 10Z + 50 = 10\left(\frac{65-50}{10}\right) + 50$$

$$= 15 + 50$$

$$= 65$$

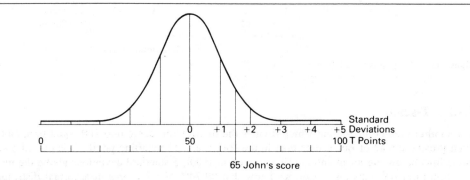

Figure 33 *T*-points calculation.

John scores 65 *T*-points for the test.

7.13   Example problem

Given the following examination results, calculate which student performed better overall (assume normal distribution).

Table 22 Hypothetical examination marks of two Business Studies students, Candidate Y and Candidate Z

	Law Mean = 60 S.D. = 10	Accounting Mean = 55 S.D. = 12	Management Mean = 50 S.D. = 5
Student Y	70	31	60
Student Z	65	43	53

First inspection of the examination results suggests that Student Y and Student Z did equally well, both scoring a total of 161 marks. This would be an incorrect conclusion however, as the distribution of marks for each subject area is different, that is, the means and standard deviations are different.

In order to compare the two sets of marks we must convert them into standardized units of measurement before adding. The relative position or achievement of each student in each subject area must be taken into account.

The standardization of marks can be achieved by methods already described earlier in the chapter. We shall use a T-scale.

Step 1 Calculate T-points gained by both students in each subject area.
(a) Law

$$\text{Student Y} \quad T\text{-points} = 10Z + 50 = 10\left(\frac{70-60}{10}\right) + 50 = 60$$

$$\text{Student Z} \quad T\text{-points} = 10Z + 50 = 10\left(\frac{65-60}{10}\right) + 50 = 55$$

(b) Accounting

$$\text{Student Y} \quad T\text{-points} = 10Z + 50 = 10\left(\frac{31-55}{12}\right) + 50 = 30$$

$$\text{Student Z} \quad T\text{-points} = 10Z + 50 = 10\left(\frac{43-55}{12}\right) + 50 = 40$$

(c) Management

$$\text{Student Y} \quad T\text{-points} = 10Z + 50 = 10\left(\frac{60-50}{5}\right) + 50 = 70$$

$$\text{Student Z} \quad T\text{-points} = 10Z + 50 = 10\left(\frac{53-50}{5}\right) + 50 = 56$$

Step 2 Total the T-points for each student

$$\text{Student Y} = 60 + 30 + 70 = 160$$

$$\text{Student Z} = 55 + 40 + 56 = 151$$

In relation to the overall distribution of scores for the whole examination Student Y achieved better results than Student Z.

63

7.14   Grading

As well as considering individual performance in relation, to that of the total group, we often categorize students into broad bands of achievement, giving those who score within a certain range of specific classification a grade. For example, after collecting test scores, we may wish to award students a grade of A, B, C, D, or E.

If we assume that the scores on our test are normally distributed, we may use the rationale behind the standardization of scores already outlined to construct a 5-grade scale.

First we must divide the distribution curve into 5 equal parts along the horizontal axis with the top of the A grade 3 standard deviations above the mean and the bottom of the E grade 3 standard deviations below the mean. That is, a six standard deviations range within which 99.74% of all the scores will lie.

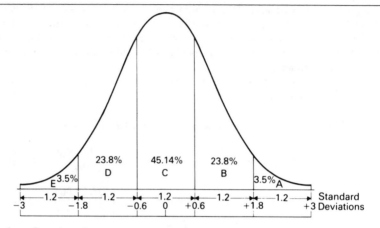

Figure 34   Constructing a 5-grade scale.

The range of each grade, that is, the difference between the upper and the lower limits, will be equal to 1.2 standard deviations (6 standard deviations divided by 5). However this information does not allow us to fix the actual positions of the grades since as yet we have no reference point. If we take the mean of the distribution as the reference point we can now place the limits of the grades accordingly.

A grade—Upper Limit = +3.0 S.D.'s
Lower Limit = +1.8 S.D.'s
B grade—Upper Limit = +1.8 S.D.'s
Lower Limit = +0.6 S.D.'s
C grade—Upper Limit = +0.6 S.D.'s
Lower Limit = −0.6 S.D.'s
D grade—Upper Limit = −0.6 S.D.'s
Lower Limit = −1.8 S.D.'s
E grade – Upper Limit = −1.8 S.D.'s
Lower Limit = −3.0 S.D.'s

Now we need only know the mean and standard deviation of our data in order to fix the upper and lower limits of the grades. The percentage of scores within each grade band can be estimated using Table 21.

7.15   Example

Suppose the mean of a distribution is 65 and the standard deviation 15. What are the limits of the grade bands?

$$
\begin{aligned}
\text{A grade—Upper Limit} &= 65 + (3 \times 15) & = 110 \\
\text{Lower Limit} &= 65 + (1.8 \times 15) & = 92 \\
\text{B grade—Upper Limit} &= & = 92 \\
\text{Lower Limit} &= 65 + (0.6 \times 15) & = 74 \\
\text{C grade—Upper Limit} &= & = 74 \\
\text{Lower Limit} &= 65 - (0.6 \times 15) & = 56 \\
\text{D grade—Upper Limit} &= & = 56 \\
\text{Lower Limit} &= 65 - (1.8 \times 15) & = 38 \\
\text{E grade—Upper Limit} &= & = 38 \\
\text{Lower Limit} &= 65 - (3 \times 15) & = 20
\end{aligned}
$$

Depending upon particular requirements, this method can be applied to give any number of grade bands.

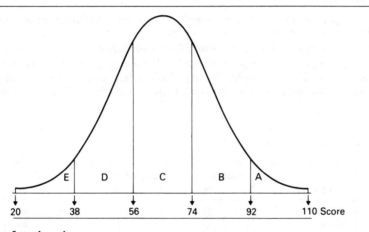

Figure 35   Using a 5-grade scale.

CHAPTER 8
MEASURING ASSOCIATION*

8.1 Introduction

In Chapters 5 'Measuring Typical Achievement', 6 'Measuring Variations in Achievement', and 7 'Measuring Relative Achievement' we have been concerned with measures of a *single quality*, be it neuroticism, numeracy, or need for achievement. Many studies, however, involve measuring *a number of qualities* because people are especially interested in identifying and interpreting *associations among qualities*. Look at Table 23 for example. The hypothetical data there refer to job satisfaction among employees of a municipal hospital. It's clear from Table 23 that job satisfaction is associated with the

Table 23   Job satisfaction in hospital employment

Position	Satisfied		Dissatisfied		Total	
	%	n	%	n	%	n
Doctors	75	30	25	10	100	40
Nurses	50	15	50	15	100	30
Ancillaries	20	6	80	24	100	30

positions that respondents occupy in their hospital employment since the distribution of job satisfaction scores differs for different positions. One definition of association (Weiss 1968) follows directly from our hypothetical example. It is this:

Two qualities are associated when the distribution of values of the one differs for different values of the other.

Look now at Table 24. The hypothetical data there refer to scores on a creativity measure among university students identified by faculty of study. It's equally apparent from Table 24 that

---

* Our outline in this section draws upon that of R. S. Weiss (1968) *Statistics in Social Research*. New York: John Wiley and Sons, Chapter 9.

Table 24 Creativity scores of university students identified by faculty of study

| Faculty of study | Creativity score | | | | Total |
| | Above average | | Below average | | |
	%	n	%	n	n
Science	52	21	48	19	40
Arts	52	16	48	14	30
Languages	52	26	48	24	50

the two qualities, *faculty of study* and *creativity,* are independent since the distribution of values of the one is the same for every value of the other. Put differently we can say:

> *When subgroups do not differ, we have independence;*
> *when subgroups do differ, we have association.*

One way of defining association, then, has to do with the proportions in the subgroups of A and B. However, there are other facets of the concept of association.

Weiss (1968) identifies five alternative ways of viewing association which we now use as a framework for the discussion that follows.

8.2   Departure from independence between two factors

This way of looking at association involves imagining what the data would look like if there were no association and then saying that association is present to the extent that the observed data depart from this. Let's illustrate the approach with an example.

Table 25 Salaried staff's pension fund scheme preferences

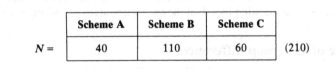

	Scheme A	Scheme B	Scheme C	
N =	40	110	60	(210)

Table 25 reports the findings of a survey by the personnel department of a large manufacturing corporation into salaried staff's preferences for three different pension fund schemes. The personnel department wishes to know whether staff preference is associated with any one particular scheme. What would the above data look like, we ask ourselves, if there were no association between staff preference and any particular pension fund scheme? Theoretically, the choices of the 210 respondents would be equally divided among the three schemes as shown in Table 26.

Table 26 The expected choices of salaried staff given no particular preference for Scheme A, B or C

	Scheme A	Scheme B	Scheme C	
N =	70	70	70	(210)

Our task then is to ascertain the extent to which the *observed* data (Table 25) depart from the *expected* data (Table 26). Chi square is appropriate to our problem.

Chi square, probably the most commonly used of all statistical techniques, allows us to ascertain the extent of the difference between observed frequencies and expected or theoretical frequencies.

Chi square is given by:

$$\chi^2 = \Sigma \frac{(O-E)^2}{E}$$

where O = the observed frequency in each category and E = the expected frequency in each category. The data from Tables 25 and 26 are summarized in Table 27.

Table 27   Observed and expected frequencies in staff preferences for pension fund schemes

Frequency	Scheme A	Scheme B	Scheme C
Observed (O)	40	110	60
Expected (E)	70	70	70

$$\chi^2 = \frac{(40-70)^2}{70} + \frac{(110-70)^2}{70} + \frac{(60-70)^2}{70} = 12.86 + 22.86 + 1.43$$

$$= 37.15$$

We can leave the interpretation of our obtained chi square value until we have read the section on Inferential Statistics that follows. For the moment, it is sufficient to note that we illustrate the use of various measures of association based on the chi square statistic on pages 133, 186, 243, and 278.

8.3   Magnitude of subgroup differences

The second way of viewing association identified by Weiss (1968) involves looking at subgroup differences. Assuming that there is some association, its degree can be measured by comparing subgroup proportions with each other. Look at the following example. What information can we glean from Table 28?

Table 28   Percentage of July school leavers employed and unemployed in December by educational level

Employment status	Educational Level	
	G.C.S.E. or N.V.Q.	No examination qualifications
Employed	87	38
Unemployed	13	62
Total	100	100

By comparing percentages in different columns of the same row we see that 49% more qualified school leavers get jobs than do unqualified school leavers.

By comparing percentages in different rows of the same column we see that 74% more qualified school leavers are employed than are unemployed.

The data suggest an association between the educational qualifications of school leavers and their success in finding jobs, the first variable (educational qualifications) implying the second (employment), *not* the other way round.

The *percentage difference,* as we have outlined above, reflects this second way of viewing association. It consists of comparing percentages in columns of the same row, or conversely, in different rows of the same column. The percentage difference is an *asymmetric* measure of association. An asymmetric measure is a measure of *one-way association.* That is to say, it measures the extent to which one phenomenon implies the other but not vice versa. Measures which are concerned with the extent to which two phenomena imply each other are referred to as *symmetric* measures.

A second way in which we can make use of the data in Table 28 is to compute a *percentage ratio* (%R). Take the information in the second row of the table for instance.

By dividing 62 by 13 (%R = 4.8) we can say that almost five times as many non-qualified youngsters are unemployed as are employed within six months of leaving school.

The percentage difference ranges from 0% when there is complete independence to 100% when there is complete association in the direction being examined. It's easy to calculate and simple to comprehend. Notice however that the percentage difference as we have defined it can only be employed when there are only two categories in the variable along which we percentage and only two categories in the variable in which we compare.

8.4 Summary of pair-by-pair comparisons

A third approach to association requires us to think of forming all possible comparisons of one member of the population with another. In these comparisons we decide whether the two qualities under study occur together or not. We then summarize the results of all these pair-by-pair comparisons, association being measured by the preponderance of one type of pair.

Look at Table 29.

Table 29 Pair-by-pair comparison procedure in measuring association

		Quality A	
		Present	Absent
Quality B	Present	a	b
	Absent	c	d

In the pair-by-pair comparison approach to association we examine every possible pair in our sample to determine the extent to which quality A is associated with quality B. Thus, if one member of a pair possesses both quality A and quality B while the other possesses neither, we count it as one pair in which there is *positive association.* In the case of one member of a pair possessing quality A but not quality B and the other possessing quality B but not quality A, we count the pair as one in which there is *negative association.* Where both members of a pair possess either quality A or quality B, we infer for present purposes, that the pair cannot help us determine whether the qualities are associated. From Table 29 then, all pairs in which one member is from Cell a and one member is from Cell d show *positive association* between qualities A and B. Similarly, all pairs in which one member is from Cell b and one member is from Cell c show *negative association* between qualities A and B. Pairs formed from other combinations of cells are alike in possessing one or both qualities and are irrelevant to our present purpose.*

This focus upon Cells a and d on the one hand, and Cells b and c on the other is the rationale of Yule's Q, a simple pair-by-pair comparison measure of association developed especially for 2 x 2 tables where both measures are dichotomous.

Q is given by the formula:

$$Q = \frac{ad - bc}{ad + bc}$$

An example of using Yule's Q in measuring association is outlined on pages 184 to 185.

What of the case, however, when data are cast into other than 2 x 2 contingency tables? Take for example, the ranking of 60 university students on two qualities, level of creativity and first-year English examination results shown in Table 30.

Table 30   60 university students ranked on level of creativity and first-year English examination results

| | | Creativity | | | |
		High	Medium	Low	Total
English	High	12	6	2	20
Examination	Medium	8	10	2	20
Results	Low	0	4	16	20
	Total	20	20	20	60

The total number of pair-by-pair comparisons is $(60 \times 59)/2 = 1770$ different pairs! Clearly listing and counting this number of pairings is both tedious and impracticable. Weiss (1968) suggests the following method of computing the number of positive and negative pairs. If we take *any* student from the high-high cell and *any* student from the medium-medium cell, then the result is a positive pair. In our table there are 12 students in the high-high cell and 10 in the medium-medium cell. It follows therefore that we can constitute 120 different pairs, all positive, of students from these two cells. Furthermore we can form only positive pairs if we take students from the high-high cell and students from the low-low cell. Together, these two cells form 192 positive pairs. In general, Weiss observes, so long as the second member of the pair comes from a cell to the right of, and below, the cell of the first pair, the result is a *positive pair.*

Similarly, if the second member of a pair comes from a cell to the left of, and below, the cell of the first member of a pair, then the result is a *negative pair.*

The computing of positive and negative pairs can be systematized by developing P (positive pairs) and Q (negative pairs) tables as we show in connection with our data on university students' rankings on creativity and English examination results.

---

* Measures of association do exist, of course, that make use of such additional information as pairs that are 'tied on A but not on B', 'tied on B but not on A', and 'tied on both A and B'. We deal with these in detail on pages 264 to 270.

Table 31 60 university students, ranked on level of creativity and first-year English examination results: calculating P and Q

		Creativity				P Table		Q Table	
		High	**Medium**	**Low**	**Total**				
English	High	12	6	2	20	384	108	48	44
Examination	Medium	8	10	2	20				
Results	Low	0	4	16	20	160	160	0	8
	Total	20	20	20	60	$P = 812$		$Q = 100$	

Cell values for the P table are computed by multiplying the number of cases in a cell by the sum of all the cases in cells to the right of, and below, that cell. Thus the P table is constituted as follows:

$$12 \times (10 + 2 + 4 + 16) = 384$$

$$6 \times (2 + 16) = 108$$

$$8 \times (4 + 16) = 160$$

$$10 \times (16) = 160$$

Cell values for the Q table are computed by multiplying the number of cases in a cell by the sum of all the cases in cells to the left of, and below, that cell. Thus the Q table is constituted as follows:

$$2 \times (8 + 10 + 0 + 4) = 44$$

$$6 \times (8 + 0) = 48$$

$$2 \times (0 + 4) = 8$$

$$10 \times (0) = 0$$

The results of the calculation of the P and the Q tables shows a preponderance of 668 positive pairs (i.e. $812 - 144 = 668$). This preponderance of positive pairs, $(P - Q)$, is referred to as S, the 'sum of all pairs'. In the case where there are more negative pairs than positive pairs, S takes a negative value.

One commonly used measure of association involving S is Goodman-Kruskal's gamma (γ) which is given by the formula:

$$\gamma = \frac{P - Q}{P + Q} \quad \text{or} \quad \frac{S}{P + Q}$$

In the case of the data on university students' creativity and English examination results,

$$\gamma = \frac{712}{812 + 100} = .78$$

Gamma is a measure of association that can range between -1.0 and $+1.0$. In our hypothetical example, we may infer that a gamma of .70 indicates a relatively strong association between creativity and success in the English examination.

A diagrammatic summary of the computation of gamma is given by Riley (1963).

To find P values, i.e., positive pairs

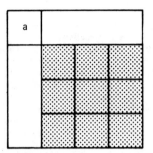

Step 1. Start with the upper left-hand cell *(a)*. Add all the frequencies in the shaded area in the diagram and multiply them by the cell *(a)* absolute number. The shaded area excludes the row and column containing the upper left-hand cell.

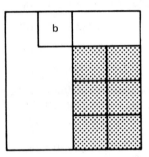

Step 2. Then take the adjacent cell in the same row *(b)*. Add all the frequencies in the shaded area, and multiply them by the cell *(b)* figure. The shaded area excludes the row and column containing cell *(b)* and the column to the left of the column containing cell *(b)*.

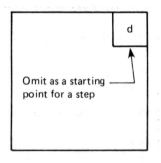

Step 3. Then go on across the top row always adding the frequencies that are simultaneously to the right of and below the cell and always excluding cells in the same row and same column, and all cells above and to the left of these rows and columns. Omit cells in the last column (for example, *d*)—since there are no cells to the right of them—from the role of starting point for a step.

72

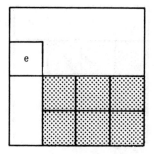

Step 4. Return to the next row and cell (*e*). Add all the frequencies in the shaded area and multiply by the cell (*e*) figure. The shaded area excludes Row 2 and Column 1 (containing cell *e*) and also the row above.

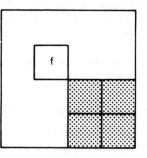

Step 5. Move to cell (*f*) and so forth. Note that what you are doing, essentially, is summing up all the chances of drawing a joint frequency of *like order* with respect to a given cell.

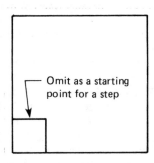

Step 6. Continue until you have used up all the cells for which you can find cells to the right and below. This means that you will omit not only the last column but also the bottom row, since there are no cells below it.

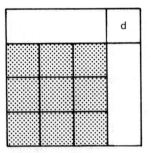

Step 7. Now add up all the products that result from multiplying cells by the sum of the frequencies below and to the right of them (Steps 1 through 6).

To find Q values, i.e. negative pairs, simply reverse the process described for finding P above. Start this time with the upper right-hand cell (here cell d), and add the frequencies to the left of it and below, as in the shaded area. Continue according to the procedure of Steps 1 to 6, simply reversing the order of the moves. Stop with the left-hand column and bottom row this time, neither of which have cells both to the left and below. Add up all the products that result from multiplying cells by the sums of the frequencies below and to the left of them, as in Step 7.

Other measures of association involving S, 'the sum of all the pairs', are Kendall's tau and Somer's d. We discuss these measures on pages 264 and 265.

8.5 Proportional reduction in error measures of association

A fourth approach to measuring association proposed by Weiss (1968) requires us to imagine that we are called upon to predict whether quality A exists or not, first without information regarding B, and then with information regarding B. The rationale of this fourth approach is as follows.

Some measures describe the association between two variables, A and B, in terms of the extent to which information about B helps in the prediction of A. When A and B are independent of each other, of course, information about B is of no help in making guesses about A. When A and B are closely connected, however, knowing about B may be very useful in guessing A.

It's been suggested that we can take as a measure of association the extent to which knowing B helps predict A. How much, we can ask, does knowledge of B *proportionally reduce the probable error* in predicting A? Such measures are commonly referred to as P.R.E. statistics, P.R.E. standing for *proportional reduction in error.*

Following Weiss (1968) let's call our P.R.E. statistic, g, for the moment and see how it can be demonstrated in a concrete example.

Table 32 gives details of the school backgrounds of commissioned and non-commissioned officers in a famous Regiment of Guards.

Looking at the data in Table 32, suppose we are told only that an individual is a member of the Regiment of Guards. How should we guess whether he is a commissioned officer (CO) or a non-commissioned officer (NCO)? Since 30 of the individals identified in the table are COs and 33 are NCOs, on balance we would be marginally better off guessing that he is an NCO. Suppose now, however, that we are given the added information that the individual about whose rank we are required to guess attended a state secondary rather than an independent public school. Clearly, our best guess now is that the person in question is an NCO because of the 36 individuals educated in the state system 31 are NCOs. The added information has enabled us to make a far more accurate guess about his regimental designation.

Table 32 Secondary school background and rank in a Regiment of Guards

School background	Commissioned officers	Non-commissioned officers	Total
Public school	25	2	27
State secondary	5	31	36
Total	30	33	63

How much better is our guess now that we know the school background of the person? Without this additional information, we would be wrong $\frac{30}{63}$ of the time—roughly half the time. With knowledge of school background we shall be mistaken in only $\frac{5}{36}$ of the cases of persons with a state secondary background and in only $\frac{2}{27}$ of the cases of persons with a public school background, that is, a total of 7 cases in 63, a reduction of probable error of 23 cases from the original 30.

We can take this proportional reduction of probable error as our measure of association. In the example we have computed the reduction in probable error in predicting regimental designation when school background is given. We refer to this as (g) regimental designation/school background or 'the g of regimental designation given school background.'

The formula for g when we are given the *row variable* (school background) and are predicting the column variable (regimental designation) is given by:

$$g \text{ (column variable/row variable)} = \frac{\sum_{\substack{\text{over all} \\ \text{rows}}} \left[\begin{array}{c} \text{largest cell} \\ \text{frequency in} \\ \text{rows} \end{array} \right] - \begin{array}{c} \text{largest column} \\ \text{total} \end{array}}{n - \text{largest column total}}$$

Substituting the values from Table 32.

$$g \text{ (regimental designation/school background)} = \frac{(25+31) - 33}{63 - 33}$$

$$g = \underline{0.77}$$

What exactly does $g = 0.77$ mean? Simply that when we know an individual's school background, there is an overall reduction of 77% in probable error in guessing whether he holds commissioned or non-commissioned rank in the Regiment of Guards.

What of the converse situation, g (school background/regimental designation)? What will be the proportional reduction in error in guessing an individual's school background once we know his regimental rank? The formula in this case is:

$$g \text{ (row variable/column variable)} = \frac{\sum_{\substack{\text{over all} \\ \text{columns}}} \left[\begin{array}{c} \text{largest} \\ \text{frequency in} \\ \text{columns} \end{array} \right] - \begin{array}{c} \text{largest} \\ \text{row} \\ \text{total} \end{array}}{n - \text{largest row total}}$$

$$g \text{ (row variable/column variable)} = \frac{(25+31) - 36}{63 - 36}$$

$$g = \underline{0.74}$$

Notice that this is a different value for g from that computed for g (regimental designation/school background). This is because in both cases our P.R.E. measures are *asymmetric*, that is to say, they produce different values for a table depending on the *direction of association* in which we are interested.

8.6 Measures involving correlation

A fifth way of measuring association identified by Weiss (1968) has to do with the extent to which increments in one factor occur together with increments in the other.

Variables are often related in such a way that an increase in one is accompanied by an increase in the other, as for example in the relationship between intelligence test scores and achievement in mathematics. Conversely, variables may be related in such a way that an increase in one is accompanied by a decrease in the other, the relationship between age and short term memory being a case in point. Where a relationship can be further specified in terms of an increase or decrease of a *certain number of units* in the one variable producing an increase or decrease of a *related number of units* of the other, then it is more apt to refer to that relationship as a *correlation* rather than association.

A correlation, then, refers to a quantifiable relationship between two variables and the statistic that provides an index of that relationship is called a *correlation coefficient.*

The correlation coefficient is a measure of the *linear relationship** between two variables. Where variables are linearly related it's possible to produce a graph of observations showing values of one variable for specific values of the other. Such graphs are known as *scatter diagrams* or *scatterplots.*

The meaning of statistical correlation

A scatter diagram is simply a graph on which points are placed to represent pairs of values for two variables. Look at Table 33 showing the Final Year Examination results of ten social work students.

Table 33 Hypothetical examination scores of ten social work students

Student	Social psychology	Counselling	Government/Law	Sociology
A	75	75	45	71
B	70	70	50	45
C	70	70	50	56
D	65	65	55	50
E	60	60	60	60
F	60	60	60	70
G	55	55	65	70
H	50	50	70	50
I	50	50	70	65
J	45	45	75	51

In scatter diagrams 1, 2, and 3, we have plotted pairs of values to show the relationship between students' scores in Social Psychology and Counselling, Counselling and Government/Law, and Counselling and Sociology, respectively.

---

*   A correlation coefficient has been developed to measure *non-linear* relationships, too. (See pages 84 to 86.)

It appears that, in the Social Psychology and Counselling examinations, each individual scores an identical mark on both papers, a truly remarkable result rarely found outside of fictitious data such as ours. Scatter diagram 1 shows a 'perfect' positive or direct correlation between the two sets of scores.

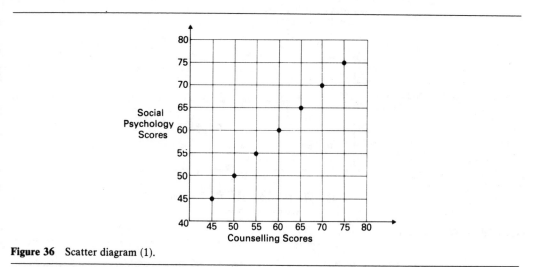

Figure 36 Scatter diagram (1).

Equally surprising (and fictitious) we find the very reverse of the Scatter diagram 1 situation when we look at the relationship between Counselling and Government/Law results. In this latter case (Scatter diagram 2), there is a 'perfect' negative or inverse correlation between the two sets of examination scores.

The reason for the term *scatter* is well illustrated when we plot the relationship between Counselling and Sociology examination results. Students who do well in Counselling are just as likely to do well or badly in Sociology. There is no obvious relationship between the two sets of scores.

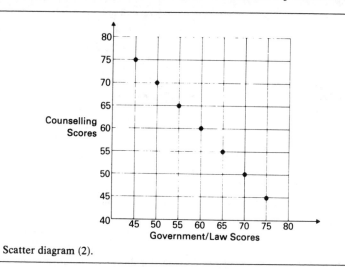

Figure 37 Scatter diagram (2).

Scatter diagram 3 shows a zero, or a near zero, relationship. We may conclude that the two variables are independent.

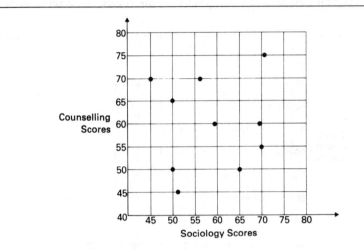

Figure 38   Scatter diagram (3).

Few if any variables in everyday life are perfectly related to the degree shown in Scatter diagrams 1 and 2. Scatter diagrams 4, 5, 6 and 7 demonstrate the sorts of relationships more usually found.* Notice in Scatter diagrams 4 and 5 that although the points do not fall along perfectly straight lines, trends are nevertheless apparent.

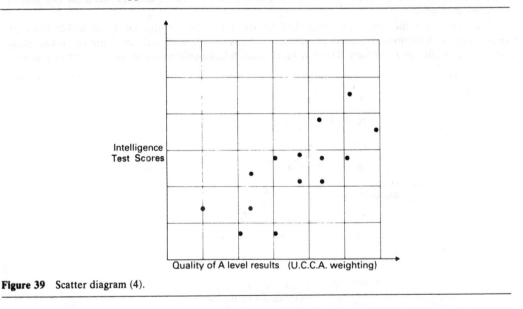

Figure 39   Scatter diagram (4).

---

*   The data in Scatter diagrams 4, 5, 6, and 7 are hypothetical and for illustrative purposes only.

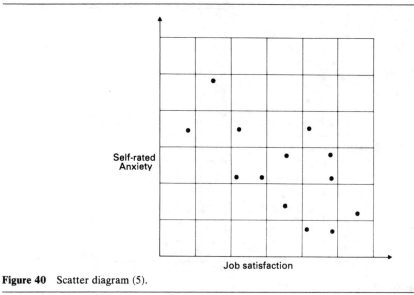

Figure 40   Scatter diagram (5).

Sometimes relationships between variables assume *curve-like shapes* when plotted on scatter diagrams. We call these *curvilinear relationships*.

Scatter diagram 6 shows a positive or direct curvilinear correlation such as might be found in a developmental psychologist's study of the growth of vocabulary in very young children.

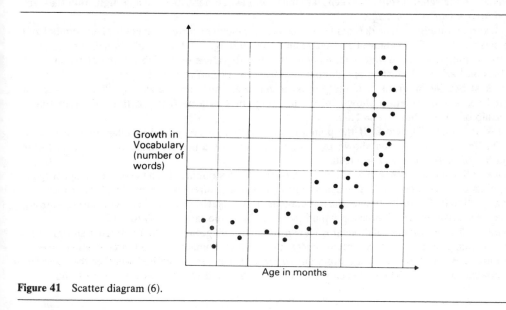

Figure 41   Scatter diagram (6).

Scatter diagram 7 illustrates a negative or inverse curvilinear correlation such as might be found in a study of the relationship between age and rate of growth.

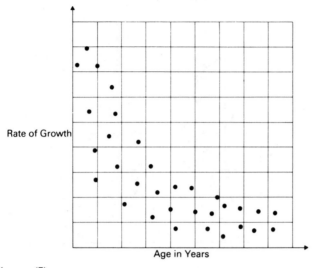

Rate of Growth

Age in Years

Figure 42 Scatter diagram (7).

We said earlier that a *correlation coefficient* provides an index of the degree to which two variables are related. When relationships between variables can best be described by a straight line they are referred to as *linear relationships.*

Linear relationships can be determined by the product moment correlation coefficient, symbolized by the small letter r. The values of r range from +1.00, through 0.00, to −1.00. In using r, we must make two assumptions about our data, namely, that for all values on Variable 1 the distribution of the values on Variable 2 must be approximately *normal* and *equal in variability.* *

Look at Scatter diagram 1. If the points in that scatterplot are joined together they form a straight line. The correlation shown in Scatter diagram 1 is a perfect positive or direct linear relationship of +1.00; that is, $r = 1.00$.

Look at Scatter diagram 2. If the points in that scatterplot are joined together they also form a straight line. The correlation shown in Scatter diagram 2 is a perfect negative or inverse linear relationship of −1.00, that is, $r = -1.00$.

Look at Scatter diagram 4. If we find the 'best fit' that a straight line can take with respect to all of the points in the scatterplot, it represents a strong positive or direct linear relationship, probably in the range 0.80 to 0.90. Similarly with Scatter diagram 5. A 'best fit' straight line representing a strong negative or inverse linear relationship would probably be in the range −0.60 to −0.70.

What of the curvilinear relationships shown in Scatter diagrams 6 and 7? Since a straight line 'best fit' is inappropriate in such circumstances, the product moment correlation coefficient, too, is inappropriate. What is needed is a correlation coefficient that can be applied whether the 'best fit' line is straight or curved. There is such a coefficient. It is called *eta* (η). (See pages 84 to 86.)

---

* The technical term for equality in variance is *homoscedasticity, (homo* meaning equal, *scedasticity* meaning scattering).

8.7 Calculating the product moment correlation coefficient, r

Method 1

There are a number of ways of calculating r, each suitable for particular types of data. The first method shown here is used when the data are in the form of raw scores and at the interval or the ratio level of measurement.

The formula for r is given as:

$$r = \frac{n \sum XY - (\sum X)(\sum Y)}{\sqrt{[n \sum X^2 - (\sum X)^2][n \sum Y^2 - (\sum Y)^2]}}$$

where

r = the product moment correlation coefficient

n = the number of pairs of scores

X = each of the scores on the first variable

Y = each of the scores on the second variable

Σ = the 'sum of'.

The data in Table 34 below show the scores of ten employees, randomly selected by a team of occupational psychologists studying work attitudes and psychological well-being. The respondents have completed two measures—a work involvement schedule and a job-satisfaction inventory.

Table 34 Calculating the product moment correlation coefficient (Method 1)

Work involvement X	Job satisfaction Y	X^2	Y^2	XY
43	4	1849	16	172
55	5	3025	25	275
67	6	4489	36	402
38	4	1444	16	152
49	5	2401	25	245
70	7	4900	49	490
80	9	6400	81	720
62	5	3844	25	310
73	6	5329	36	438
83	9	6889	81	747
$\sum X = 620$	$\sum Y = 60$	$\sum X^2 = 40570$	$\sum Y^2 = 390$	$\sum XY = 3951$

$$r = \frac{10\,(3951) - (620)(60)}{\sqrt{[10\,(40{,}570) - (620)^2][10\,(390) - (60)^2]}}$$

$$= \frac{39{,}510 - 37{,}200}{\sqrt{[405{,}700 - 384{,}400][3900 - 3600]}} = \frac{2310}{\sqrt{6{,}390{,}000}} = \frac{2310}{2528} = 0.91$$

8.8 Calculating the product moment correlation coefficient, r

Method 2

The second method of calculating r is short and simple, employing deviations from the means of all the X and Y values respectively. The formula is given as:

$$r = \frac{\Sigma xy}{\sqrt{(\Sigma x^2)(\Sigma y^2)}}$$

where

r = the product moment correlation coefficient

x ≐ the deviation of any X value from the mean of all the X values

y = the deviation of any Y value from the mean of all the Y values

Σ = the 'sum of'.

Table 35 shows the computation of the correlation between the work involvement and the job satisfaction scores of ten employees as in Table 34 above.

Table 35 Calculating the product moment correlation coefficient (Method 2)

Work involvement X	Job satisfaction Y	x	y	x^2	y^2	xy
43	4	−19	−2	361	4	38
55	5	−7	−1	49	1	7
67	6	+5	0	25	0	0
38	4	−24	−2	576	4	48
49	5	−13	−1	169	1	13
70	7	+8	+1	64	1	8
80	9	+18	+3	324	9	54
62	5	0	−1	0	1	0
73	6	+11	0	121	0	0
83	9	+21	+3	441	9	63
$\Sigma X = 620$ $M_X = \frac{620}{10} = 62$	$\Sigma Y = 60$ $M_Y = \frac{60}{10} = 6$	$\Sigma x = 0$	$\Sigma y = 0$	$\Sigma x^2 = 2130$	$\Sigma y^2 = 30$	$\Sigma xy = 231$

$$r = \frac{\Sigma xy}{\sqrt{(\Sigma x^2)(\Sigma y^2)}} = \frac{231}{\sqrt{(2130)(30)}} = \frac{231}{\sqrt{63,900}} = \frac{231}{252.8} = 0.91$$

As a rough and ready guide to the meaning of r, the following table offers a descriptive interpretation.

r	meaning
0.00 to 0.19	a very low correlation
0.20 to 0.39	a low correlation
0.40 to 0.69	a modest correlation
0.70 to 0.89	a high correlation
0.90 to 1.00	a very high correlation

8.9 Rank order correlation coefficients

The product moment correlation coefficient, r, is appropriate in describing the degree of association between two variables when the data are at the interval or the ratio level of measurement.

Very often, however, variables cannot be described with sufficient precision as to warrant the interval or the ratio level of measurement.

Suppose that instead of computing scores for employees' job satisfaction from a schedule, two of the research psychologists use focused interviews to put the ten employees into *rank order* in respect of their perceived level of job satisfaction.

Set out below are the hypothetical rank orderings of Psychologist 1 and Psychologist 2 in respect of the ten employees' perceived job satisfaction levels.

Is there a significant relationship between the two sets of rankings? Spearman's rank order correlation coefficient, r_s, is appropriate in describing the degree of association in the judgements of the two psychologists.

Spearman's rank order correlation coefficient (r_s)

Spearman's correlation coefficient (r_s) is given by:

$$r_s = 1 - \frac{6 \sum d^2}{n(n-1)(n+1)}$$

where

d = the difference in rank between the items in a pair

n = the number of items

$\sum$ = 'the sum of'.

Table 36   Rank ordering on job satisfaction by two psychologists

	Psychologist 1	Psychologist 2	d	d^2
A	10	8	2	4
B	7	9	2	4
C	4	4	0	0
D	1	1	0	0
E	3	5	2	4
F	2	2	0	0
G	9	10	1	1
H	5	6	1	1
J	8	7	1	1
K	6	3	3	9
			$\sum d^2 = 24$	

$$r_s = 1 - \frac{6 \sum (d^2)}{n(n-1)(n+1)}$$

$$= 1 - \frac{144}{990}$$

$$= 0.855$$

From the rough and ready guide to the meaning of correlation shown on page 83 a value of 0.855 indicates a high correlation between the judgements of the two psychologists. Whether or not that value is *significant* we must leave until the concept of significance is discussed later in the text. In Appendix 9 we test the significance of the results shown in Table 36.

8.10   Kendall's rank order correlation coefficient (τ, tau)

The Kendall rank order correlation coefficient, τ, is a suitable alternative measure to Spearman's r, when the data on both measures are at the ordinal level of measurement, or above. For an outline of Kendall's τ, see the example on pages 148 to 150.

8.11   The correlation coefficient eta (η)

On page 80 we note that the product moment correlation coefficient is inappropriate in the case of curvilinear relationships. We now add to that brief comment by way of introducing the correlation coefficient *eta*. Whilst a product moment correlation coefficient can be computed for any set of paired X, Y values, wherever the relationship between those values departs from *linearity* in any way, the correlation coefficient underestimates the degree of relationship present. To that extent, the product moment correlation coefficient is an adequate measure of relationship only when variables are linearly related. The correlation coefficient *eta* (η) is an appropriate measure of relationship between two variables when the relationship is not linear.

Suppose that a researcher obtains general knowledge quiz scores from 130 people aged between 5 and 85. Suspecting that the relationship between general knowledge and age may be such that both young and old score less well than those of 'middle' years, the investigator decides to set up a scatterplot of scores as shown in Table 36.1. The data are plotted with chronological age along the X axis and quiz scores along the Y axis.

It is obvious from an inspection of Table 36.1 that the trend of the plotted points cannot very well be represented by any straight line. An appropriate measure of non-linear relationship is called for. *Eta* is just such a measure. The usual way to compute *eta* is to define eta squared as the ratio of the sum of the squares of the 'between' columns ($\sum y_b^2$) for variable Y to the total sum of squares ($\sum y_t^2$) for variable Y:

$$\eta_{yx}^2 = \frac{\sum y_b^2}{\sum y_t^2}$$

Table 36.1   Scatterplot for general knowledge quiz scores and ages for 130 people.

Y axis quiz scores	X axis: age									f	y'	fy'	$f(y')^2$
	1–10	11–20	21–30	31–40	41–50	51–60	61–70	71–80	81–90				
85–94			1	1	1	2				5	8	40	320
75–84		1	1	1	5	4	4	3		19	7	133	931
65–74		5	4	1	7	4	2	7	2	32	6	192	1152
55–64		2	3		3	2	1	2		13	5	65	325
45–54	4	2	1					3	4	14	4	56	224
35–44	9	7			1				4	21	3	63	189
25–34	7								8	15	2	30	60
15–24	2								8	10	1	10	10
5–14	1									1	0	0	0
f	23	17	10	3	17	12	7	15	26	130		589	3211
x'	0	1	2	3	4	5	6	7	8				

From the scatterplot data in Table 36.1, the *total sum of squares for Y* is given by:

$$\Sigma y_t^2 = \Sigma f(y')^2 - \frac{(\Sigma f y')^2}{N}$$

$$\Sigma y_t^2 = 3211 - \frac{(589)^2}{130} = 542.38$$

The *'between' sum of squares for variable Y* is determined from the means of the columns taken from the means of the entire distribution. The calculation is set out in Table 36.2. There, Column (1) indicates the 9 columns and is composed of the x' values from the bottom of the scatterplot in Table 36.1. The frequencies (fx) of the various columns are set out in Column (2). Column (3) contains the sum of the y' for the frequencies of each of the columns. These values are determined as follows. Looking at Table 36.1, take, for example, the line representing column 1 in the x' values in the table and read 'up' the column. There are 7 frequencies with y' values of 3, 2 frequencies with y' values of 4, 2 frequencies with y' values of 5, 5 frequencies with y' values of 6 and 1 frequency with a y' value of 7, summing to 76. All of the values in Column (3) are thus determined. Each of the Column (3) values is then squared and that squared value is located in Column (4). In Column (5) each of the squared values is divided by the appropriate column frequency. These are then summed and the sum of the squares for 'between' columns is given by:

$$\Sigma y_b^2 = \sum \left[\frac{\Sigma (y')^2}{fx} \right] - \frac{[\Sigma(\Sigma y')]^2}{N}$$

Table 36.2 Calculation of the 'between' sum of squares for Y

Column (1)	Column (2) fx	Column (3) $\Sigma y'$	Column (4) $(\Sigma y')^2$	Column (5) $(\Sigma y')^2/fx$
0	23	59	3481	151.35
1	17	76	5776	339.76
2	10	58	3364	336.40
3	3	21	441	147.00
4	17	103	10609	624.06
5	12	78	6084	507.00
6	7	45	2025	289.29
7	15	85	7225	481.67
8	26	64	4096	157.54
	$\Sigma fx = 130$	$\Sigma(\Sigma y') = 589$		$\Sigma(\Sigma y')^2/fx = 3034.07$

$$= 3034.07 - \frac{(589)^2}{130} = 3034.07 - 2668.62 = 365.45$$

$$eta\ (\eta_{yx}) = \frac{365.45}{542.38} = 0.82$$

Note:
In the case of the product moment correlation coefficient, $r_{xy} = r_{yx}$ and, in consequence, the subscripts are omitted. With the correlation ratio *eta*, however, there are two coefficients, one between X and Y and one between Y and X. The subscripts are therefore important in signalling that in the case of η_{xy} we are concerned with the relation of X to Y and in the case of η_{yx} with the relation of Y to X. The formulae for η_{yx} outlined above may be used to find η_{xy} by replacing the sum of the squares for Y variable by the sum of the squares for X variable and inserting the word *rows* for *columns* in the formulae.

Significance of eta (η)

The significance of the correlation ratio may be tested by the F test. Briefly, F is the ratio of the *'between' mean square* divided by the *'within' mean square*.

The *mean square between columns* is derived by dividing the sum of the squares for the 'between' columns by the number of columns in the scatter-plot minus 1 (ie. $c - 1$). from the data on page 84:

$$\text{mean square 'between' columns} = \frac{365.45}{c - 1} = \frac{365.45}{8} = 45.68$$

The *mean square within columns* is obtained as follows:

(a) first, the 'within' sum of squares is calculated:

$$\Sigma y_w^2 = \Sigma y_t^2 - \Sigma y_b^2 = 542.38 - 365.45 = 176.93$$

(b) the 'within' sum of squares is then divided by the *number of pairs in the sample* minus the *number of columns*.

$$\frac{176.93}{N - c} = \frac{176.93}{130 - 9} = 1.46$$

(c) the *F* ratio is then computed:

$$F = \frac{\text{mean square 'between' columns}}{\text{mean square 'within' columns}} = \frac{45.68}{1.46} = 31.28$$

The table in Appendix 6 is interpolated at $(c - 1)$ degrees of freedom for the 'between' mean square and at $(N - c)$ degrees of freedom for the 'within' mean square; that is to say, at 8 df. and at 121 df. respectively. The obtained value for F far exceeds the critical value of 2.6629. We conclude that the *eta* correlation coefficient of .82 is highly significant.

8.12 Some further thoughts on relationships

Let's reconsider for a moment the basic concept of variability that we touched upon in Section 6.1. It will help us grasp the meaning of correlation more fully.

As has already been shown, measures of human variables such as intelligence, anxiety, job satisfaction, need for achievement, etc., result in a distribution of scores which vary from one individual to another. The factors known to account for this variability are often numerous and, in many cases, unknown. Whether known or not, they are generally categorized as (a) systematic factors and (b) error factors.

Systematic variability, that is, variability due to systematic factors and *error variability,* that is, variability due to error factors, together constitute TOTAL VARIABILITY. When 'parcelling up' (the correct term is *partitioning)* variability in this way, the best statistical measure of total variability is the VARIANCE. The variance is simply the standard deviation squared (S.D.^{2}). Because of its non-linear units   of measurement (S.D.^{2}) , variance can be thought of as an *amount* of variability, made up of different components.

The *total* variance is made up of *systematic* variance and *error* variance. Kerlinger (1970) describes these components as follows:

'Systematic variance is the variance in measures due to some known or unknown influences that "cause" scores to lean in one direction more than another. Any natural or man-made influences that cause events to happen in a certain predictable way are systematic influences'.

and:

'Error variance is the fluctuation or varying of measures due to chance . . . It is the variation in measures due to the unusually small and self-compensating fluctuations of measures—now here, now there, now up, now down'.

8.13 The coefficient of determination

A correlation coefficient gives a measure of the relationship between two variables. It tells us very little however about the nature of that relationship, only that it exists and that it is either relatively high or low.

A fuller grasp of correlation is gained if we consider the COEFFICIENT OF DETERMINATION. This coefficient ($r^2 \times 100$), determines what percentage of the total variance of variable X is due to the variance of variable Y.

For example:

(a) If the correlation (r) between variable X and variable $Y = 0$, then the coefficient of determination $= 0^2 \times 100 = 0\%$

None of the factors accounting for variability are common to both variables.

(b) If the correlation (r) between variable X and variable $Y = 0.8$, then the coefficient of determination $= 0.8^2 \times 100 = 64\%$. 64% of the factors accounting for variability are common to both variables.

(c) If the correlation (r) between variable X and variable $Y = 1$, then the coefficient of determination $= 1^2 \times 100 = 100\%$. 100% of the factors accounting for variability are common to both variables.

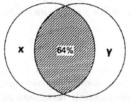

The coefficient of determination may be used as a measure of 'generality' between skills. If, for example, the correlation between two skills is +1 then the abilities needed to perform one skill are identical with the abilities needed to perform the other. On the other hand, if the correlation between two skills is 0, then the abilities needed to perform one skill are completely different from the abilities needed to perform the other, that is to say, the skills are 100% specific.

Notice that one cannot obtain a negative coefficient of determination, since the correlation value has to be squared.

If the correlation between two skills is –1 then the abilities needed to increase performance in one skill will decrease performance in the other.

87

CHAPTER 9
REGRESSION ANALYSIS

9.1   Introduction

In the previous section we looked at how the strength of a relationship between two variables can be determined by means of the correlation coefficient r. This coefficient gives us a measure of the *general* relationship between two variables.

We now turn our attention to a technique closely related to correlation which provides a means of predicting the *specific* values of one variable when we know or assume values of the other variable(s). This technique is known as REGRESSION ANALYSIS. For the purpose of this text our account will be restricted to the analysis of variables having linear, not curvilinear, relationships.

9.2   Simple linear regression

In our discussion of the meaning of statistical correlation (Section 8.6) we illustrate the concept of correlation by reference to scatter diagrams plotted for hypothetical values of certain variables. These scatter pictures give an indication of the possible relationships between the two variables.

Figure 43 shows the plotted pairs of values of Social Psychology and Counselling scores from Table 33 (page 76).

As before we see that all the points are situated on a straight line, indicating a perfect positive relationship between the two variables.

From our knowledge of elementary geometry we know that any straight line through a set of points on a graph can be defined by a mathematical equation in the form of $Y = a + bX$, where a is the intercept on the Y axis (Y value when $X = 0$) and b is the slope of the line (measure of the angle of the line to the X axis).

Given that the equation of the straight line is $Y = a + bX$ it follows that when a and b are known we can predict the Y value for any particular X value. In our example (Figure 43) we see that the graph intercepts the Y axis at point zero ($a = 0$) and the slope is 1 ($b = 1$) since the line is at 45°. Our equation therefore becomes $Y = 0 + 1X$ or $Y = X$. Substituting in this equation for various values of X we can predict the corresponding values of Y. For example if we obtained an X score of 60 we could predict that the Y score would be 60 ($Y = 60$). This is shown diagrammatically in Figure 43.

The example above is rather unusual. In real life it's extremely rare to get a perfect positive relationship between variables. It's more than likely that we would obtain data pairs that do not fall on a straight line, as we show in the following example.

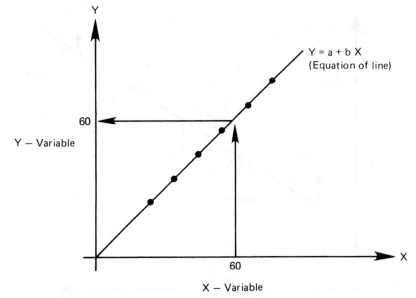

Figure 43   Equation of Line of Best Fit.

Let's assume that in investigating the relationship between two variables (X and Y) we have collected the following data.

Table 37   X and Y Scores for 8 observations

Observation	Variable X	Variable Y
1	8	20
2	7	16
3	6	18
4	5	14
5	5	12
6	4	14
7	3	8
8	1	10

When we plot these data on a scatter diagram (Figure 44) we see that although the relationship appears to be linear we cannot draw a straight line through all the points. However, using what is called the 'method of least squares', we can draw a line which describes the average relationship between X and Y.

This line is called the LINE OF BEST FIT or REGRESSION LINE of Y on X, the equation of which can be used to predict *average* values of Y for given values of X. The equation, known as *the regression equation for Y on X*, is given by:

$$Y_c = a_y + b_y X$$

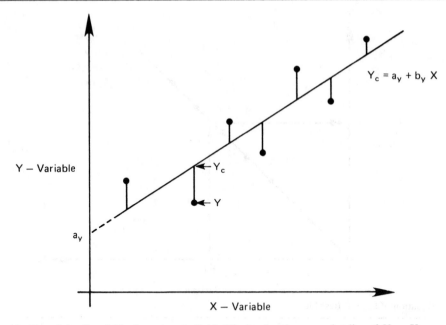

$$Y_c = a_y + b_y \, X$$

Y — Variable

Y_c

Y

a_y

X — Variable

Figure 44 Plot of the X and Y values given in Table 37, showing the regression line of Y on X.

where Y_c is the computed value of Y from X, a_y is the intercept on the Y axis when $X = 0$ and b_y, or the regression coefficient as it is known, is the slope of the line relative to the X axis.

The vertical lines in Figure 44 $(Y_c - Y)$, indicate the errors in prediction resulting from the regression equation of Y on X. The regression line is drawn in such a way that the sum of the squares of these vertical distances is minimized, i.e. $\Sigma (Y_c - Y)^2$ is minimized and $\Sigma (Y_c - Y) = 0$.

Before showing how the regression equation $Y_c = a_y + b_y X$, is computed and used predictively, it's important to note that this equation can only be used to estimate Y values from X values.

If we want to predict X values from Y values we must use another regression line, the regression line of X on Y. Figure 45 shows this regression line drawn for the same data points.

The line in Figure 45 is not the same as that shown in Figure 44. The regression line of X on Y has been drawn in such a way that the sum of the squares of the horizontal distances $(X_c - X)$ is minimized. I.e. $\Sigma (X_c - X)^2$ is minimized and $\Sigma (X_c - X) = 0$.

It is defined by the equation:

$$X_c = a_x + b_x Y$$

where a_x is the intercept on the X axis when $Y = 0$ and b_x, or regression coefficient for X on Y, is the slope of the line relative to the Y axis.

If we now look at Figure 46 we see both regression lines drawn on the same scatter diagram. Only if the correlation between X and Y is perfect (i.e. $r = 1$) do the lines coincide. The greater the angle between the regression lines, the less perfect the correlation. For a zero correlation, the regression lines would be at right angles.

Using our normal step-by-step procedure, let's now illustrate the total regression operation by developing the actual regression equations for the data in Table 37.

90

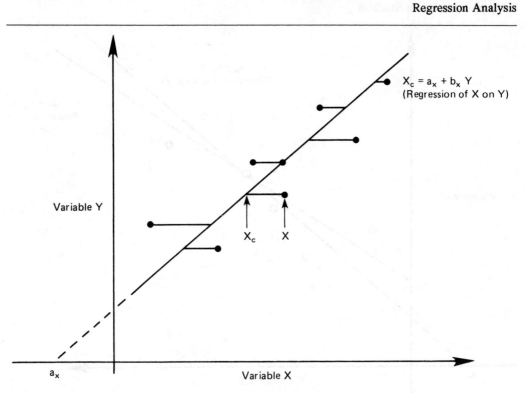

Figure 45 Plot of the X and Y values given in Table 37, showing the regression line of X on Y.

(A) Regression of Y on X

1 Expand the data in Table 37 by computing XY, X^2, Y^2, ΣXY, ΣX^2, ΣY^2 and include these in Table 38.

Table 38 Computing the regression equations

Observations	X	Y	XY	X^2	Y^2
1	8	20	160	64	400
2	7	16	112	49	256
3	6	18	108	36	324
4	5	14	70	25	196
5	5	12	60	25	144
6	4	14	56	16	196
7	3	8	24	9	64
8	1	10	10	1	100
Σ	39	112	600	225	1680

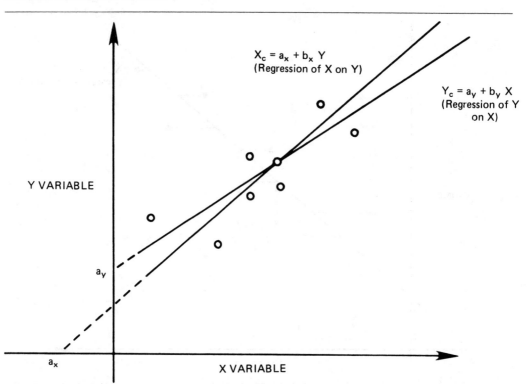

Figure 46 Plot of X and Y values of Table 37, showing the regression line of Y on X and regression line of X on Y.

The regression of Y on X is given by:

$$Y_c = a_y + b_y X$$

where

$$b_y = \frac{n \sum XY - (\sum X)(\sum Y)}{n \sum X^2 - (\sum X)^2} \tag{1}$$

and

$$a_y = \frac{\sum Y}{n} - (b_y)\frac{(\sum X)}{n} \tag{2}$$

2 Calculate (b_y) using formula (1)

$$b_y = \frac{(8)600 - (39)(112)}{(8)225 - (39)^2}$$

$$= \frac{4800 - 4368}{1800 - 1521} = \frac{432}{279}$$

$$= 1.55$$

3 Calculate (a_y) using formula (2)

$$a_y = \frac{112}{8} - (1.55)\frac{39}{8} = 14 - 7.56$$

$$= 6.44$$

In the present example, the regression equation for Y on X can now be defined as:

$$Y_c = 6.44 + 1.55(X)$$

We can use this to predict the value of Y_c given a value of X. For example, for $X = 5.5$

$$Y_c = 6.44 + 1.55(5.5)$$

$$= 14.97$$

When the raw scores are large and the arithmetical computation becomes unwieldy, the formula for b_y can be modified to give:

$$b_y = \frac{\Sigma xy}{\Sigma x^2}$$

where

x = the difference between the raw score X and the mean of the X scores.

y = the difference between the raw score Y and the mean of the Y scores.

To illustrate, let's determine the value of b_y for the above data using this modified formula. Draw up Table 39, to include X, Y, Mean X, Mean Y, x, y, xy, and x^2

Table 39 Computing the regression equation, when raw scores are large

Observations	X	Y	x $(X - M_x)$	y $(Y - M_y)$	xy	x^2
1	8	20	3.12	6	18.72	9.73
2	7	16	2.12	2	4.24	4.49
3	6	18	1.12	4	4.48	1.25
4	5	14	0.12	0	0	0.01
5	5	12	0.12	-2	-0.24	0.01
6	4	14	-0.88	0	0	0.77
7.	3	8	-1.88	-6	11.28	3.53
8	1	10	-3.88	-4	15.52	15.05
Σ	39	112			54	34.84
Mean	4.88	14				

Substituting these values into

$$b_y = \frac{\Sigma xy}{\Sigma x^2}$$

93

we get

$$b_y = \frac{54}{34.84}$$

$$= 1.55$$

which is the same value obtained by using the formula (1).

(B)   Regression of X on Y

The regression of X on Y is given by

$$X_c = a_x + b_x Y$$

where

$$b_x = \frac{n \sum XY - (\sum X)(\sum Y)}{n \sum Y^2 - (\sum Y)^2}$$

and

$$a_x = \frac{\sum X}{n} - (b_x)\frac{(\sum Y)}{n}$$

Substituting the values from Table 39 we obtain:

$$b_x = \frac{(8)600 - (39)(112)}{(8)1680 - (112)^2}$$

$$= \frac{4800 - 4368}{13440 - 12544} = \frac{432}{896}$$

$$= 0.48$$

and

$$a_x = \frac{39}{8} - 0.48\frac{(112)}{8}$$

$$= 4.88 - 6.72$$

$$= -1.84$$

Our regression equation for X on Y, in this example, can now be defined as:

$$X_c = -1.84 + 0.48(Y)$$

which we can use to predict the value of X_c given a value of Y. For example, for $Y = 15$

$$X_c = -1.84 + 0.48(15)$$

$$= 5.36$$

As in our previous example, if the raw scores are large, we can modify the formula for b_x to give:

$$b_x = \frac{\sum xy}{\sum y^2}$$

Throughout this section we have concerned ourselves with the way in which values of one variable can be predicted from known values of another when the variables are linearly related. Generally the predicted variable (Y) is known as the dependent variable and the predictor variable (X) as the independent variable, giving the regression equation

$$Y_c = a + bX \quad \text{where } a = a_y \quad \text{and} \quad b = b_y$$

Predicting values of one variable from values of another is a risky business unless the correlation coefficient is high. The risk arises out of the fact that there may be other sources of influence on the dependent variable. In other words, the varying of the Y scores may be affected by more than just the X variable.

If the requisite assumptions of homoscedasticity and normality are met however, we can gauge the accuracy of our predictions by estimating the possible error. How this is done can be seen using our specific research example on page 159.

One of the ways of reducing the prediction error in regression analysis is to identify and analyse the influence of more than one independent variable on the dependent variable. It is to this that we now turn our attention.

9.3 Multiple regression

Multiple regression enables us to predict values of the dependent variable (Y) from our knowledge of the values of several related variables $(X_1, X_2 \ldots X_k)$. The process involved is a logical extension of the simple (two-variable) linear regression.

For present purposes, it's sufficient to say that once again the *method of least squares* is used to obtain the regression equation. Because the scatterplot now becomes multi-dimensional however, a *line* of best fit is no longer appropriate. In a three variable case the points are plotted in three dimensions along the Y, X_1 and X_2 axes and the method of least squares serves to identify the *plane* of best fit.

The regression equation relating to more than two variables is given by:

$$Y_c = a + b_1X_1 + b_2X_2 + \ldots b_kX_k$$

where the symbols have similar meaning to those in the simple regression equation except that there are now k variables and k regression coefficients. The constants $b_1, b_2 \ldots b_k$ are known as the *partial regression coefficients*.

As an example, let's consider the following problem. Suppose that during a survey carried out at a local Leisure Centre we find that members' attendance is influenced, in part, both by their personal incomes and by the distance they live away from the Centre. We can use the information collected from our sample to predict the possible use of the Centre by future new members. The raw data are set out in Table 40.

1   For each subject in the sample compute X_1Y, X_2Y, X_1X_2, Y^2, $(X_1)^2$, $(X_2)^2$ and include in the table.

2   Compute the totals (Σ) for each column of the table.

3   Substitute the appropriate values into the following equations. (For *three* variable regression.)

$$\Sigma Y = na + b_1 \Sigma X_1 + b_2 \Sigma X_2 \qquad (1)$$

$$\Sigma X_1 Y = a \Sigma X_1 + b_1 \Sigma (X_1)^2 + b_2 \Sigma X_1 X_2 \qquad (2)$$

$$\Sigma X_2 Y = a \Sigma X_2 + b_1 \Sigma X_1 X_2 + b_2 \Sigma (X_2)^2 \qquad (3)$$

where n = number of subjects in sample.

Table 40 Computing the multiple linear regression equation

Subjects	Attend (per mth)	Personal income (£×1000)	Travel distance (miles)						
	Y	X_1	X_2	X_1Y	X_2Y	X_1X_2	Y^2	$(X_1)^2$	$(X_2)^2$
A	15	6	4	90	60	24	225	36	16
B	24	8.5	2	204	48	17	576	72.25	4
C	22	7	3	154	66	21	484	49	9
D	16	4.5	7	72	112	31.5	256	20.25	49
E	7	3.0	6	21	42	18	49	9	36
F	20	6.5	3	130	60	19.5	400	42.25	9
G	21	7.0	3	147	63	21	441	49	9
H	24	10	3	240	72	30	576	100	9
I	18	6	4	108	72	24	324	36	16
J	10	4.5	5	45	50	22.5	100	20.25	25
Totals (Σ)	177	63	40	1211	645	228.5	3431	434	182

Therefore

$$177 = 10a + 63b_1 + 40b_2 \tag{1}$$

$$1211 = 63a + 434b_1 + 228.5b_2 \tag{2}$$

$$645 = 40a + 228.5b_1 + 182b_2 \tag{3}$$

4 Solve these three linear equations simultaneously. A step-by-step explanation follows for the reader unfamiliar with this process.

(i) Divide each equation by its respective coefficient of a: i.e. divide equation (1) by 10; equation (2) by 63; and equation (3) by 40. Use as many decimal places as your calculator will allow.

For equation (1)

$$\frac{177}{10} = \frac{10}{10}a + \frac{63}{10}b_1 + \frac{40}{10}b_2$$

$$17.7 = a + 6.3b_1 + 4.0b_2 \tag{4}$$

For equation (2)

$$\frac{1211}{63} = \frac{63}{63}a + \frac{434}{63}b_1 + \frac{228.5}{63}b_2$$

$$19.2222 = a + 6.8889b_1 + 3.62698b_2 \tag{5}$$

For equation (3)

$$\frac{645}{40} = \frac{40}{40}a + \frac{228.5}{40}b_1 + \frac{182}{40}b_2$$

$$16.125 = a + 5.7125b_1 + 4.55b_2 \tag{6}$$

(ii) Remove a by subtracting (or adding if the signs of (a) are unlike) equation (6) from equation (5) and (4) respectively.

(5) *minus* (6)

$$19.2222 - 16.125 = (6.8889b_1 - 5.7125b_1) + (3.62698b_2 - 4.55b_2)$$

$$3.097 = 1.1764b_1 - 0.92302b_2 \tag{7}$$

(4) *minus* (6)

$$17.7 - 16.125 = (6.3b_1 - 5.7125b_1) + (4.0b_2 - 4.55b_2)$$

$$1.575 = 0.5875b_1 - 0.55b_2 \tag{8}$$

(iii) Divide equations (7) and (8) by their respective b_2 coefficients

For equation (7)

$$\frac{3.097}{0.92302} = \frac{1.1764}{0.92302}b_1 - \frac{0.92302}{0.92302}b_2$$

$$3.35529 = 1.2745b_1 - b_2 \tag{9}$$

For equation (8)

$$\frac{1.575}{0.55} = \frac{0.5875}{0.55}b_1 - \frac{0.55}{0.55}b_2$$

$$2.86364 = 1.06818b_1 - b_2 \tag{10}$$

(iv) Remove b_2 by subtracting (or adding if signs of b_2 are unlike) equation (10) from equation (9).

$$3.35529 - 2.86364 = 1.2745b_1 - 1.06818b_1$$

$$0.49165 = 0.20632b_1$$

$$b_1 = 2.38295$$

(v) Substitute the value of b_1 into equation (10) (or (9)) and solve for b_2.

$$2.86364 = 1.06818(2.38295) - b_2$$

$$b_2 = -0.31822$$

(vi) Substitute the value of b_1 and b_2 into equation (6) (or (5) or (4)) and solve for a.

$$16.125 = a + (5.7125)(2.38295) + (4.55)(-0.31822)$$

$$16.125 = a + 13.61260 - 1.44790$$

$$a = 3.96030$$

5 Substitute the value of a, b_1 and b_2 into equation (5) as an arithmetical check.

$$19.2222 = 3.96030 + (6.8889)(2.38295) + (3.62698)(-0.31822)$$

$$19.2222 = 19.22202 = \text{CHECKED}$$

6 Complete the computed multiple regression equation.

$$Y_e = 3.96030 + 2.38295X_1 - 0.31822X_2$$

We are now in a position to predict the value of Y_c for given values of X_1 and X_2. For example, if a new member of the Leisure Centre with a salary of £5000 lives 4 miles away, how many visits might we expect him to make in one month?

$$Y_c = 3.96030 + 2.38295(5) - 0.31822(4)$$

$$= 3.96030 + 11.91475 - 1.27288$$

$$= 14.60217$$

The new member could be expected to make approximately 14 or 15 visits in one month.

9.4 Using the coefficient of determination in multiple regression analysis

How good a multiple regression equation is as a predictor of the dependent variable is contingent upon how much of the variance of that dependent variable (Y) is explained by the effects of the independent variables (X_1, X_2 . . . X_k) working together. If, for example, we find that the Y scores are more influenced by factors other than the independent variables under consideration, it follows that a regression equation using just those variables will not be a good predictor.

In order to identify what effect the independent variables have on the dependent variable we have to consider the coefficient of determination. In multiple regression this effect is measured by the COEFFICIENT OF MULTIPLE DETERMINATION (R^2), which is the square of the multiple correlation coefficient.

The coefficient of multiple determination is used to compute the proportion of the variance of the dependent variable that is due to the combined effects of the independent variables.

9.5 Calculating the coefficient of multiple determination; method 1

Since we have already calculated a multiple regression equation for our hypothetical Leisure Centre data let's now compute the coefficient of multiple determination for these data and use it to estimate the value of the equation as a predictor. The procedure for computation is as follows.

The method is based upon the calculation of BETA WEIGHTS (β), which are standard partial regression coefficients that place the weightings in the multiple regression equation in the form of Z's or standard scores.

Central to the calculation is the equation:

$$R^2_{123...k} = \beta_1(ry_{.1}) + \beta_2(ry_{.2}) + . . . \beta k(ry_{.k})$$

where

$$r = \text{product moment correlation coefficient}$$

$$R^2 = \text{coefficient of multiple determination.}$$

1   Calculate the standard deviations of all the variables, using

$$\text{S.D.} = \sqrt{\frac{\sum X^2 - \frac{(\sum X)^2}{n}}{n-1}} \quad \text{(for variable } X\text{)}$$

For variable Y (Attendance)

$$S.D. = \sqrt{\frac{3431 - \frac{(177)^2}{10}}{10 - 1}}$$

$$= 5.75519$$

For variable X_1 (Personal income)

$$S.D. = \sqrt{\frac{434 - \frac{(63)^2}{10}}{10 - 1}}$$

$$= 2.03033$$

For variable X_2 (Distance travelled)

$$S.D. = \sqrt{\frac{182 - \frac{(40)^2}{10}}{10 - 1}}$$

$$= 1.56347$$

2 Compute beta weights (β) for each independent variable where

$$\beta_k = b_k \left(\frac{S_k}{S_Y}\right)$$

where

b_k = regression coefficient for variable k

S_k = standard deviation of variable k

S_Y = standard deviation of variable Y

$$\beta_1 = b_1 \left(\frac{S_1}{S_Y}\right)$$

$$= 2.38295 \left(\frac{2.03033}{5.75519}\right)$$

$$= 0.84066$$

$$\beta_2 = b_2 \left(\frac{S_2}{S_Y}\right)$$

$$= -0.31822 \left(\frac{1.56347}{5.75519}\right)$$

$$= -0.08645$$

3 Determine the product moment correlation coefficients among all the variables, namely, attendance and personal income (r_{YX_1}); attendance and distance travelled (r_{YX_2}); and personal income and distance travelled ($r_{X_1X_2}$). Refer to the data in Table 40.

(a) For attendance and personal income

$$r_{YX_1} = \frac{n\sum X_1 Y - (\sum X_1)(\sum Y)}{\sqrt{[n\sum X_1^2 - (\sum X_1)^2][n\sum Y^2 - (\sum Y)^2]}}$$

$$r_{YX_1} = \frac{10(1211) - (63)(177)}{\sqrt{[10(434) - (63)^2][10(3431) - (177)^2]}}$$

$$= \frac{12110 - 11151}{\sqrt{[4340 - 3969][34310 - 31329]}}$$

$$= 0.91191$$

(b) For attendance and distance travelled.

$$r_{YX_2} = \frac{n\sum X_2 Y - (\sum X_2)(\sum Y)}{\sqrt{[n\sum X_2^2 - (\sum X_2)^2][n\sum Y^2 - (\sum Y)^2]}}$$

$$= \frac{10(645) - (40)(177)}{\sqrt{[10(182) - (40)^2][10(3431) - 177)^2]}}$$

$$= \frac{6450 - 7080}{\sqrt{[1820 - 1600][34310 - 31329]}}$$

$$= -0.77794$$

(c) For personal income and distance travelled

$$r_{X_1 X_2} = \frac{n\sum X_1 X_2 - (\sum X_1)(\sum X_2)}{\sqrt{[n\sum X_1^2 - (\sum X_1)^2][n\sum X_2^2 - (\sum X_2)^2]}}$$

$$= \frac{10(228.5) - (63)(40)}{\sqrt{[10(434) - (63)^2][10(182) - (40)^2]}}$$

$$= \frac{2285 - 2520}{\sqrt{[4340 - 3969][1820 - 1600]}}$$

$$= -0.82256$$

4   Substitute the appropriate values into

$$R^2_{Y.X_1 X_2} = \beta_{x1}(r_{yx1}) + \beta_{x2}(r_{yx2})$$

$$= 0.84066(0.91191) - 0.08645(-0.77794)$$

$$R^2 = 0.834$$

9.6   Calculating the coefficient of multiple determination; method 2

The above procedure is applicable for any number of independent variables. However, when there are just two independent variables the coefficient of multiple determination can be more simply calculated by using the coefficient of multiple correlation formula as shown below.

$$R_{(Y.X_1 X_2)} = \sqrt{\frac{(r_{Y.X_1})^2 + (r_{Y.X_2})^2 - 2(r_{Y.X_1})(r_{Y.X_2})(r_{X_1 X_2})}{1 - (r_{X_1 X_2})^2}}$$

where $r_{(1,2)}$ = product moment correlation coefficient between variable 1 and variable 2.

Therefore

$$R^2_{Y.X_1X_2} = \frac{(r_{Y.X_1})^2 + (r_{Y.X_2})^2 - 2(r_{Y.X_1})(r_{Y.X_2})(r_{X_1X_2})}{1 - (r_{X_1X_2})^2}$$

Substituting into this formula we obtain:

$$R^2_{(Y.X_1X_2)} = \frac{(0.91191)^2 + (-0.77794)^2 - 2(0.91191)(-0.77794)(-0.82256)}{1 - (-0.82256)^2}$$

$$= \frac{0.83158 + 0.60519 - 1.167}{0.32340}$$

$$= .834.$$

A coefficient of multiple determination of .834 indicates that 83.4% of the variance of the attendance scores is due to the combined effects of *personal income* and *distance travelled*. As we can see, the greater proportion of the variance in attendance scores is accounted for by the combined effects of the independent variables.

We may therefore conclude that these two independent variables, taken together in a multiple regression equation, give us a good predictive measure of attendance.

In general, a multiple correlation coefficient of 0.7 or above (i.e. $R^2 \geq .49$) is considered a high relationship.

At this stage it is important to remember that the coefficient of multiple determination only tells us about the *combined effects* of the independent variables, not what the *relative contribution* of each of the independent variables is.

To a certain extent the multiple regression equation itself indicates the relative contribution of each of the independent variables by weighting them with regression coefficients. However it's not as simple as this. The regression coefficients can be misleading in an interpretation of regression particularly when the independent variables themselves are correlated. For highly intercorrelated independent variables the regression coefficients tend to be unreliable. Conversely, if the correlations among the independent variables are zero or near zero then the interpretation is greatly simplified. A detailed explanation of this problem is complex and beyond the scope of the present text.

CHAPTER 10
INFERENTIAL STATISTICS

10.1   Introduction

Up to this point in the text we have discussed *descriptive statistics,* outlining the central tendencies, the variability, and the relationships in data that are readily to hand. It is time now to move from description to an examination of statistical techniques that enable us to go from *known* to *unknown* data, that is, to make inferences about wider populations from which our 'known' data are drawn. These techniques are called *inferential statistics.* Inferential statistics deal with two different types of problems, the first to do with making estimates, the second with testing hypotheses. Both tasks involve making inferences about population parameters from sample measures. It is to samples and to sampling methods that we first turn our attention.

We said earlier that it is often impossible to obtain measures of characteristics of a total population. Population characteristics have to be inferred from measures taken from samples. Statistics are taken from samples and, using appropriate inferential techniques, population parameters are estimated.

10.2   Sampling methods

Regardless of which inferential statistical technique we intend to use, the predictive or inferential power of the test will be governed to a certain extent by the procedure used in selecting the sample. If the sample is not truly representative of the population from which it is drawn, that is, if it is a biased sample, then it becomes virtually impossible to make an accurate prediction about the population.

Bias in sample selection is reduced when methods incorporating at least an element of random selection of subjects are employed. More importantly, the predictions made from such samples have greater validity, because the principles of randomness of selection are fundamental to theories of statistical inference. These theories are based upon the laws of probability and chance.

The methods outlined below all incorporate, in part at least, some random selection of subjects.

10.3   Simple random sampling

In simple random sampling, each member of the population under study has an equal chance of being selected. The method involves selecting at random from a list of the population (a sampling frame) the required number of subjects for the sample. Because of probability and chance, the sample should contain subjects with characteristics similar to the population as a whole, i.e. some old, some young, some tall, some short, some fit, some unfit, some rich, some poor, etc. One problem associated with this particular sampling method is that a complete list of the population is needed and this is not always readily available.

10.4   Systematic sampling

This method is a modified form of simple random sampling. It involves selecting subjects from a population list in a systematic rather than a random fashion. For example, if from a population of say 2000, a sample of 100 is required, then every 20th person can be selected. The starting point for the selection is chosen at random.

10.5   Stratified sampling

Stratified sampling involves dividing the population into homogeneous groups, each group containing subjects with similar characteristics. For example, group A might contain males and group B females. In order to obtain a sample representative of the whole population in terms of sex, a random selection of subjects from both group A and B must be taken. If needed, the exact proportion of males to females in the whole population can be reflected in the sample.

10.6   Cluster sampling

When the population is large and widely dispersed, gathering a simple random sample poses administrative problems. Suppose, for example, a psychologist wishes to examine levels of neuroticism in British undergraduate students. It would be quite impractical to select undergraduates randomly and spend an inordinate amount of time travelling to and fro between institutions of higher education in order to test them. By cluster sampling, the psychologist could randomly select a specific number of institutions and test all the students in those selected venues.

10.7   Stage sampling

Stage sampling is an extension of cluster sampling. It involves selecting the sample in stages, that is, taking samples from samples. Using institutions of higher education in Great Britain as the population for example, the psychologist might employ a type of stage sampling that involves selecting a number of institutions at random and from within each of these selecting a number of faculties at random and from within each of these, randomly selecting a number of undergraduate students.

10.8   Sampling error

If many samples are taken from the same population it is unlikely that they will all have characteristics identical, either to each other or to the population. In short, there will be *sampling error.*

Sampling error is not necessarily the result of mistakes made in sampling procedures. Rather, variations may occur due to the chance selection of different individuals. For example, if we take a large number of samples from the population and measure the mean value of each sample, then the sample means will not be identical. Some means will be relatively high, some relatively low, and many will cluster around an average or mean value for the samples. Why should this occur?

We can explain this phenomenon by reference to the Central Limit Theorem which is derived from the laws of probability. The Central Limit Theorem states that if random, large samples of equal size are repeatedly drawn from any population, then the means of those samples will be approximately normally distributed. Moreover, the average or mean of the sample means will be approximately the same as the population mean. We can show this diagrammatically as follows:

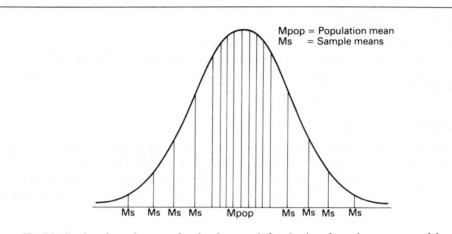

Mpop = Population mean
Ms   = Sample means

Ms   Ms Ms Ms   Mpop   Ms Ms Ms   Ms

Figure 47 Distribution of sample means showing the spread of a selection of sample means around the population mean.

With the Central Limit Theorem in mind, we can see that when a sample is selected and the mean value of a particular characteristic of that sample is calculated, that sample mean is but one of a theoretical distribution of sample means. Moreover, because of sampling error, that particular sample mean is unlikely to be the same as the mean of the population. From our knowledge of the characteristics of a normal distribution however, we know that 68.26% of scores lie between ±1 standard deviations from the mean (i.e. $Z = 1$). It follows then that in the theoretical distribution of sample means, 68.26% of all sample means will lie between ±1 standard deviations from the population mean. Put differently, we know that our particular sample mean has a 68.26% chance of falling between ±1 standard deviations from the population mean.

Furthermore, from Figure 49 we can see that our sample mean has a 95% chance of lying between ±1.96 standard deviations from the population mean (i.e. $Z = ±1.96$); a 99% chance of lying between ±2.58 standard deviations from the population mean (i.e. $Z = ±2.58$), and a 99.73% chance of lying between ±3 standard deviations from the population mean (i.e. $Z = ±3$).

Clearly, if we can calculate the standard deviation of the theoretical distribution of sample means, we are then able to predict, with varying degrees of certainty or confidence, how far our sample mean lies from the population mean. That is to say we can be:

68.26%   certain or confident that M_s is ±1 S.D.$_M$ from M_{pop}.

95%   certain or confident that M_s is ±1.96 S.D.$_M$ from M_{pop}.

99%   certain or confident that M_s is ±2.58 S.D.$_M$ from M_{pop}.

99.73%   certain or confident that M_s is ±3.00 S.D.$_M$ from M_{pop}.

The standard deviation of the theoretical distribution of sample means is a measure of sampling error and is called THE STANDARD ERROR OF THE MEAN.

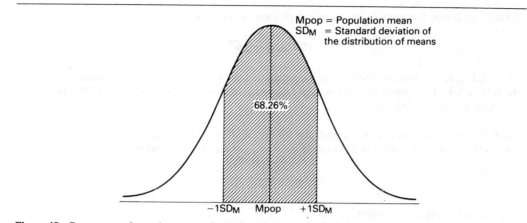

Mpop = Population mean
SD_M = Standard deviation of
the distribution of means

68.26%

−1SD_M   Mpop   +1SD_M

Figure 48   Percentage of sample means ±1 standard deviation from the population mean.

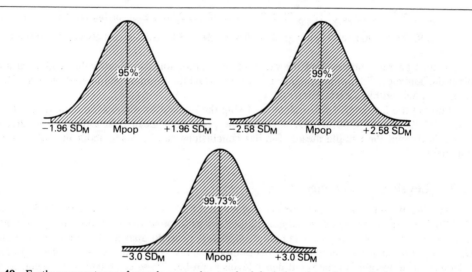

95%

−1.96 SD_M   Mpop   +1.96 SD_M

99%

−2.58 SD_M   Mpop   +2.58 SD_M

99.73%

−3.0 SD_M   Mpop   +3.0 SD_M

Figure 49   Further percentages of sample means by standard deviations from the population mean.

So far, we have referred to this standard deviation as S.D._M. More usually however, it is given the symbols S.E._M or σ_M and is calculated as follows:

$$S.E._M = \frac{S.D._s}{\sqrt{N}}$$

where

S.D._s = the standard deviation of the sample

N = the number in the sample.

105

Strictly speaking, the formula for the standard error of the mean is:

$$S.E._M = \frac{S.D._{pop}}{\sqrt{N}}$$

where $S.D._{pop}$ is the standard deviation of the population. However, as we are usually unable to ascertain the S.D. of the whole population, the standard deviation of the sample is used instead. We can illustrate this with a worked example.

Example How far away do we expect the true population mean to lie from a sample mean of 75 when the standard deviation of the sample is 18 and the size of the sample is 36?

$$S.E._M = \frac{S.D._s}{\sqrt{N}} = \frac{18}{\sqrt{36}} = 3$$

Therefore, the population mean has:

      a 68.26% chance of lying 75 ± 3, (M_s ± 1 S.E.$_M$), or between 72 and 78

      a 95% chance of lying 75 ± 1.96 x 3, (M_s ± 1.96 S.E.$_M$), or between 69.12 and 80.88

      a 99% chance of lying 75 ± 2.58 x 3, (M_s ± 2.58 S.E.$_M$), or between 67.26 and 82.74

      a 99.73% chance of lying 75 ± 3.00 x 3, (M_s ± 3.00 S.E.$_M$), or between 66 and 84.

It must be remembered that the mean of the population is a fixed value and does not vary. What the above confidence statements are saying is that the mean of the population is at a fixed but unknown point within a certain interval.

Look at the last confidence statement. Using the interval 66–84, we can conclude that there is only a 0.27% chance (100% – 99.73%) that the population mean is at a distance greater than 9, (3 × S.E.$_M$) from the sample mean. Similar conclusions can be made about the other confidence statements.

10.9 Levels of confidence

The varying degrees of certainty or confidence to which we have referred (68.26%, 95%, 99%, 99.73%) are called the *levels of confidence*. As we now know from the preceding sections, these levels tell us the probability of a sample mean being a certain distance from the population mean. Up to this point in our discussion we have described the levels of confidence in terms of percentages. It's more usual however to state confidence levels as probabilities, as the table of selected confidence levels shows below.

Table 41 Summary of selected confidence levels

Level	Interval range	Probability of sample mean lying within interval (P_B)	Probability of sample mean lying *outside* interval (P_A)
99.73%	M_{pop} ± 3.00 S.E.$_M$	0.9973	0.0027
99%	M_{pop} ± 2.58 S.E.$_M$	0.99	0.01
95%	M_{pop} ± 1.96 S.E.$_M$	0.95	0.05
68.26%	M_{pop} ± 1.00 S.E.$_M$	0.6826	0.3174

The information in the summary Table 41 can be represented diagrammatically as follows:

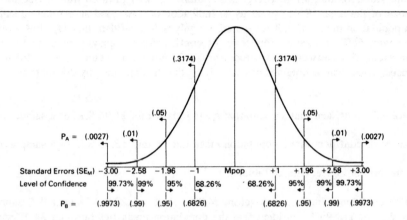

Figure 50 Diagrammatic representation of selected confidence levels.

When we define a confidence level in terms of probability we normally use the P_A value. In Figure 50 above, the P_A value is shown above the horizontal axis of the normal curve. The P_A value states the probability of a sample mean lying *outside* the confidence interval. For example, the 0.05 confidence interval is the one *outside* of which a sample mean has only a 5 in 100 chance of lying.

Let's return to the problem of whether a sample mean can be considered to be a representative or reliable estimate of the true population mean.

From our discussion we can now see that the further away a sample mean is from the population mean, the less trust we can place in it. For example, suppose the sample mean lies outside the 0.01 interval. What exactly does this tell us about it? It says, in effect, that it only has a 1 in 100 chance of being representative of the population mean. Put differently, that particular sample mean has a 99 in 100 chance of being non-representative.

The confidence interval of the mean of a large sample

Table 41.1 sets out the wing length measurements of 100 robins in the order that an ornithologist captured and measured the birds.

Table 41.1 Wing length measurements (mm)

76	73	75	73	74	74	74	74	74	77
74	72	75	76	73	71	73	80	75	75
68	72	78	74	75	74	69	77	77	72
72	76	76	77	70	77	72	74	77	76
78	72	70	74	76	72	73	71	74	74
75	79	75	74	75	74	71	73	75	73
75	70	73	75	70	72	72	71	76	73
74	76	74	75	74	76	75	75	73	73
78	74	73	75	74	73	72	76	73	76
74	71	72	71	79	78	69	77	73	71

By reference to Sections 5.3 and 6.6 we determine that the mean wing length, M_s, is 74.00mm and that the standard deviation, S.D.$_s$, is 2.34mm. As was pointed out in Section 10.8, the standard error of the mean, S.E.$_M$, gives us an indication of how good an estimate a sample mean, M_s, is of a population mean, M_{pop}. Thus we are roughly 68% confident that M_{pop} lies within ±1 S.E. of M_s. However, 68% is a rather low level of confidence; we usually want to be surer that a population mean lies between indicated limits. To meet this need, 95% or 99% limits are generally used. These can be obtained by multiplying the standard error by the appropriate Z score as follows:

We are 95% confident that a population mean falls within ±1.96 S.E. of a sample mean.

We are 99% confident that a population mean falls within ±2.58 S.E. of a sample mean.

From the ornithological data in Table 41.1:

The 95% confidence interval is therefore $74.00 \pm (1.96 \times 0.234) = 74.00 \pm 0.459$mm. This means that we are 95% confident that the population mean lies between 74.459mm and 73.541mm.

Notice that because N, the number of observations, is the denominator of the equation for estimating the standard error (and hence confidence interval) in Section 10.8 above, the value of the standard error (and the breadth of the confidence interval) gets *smaller* as N gets larger. This is a mathematical expression of the statistical axiom that, the larger the sample size, the greater the reliability of an estimate of population parameter.

The confidence interval of the mean of a small sample

In calculating the standard error of a mean (and hence a confidence interval) the standard deviation, S.D.$_s$, is used. Strictly, as we saw in Section 10.8, the standard deviation of the population, S.D.$_{pop}$, should be employed. In large samples we are confident that the standard deviation of the sample is a reliable estimate of the standard deviation of the population; we are less certain, however, in the case of small samples. We need therefore to apply a correction factor to compensate for the uncertainty in smaller samples. That factor should become larger as the sample becomes smaller. A suitable correction factor is t, as described in Section 10.10.

Confidence limits of a proportion or percentage

We sometimes wish to assign confidence limits to a proportion or percentage. Suppose, for example, an ornithologist catches 80 birds in July, of which 60 are juveniles. In that *sample* the proportion of juveniles is 60/80 = 0.75 (i.e. 75%). How sure can the ornithologist be that the true proportion of juveniles in the *population* lies within specified limits of that estimated from the *sample*?

The calculation of confidence limits is accomplished by multiplying the standard error of the sample proportion by 1.96 for 95% confidence limits or by 2.58 for 99% confidence limits. For practical purposes a satisfactory estimate of the standard error of a proportion is given by:

$$S.E. = \sqrt{\frac{p\ (1-p)}{N-1}}$$

where

p = sample proportion

N = number of sampling units.

Thus in the ornithological example above,

$$S.E. = \sqrt{\frac{0.74\ (1-0.75)}{80-1}} = 0.049$$

The 95% confidence limits are therefore 0.75 ± (1.96 × 0.049) = 0.75 ± 0.096. That is to say, 0.846 (84.6%) upper limit and 0.654 (65.4%) lower limit. We are therefore 95% confident that the true percentage of juveniles in the *population* from which the *sample* is drawn lies between 84.6% and 65.4%.

10.10   *t* distributions

In our outline of confidence intervals and limits, we used Table 21 to estimate, for example, that the 0.05 level of confidence is situated at +1. 96 standard deviations or standard errors from the mean of the population (i.e. $Z = 1.96$). To find the exact limits in actual scores for a particular sample mean we multiplied the calculated $S.E._M$ by 1.96.

Table 21 can only be used, of course, when the distribution of sample means is normal. If the distribution is not normal then the 0.05 level will not be located at a Z score of ±1.96. As samples become smaller, their distributions become flatter and more spread out as we show in Figure 51 below. These non-normal distributions are called *t* distributions.

Notice that there is a different *t* distribution for each size of sample.

To obtain a confidence level for a *t* distribution within which 95% of the theoretical sample means lie (i.e. the 0.05 level) we have to move further out in standard deviation units from the mean. We show this diagrammatically in Figure 52 below.

In a normal distribution, the distance between the mean and successive points, measured in standard deviation units, is expressed as a Z value and this value is used to estimate the area between a particular point and the mean (see Table 21).

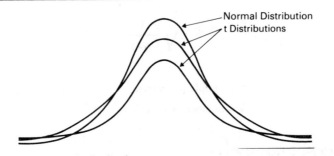

Figure 51 *t* distributions and normal distribution.

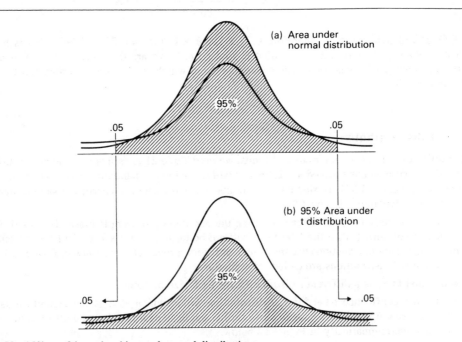

Figure 52 95% confidence level in *t* and normal distributions.

In a *t* distribution however, the distance between the mean and successive points, measured in standard deviation units, is expressed as a *t* value. For a given sample size this *t* value can be used to estimate the area between a particular point and the mean. A different table must be consulted however. An abridged *t* distribution table is shown below. It gives the *t* value needed to achieve particular levels of confidence.

Look at Table 42. The *t* values are shown for the 0.05 and the 0.01 levels of confidence. Notice that they are not set as *t* values against *N*, but as *t* values against *degrees of freedom*.

In estimating the population mean from a sample mean the degrees of freedom available are *N* – 1 where *N* = the number of observations. For example, if we want to establish the 0.05 and the 0.01 levels of confidence for the mean of a population when the sample size is 15, we enter the table at

Table 42* An abridged *t* distribution table

Degrees of Freedom	P = 0.05	P = 0.01
1	t = 12.706	t = 63.657
2	t = 4.303	t = 9.925
5	t = 2.571	t = 4.032
8	t = 2.306	t = 3.355
10	t = 2.228	t = 3.169
14	t = 2.145	t = 2.977
16	t = 2.120	t = 2.921
20	t = 2.086	t = 2.845
30	t = 2.042	t = 2.750
60	t = 2.000	t = 2.660
120	t = 1.980	t = 2.617
∝	t = 1.960	t = 2.576

* A comprehensive *t* table for use at later stages in the text is presented in Appendix 5.

degrees of freedom 14, since d.f. = $N - 1 = 15 - 1 = 14$. The appropriate *t* values are 2.145 and 2.977 for the 0.05 and the 0.01 levels respectively. We conclude that when sample sizes are 15, 95% of the sample means will lie between ±2.145 (2.15) standard deviations from the population mean. Moreover, 99% of the sample means will lie between ±2.977 (2.98) standard deviations from the population mean.

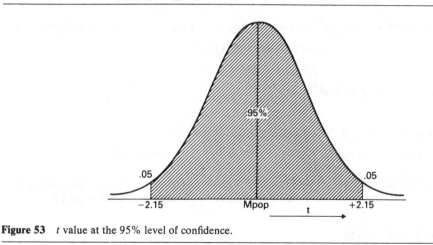

Figure 53   *t* value at the 95% level of confidence.

Recall that the standard deviation of a distribution of means (S.D.$_M$) is equal to the standard error of the mean (S.E.$_M$). We can see that once the standard error has been determined, the exact confidence limits, measured in actual scores, can be estimated.

111

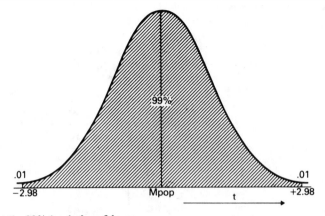

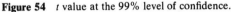

Figure 54 t value at the 99% level of confidence.

Take for example, sample size 15,

$$0.05 \text{ (95\%) confidence interval} = M_s \pm (2.15 \times \text{S.E.}_M)$$

$$0.01 \text{ (99\%) confidence interval} = M_s \pm (2.98 \times \text{S.E.}_M)$$

Let's see how this works out with an actual problem. For a sample size 17, determine the 0.05 (95%) and the 0.01 (99%) confidence intervals for the M_{pop} when the sample mean (M_s) is 50 and the sample standard deviation (S.D.$_s$) is 10.

Step 1 Calculate the standard error of the mean (S.E.$_M$)

$$\text{For small samples} \quad \text{S.E.}_M = \frac{\text{S.D.}_s}{\sqrt{N}} = \frac{10}{\sqrt{17}} = \frac{10}{4.12} = 2.42$$

Step 2 Enter Table 42 for 16 d.f. $(N - 1)$ and find t values for the 0.05 and the 0.01 levels

$$t \text{ for } 0.05 = 2.120, \ (2.12)$$

$$t \text{ for } 0.01 = 2.921, \ (2.92)$$

Step 3 Establish the confidence limits using:

$$M_s \pm t \times \text{S.E.}_M$$

For the 0.05 level, $50 \pm (2.12 \times 2.42) = 50 \pm 5.13 = 44.87$ to 55.13

For the 0.01 level, $50 \pm (2.92 \times 2.42) = 50 \pm 7.07 = 42.93$ to 57.07

We conclude that the population mean has a 95% chance of lying at a fixed value somewhere between 44.87 and 55.13. Furthermore, it has a 99% chance of lying between 42.93 and 57.07.

10.11 Degrees of freedom

Having introduced *degrees of freedom* (d.f.) in the previous section, it's appropriate at this point to deal more fully with this rather difficult concept.

Degrees of freedom is involved in many of the inferential statistical techniques we cover in later sections of the text.

In sampling statistics, the degrees of freedom can loosely be considered equal to the number of observations or scores minus the number of parameters that are being estimated. Thus, in using N sample observations to estimate the mean of the population (the mean being one parameter), the degrees of freedom will be $N - 1$.

The rationale behind this is that if observations or numbers are used to estimate a score about which we can draw a conclusion, then the establishment of that score causes the original observations or numbers to lose a certain amount of freedom.

Look at it this way. Suppose we have to select any five numbers. We have complete freedom of choice as to what the numbers are. So, we have 5 degrees of freedom. Suppose however we are then told that the five numbers must have a total value of 25. We will have complete freedom of choice to select four of the numbers but the fifth will be dependent on the other four. Let's say that the first four numbers we select are 7, 8, 9, and 10 which total 34, then if the total value of the five numbers is to be 25, the fifth number must be –9.

$$7 + 8 + 9 + 10 - 9 = 25$$

A restriction has been placed on one of the observations; only four are free to vary; the fifth has lost its freedom. In our example then d.f. = 4, that is, $N - 1 = 5 - 1 = 4$.

Suppose now we are told to select any five numbers, the first two of which have to total 9 and the total value of all five has to be 25.

One restriction is apparent when we wish the total of the first two numbers to be 9. Another restriction is apparent in the requirement that all five numbers must total 25. In other words we have lost two degrees of freedom in our example. It leaves us with d.f. = 3, that is, $N - 2 = 5 - 2 = 3$.

Notice that the degrees of freedom available are not always $N - 1$. They depend upon the particular estimation at hand. In each of the inferential statistical techniques outlined later in the text, we set out the specific method for obtaining the appropriate degrees of freedom.

10.12 Hypothesis formulation and testing

Hypotheses are hunches that the researcher has about the existence of relationships between variables. Testing hypotheses is to do with accepting or rejecting explanations of those relationships within known degrees of certainty.

This way of looking at hypothesis formulation and testing may seem somewhat strange to the reader. 'Isn't hypothesis testing', the reader might ask, 'concerned with finding *differences* rather than *relationships*'?

The apparent contradiction can be explained by the following example:

'. . . suppose a researcher tests the hypothesis that boys will out-perform girls on certain tests of geometrical aptitude. He might divide his sample into a group of boys and a group of girls, administer a geometry aptitude test and see if the two groups perform differently on the test. Now although it may appear that he is looking for a difference between the groups, deeper consideration will reveal that the researcher is really attempting to see if there is a *relationship* between the *variable* of sex, on the one hand, and the *variable of geometry aptitude* on the other. Ultimately, almost all hypotheses in educational research are suppositions about relationships between variables'.
(Popham and Sirotnik, 1973, p. 46)

Nevertheless, as Popham and Sirotnik advise, it is useful for the student researcher to think of the various methods he comes across in inferential statistics in terms of whether they are essentially *difference-testing* or *relationship-testing* techniques.

We deal first with the concept of statistical significance before looking at the formulation and testing of a hypothesis.

Suppose that in the example above, the mean score on a geometric aptitude test in a group of 11 boys is 14 with a standard deviation of 3, and in a group of 11 girls, the mean score is 10 with a standard deviation of 2. The researcher wishes to know whether or not the observed difference in the means of the two groups is *significant*. What exactly is meant by *significant*?*

10.13   Statistical significance

Recall that in previous sections, we showed that by marking off standard units of distance along the base line of the normal curve, we were able to determine the percentage of a population under specific parts of that curve. The standard scores used to mark off standard units of distance, we remember, are called Z scores.

A Z score of 1.96 taken at each end of the normal curve cuts off 5% of the total area of the curve as we illustrate in Figure 55. Similarly, a Z score of 2.58 taken at each end of the normal curve cuts off 1% of the total area of the curve as shown in Figure 56.

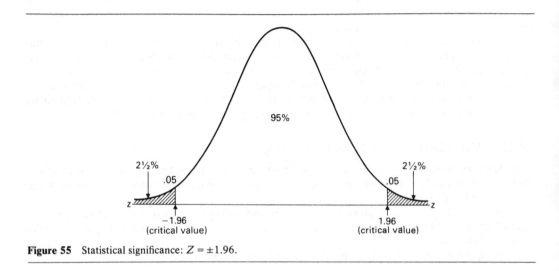

Figure 55   Statistical significance: $Z = \pm 1.96$.

Much of the research that is undertaken uses what is called the *hypothetico-deductive method*. This involves the statistical testing of a formulated hypothesis.

As stated earlier, a hypothesis is simply a 'hunch', or a statement about an expected relationship, or difference between variables or groups. In our geometrical aptitude example we hypothesize that:

there is a difference between the means of the boys' and the girls' groups.

In order to find out whether there is a statistically significant difference between the two groups, or whether the difference is caused by sampling error, we first have to state our hypothesis in the form of a NULL HYPOTHESIS, generally denoted by the shorthand (H_0).

---

*   We are solely concerned here with the concept of *statistical* significance. What may be highly significant *statistically* may be of no *educational* significance whatsoever!

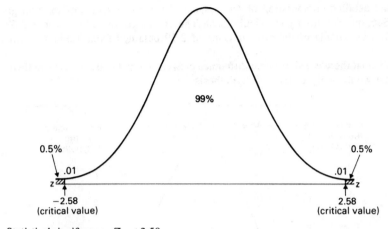

Figure 56 Statistical significance: $Z = \pm2.58$.

A null hypothesis (H_0) is *a hypothesis of no difference*. In our example, the null hypothesis would be stated as follows:

there is no difference between the means of the boys' and the girls' groups.

The reason why we have to state our hypothesis in the null form is that inferential statistical techniques are designed to allow us to estimate how far above or below zero a difference or relationship can be expected to lie due to random sampling error. The further a difference or relationship is above or below zero, the less chance it has of occurring as a result of random sampling error and the greater chance it has of being statistically significant.

Statistical tests tell us of the probability of a difference or a relationship occurring as a result of chance sampling errors. We generally accept that if a difference only has a 5 in 100 ($p = 0.05$) chance of being due to sampling error, or 95 in 100 ($p = 0.95$) chance of *not* being due to sampling error, then we take it to be significantly different and we reject H_0, the null hypothesis.

Certain statistical tests give us values in the form of Z scores, the chance probability of which we can determine along the base line of the normal curve. Thus, from a Z score of 1.96 or more, we can infer that the chance probability of the value occurring is 5 times in 100, written as the probability (p) is less than ($<$) 0.05 that is,

$$p < 0.05.$$

Similarly, from a Z score of 2.58 or more, we infer

$$p < 0.01$$

Other statistical tests give us values in the form of t scores, the chance probability of which we can, again, determine by reference to the base line of a curve which, like the normal curve, is symmetrical and has a mean of zero. The shapes of t distributions alter with the size of samples as shown in Figure 52 (page 110). As a general rule of thumb however, when a sample size is greater than 30 ($n > 30$), the t distribution and the Z distribution are taken to be approximately equal.

In our example of differences in geometrical aptitude, if a t value exceeds 2.09 then we can infer that only 5 times in 100 is the difference between the means likely to occur by chance. Put another way, 95 times out of 100, a difference of that size is *statistically significant*. It's not appropriate

115

here to show the detailed computation of t in respect of the mean scores of the two groups in our example (see Section 16.1 for t test calculations). The computed value however is 3.67. Figure 57 shows that 3.67 lies outside of the critical value of 2.09 obtained from Table 42, which indicates $p < 0.05$.

We conclude that there *is* a significant difference between the boys and the girls in their geometrical aptitude and that we must *reject* the null hypothesis.

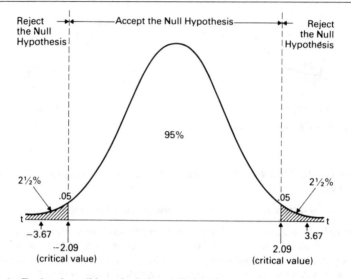

Figure 57 Example: Testing the null hypothesis (two-tailed test).

10.14   One-tailed and two-tailed tests

Notice that in formulating the null hypothesis in our example above we simply stated that there would be no difference between the mean scores on geometrical aptitude. The test of the null hypothesis that we employed was a *non-directional* test; non-directional because we did not specify the direction in which we believed the means would differ. In Figure 57 we established a critical value of 2.09 on either side of the curve, thus allowing for the possibility that our statistical test would give us either a positive or a negative t value. Because a non-directional test locates critical values at both 'tails' of the distribution, it is referred to as a *two-tailed* test.

Suppose, however, that prior to giving the test of geometrical aptitude we have good reasons to predict that the boys' mean performance will be significantly superior to that of the girls. Here we are predicting the *direction* in which we expect the difference to lie and we are, in consequence, concerned with only one 'tail' of the distribution. In this event a *one-tailed* test is appropriate. Figure 58 shows the critical value (1.73) of t in respect of our sample of 22 boys and girls.

10.15   Type 1 and Type 2 errors

As we said earlier, it's common practice for the researcher to 'accept the 0.05 level of significance' or to 'accept the 0.01 level of significance' in his investigation.

What exactly does this mean?

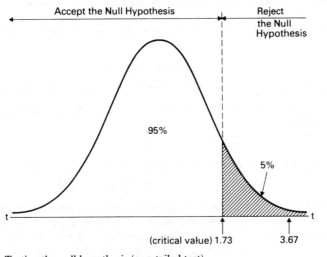

Accept the Null Hypothesis

Reject the Null Hypothesis

95%

5%

t

t

(critical value) 1.73   3.67

Figure 58   Example: Testing the null hypothesis (one-tailed test).

'Accepting the 0.05 level of significance' in rejecting the null hypothesis means that 95 times out of a 100 we are probably correct in our decision, *but,* 5 times out of 100, we run the risk of rejecting the null hypothesis when in fact it is *true.*

When the null hypothesis is rejected and it is actually true, we refer to a TYPE 1 error having been committed. How can the risk of committing a Type 1 error be reduced? Simply by setting our level of acceptance at a more rigorous standard, for example, at the 1 in a 1000 times level of significance ($p < 0.001$), or by increasing the sample size.

The reader will appreciate however that the researcher faces a 'swings and roundabouts' situation. The opposite of the case we have just outlined is referred to as a TYPE 2 error, that is, *not* rejecting the null hypothesis when, in fact, it should be rejected.

Thus, as we decrease the possibility of making a TYPE 1 error, we increase the probability of making a TYPE 2 error.

Most researchers are by nature cautious characters. They aim to limit the probability of committing Type 1 errors.

10.16   Independent and dependent variables

Recall that in Section 10.12 we gave an example to do with geometric skills among boys and girls and proposed that formulating a hypothesis involved making a supposition about the relationship between *sex* on the one hand, and *geometric aptitude* on the other. We went on to say that almost all educational hypotheses have to do with hunches about relationships between variables.

What sort of relationship might we propose about *sex* and *aptitude for geometry?* That sex is, perhaps, one of a number of variables that are *presumed to cause** geometrical aptitude? The other way round is clearly absurd—that geometrical aptitude is *presumed to cause* sex!

---

* Notice how careful we are to use the phrase 'presumed to cause' rather than 'cause'!

A fundamental way of classifying variables in the social sciences is to do with their *presumed causal* relationships.

An independent variable is the presumed cause of a *dependent* variable. What we are saying, in effect, is:

If A, then B

In research, we attempt to predict *from* independent variables *to* dependent variables. Thus:

the sex of the child (independent variable) is the presumed cause of geometrical aptitude (dependent variable),

and

coaching technique A (independent variable) is the presumed cause of skill rating B (dependent variable).

A variable can, of course, serve both as an independent and as a dependent variable. For example, a researcher may wish to examine the presumed causes of *anxiety* (dependent variable) from among the following independent variables—amount of time in revision, clarity of written instructions, and the difficulty of the questions. On the other hand, his research might feature *anxiety* as an independent variable in a study of the presumed causes of examination failure.

10.17 Correlated and uncorrelated data

Recall that in Section 10.3 (page 102) in our discussion of simple random sampling, we said that this method ensures each member of a group an equal chance of being selected. Suppose we draw two samples completely randomly from a population of 1000 pupils. We should rightly expect that the two groups would be unrelated to various characteristics that we might be interested in. Paraphrasing what we said in Section 10.3 pupils in both groups should be both old and young, tall and short, intelligent and dull, swimmers and non-swimmers, and so on. In a word, the data to do with the two groups are unrelated or *uncorrelated*.

Suppose now that we wish to examine the difference between two groups of pupils in respect of *staying on at school after* 16+ and that we have matched the groups by *intelligence, social class background* and the *subjects* they are studying. On only one variable are they different, and that is *sex*. Our two groups then, one consisting of boys, the other girls, have been matched or related in respect of three variables. Our data in this case are *correlated*.

Suppose further that in one of the two groups, say the girls, we are interested in the decision to *stay on at school after* 16+ both *prior to* and *immediately after* 'O' level results. Here again, we have an example of correlated data since our two sets of information refer to the same subjects.

Certain statistical techniques, as we shall see, are appropriate in the case of correlated data whilst other techniques are suitable for uncorrelated data.

10.18 Parametric and non-parametric statistics: some further observations

In Section 2.4 we were content simply to distinguish between parametric and non-parametric methods and to note that unlike parametric techniques which involve assumptions about the distributions of the populations that are sampled, non-parametric methods require far fewer assumptions about population data; that is, non-parametric tests, as their name implies, may be regarded as 'distribution-free' procedures.

What, then, are the advantages of using non-parametric statistics? In the first place, they employ simple formulae! In the second, they are easy and quick to apply and in comparison with parametric tests they can be used under a wide range of conditions. Many of the non-parametric tests that we employ in subsequent sections of the text are concerned with the rank ordering, rather than the numerical value of variables or observations. Indeed, as we shall see, sometimes not even rank order is required. In brief, non-parametric procedures can be used when data are at the nominal and the ordinal level of measurement.

What are the disadvantages of non-parametric tests? Perhaps the most important is to do with what is termed the *power* of a test. The power of a test refers to the probability of rejecting the null hypothesis when, in fact, it is false. The greater the ability of a test to reject a false hypothesis, the greater its relative power.

$$\text{Power} = 1 - \beta$$

where β is the probability of a Type 2 error.

Although, in general, non-parametric tests are not as powerful as parametric tests, for large samples they are often more *robust*. This means that often inferences based on non-parametric tests are valid, despite the strict assumptions of the equivalent parametric tests not being adhered to.

As you will see later in the text, many parametric tests have non-parametric equivalents. In order to be more discerning in our selection it is important to have some idea of their relative power. This is done by estimating the *power efficiency* or *asymptotic relative efficiency* (ARE), as it is sometimes called. The power efficiency gives us a single measure of relative performance of two tests, both using large sample sizes. It tells us how much we must increase sample size N in test I (non-parametric test) to make it as powerful as test II (parametric test). It is calculated from the following formula:

$$\text{Power efficiency of test I} = (100) \frac{N_{\text{II}}}{N_{\text{I}}} \text{ per cent}$$

where N_{I} is the number of cases needed by test I to achieve the same power as test II when it (test II) has N_{II} cases.

For example, if test I needs 30 cases to be as powerful as test II when it has 21 cases, then test I has a power efficiency of $(100)\frac{21}{30} = 70\%$. This indicates that the power of test I using 100 cases is the same as that of test II using 70 cases.

As we have said, non-parametric tests are generally less powerful than parametric tests, due in part to dealing in rank orders rather than actual numerical values and therefore 'wasting' a certain amount of available information. However, if the assumptions of a particular parametric test are not met then it is often more appropriate to use an equivalent non-parametric test. The advantage here is that since the non-parametric test is not based upon parametric assumptions it does not lose its validity when those assumptions are not met. Indeed, it would be more valid than its parametric equivalent, particularly when the sample size is small.

For a more detailed discussion of the merits and limitations of non-parametric tests see Gibbons (1976, pages 26-30).

Finally, in Chapter 18 onwards, in which we set out what are termed *Factorial Designs,* readers will grasp an important advantage of parametric over non-parametric techniques. When the researcher is interested in interaction effects between variables, it is parametric techniques, such as two-way analysis of variance, which permit the assessment of interaction far more readily than non-parametric methods.

CHAPTER 11
CHOOSING AN APPROPRIATE TEST

11.1 Advice

In choosing an appropriate test readers may find the following advice helpful.

1 Enter the Tables at the appropriate *RESEARCH DESIGN*. That is to say, is your research concerned with ONE GROUP, TWO GROUPS, SEVERAL GROUPS, or is it FACTORIAL in design?

2 Which LEVEL OF MEASUREMENT is involved? That is to say, are your data at the INTERVAL/RATIO LEVEL? If this is so and your sample is large, then a PARAMETRIC Test is appropriate. If your data are at the NOMINAL or the ORDINAL LEVEL, then you should select your test from those in the NON-PARAMETRIC boxes.

3 What is the NATURE OF YOUR DATA? That is to say, are they CORRELATED or UNCORRELATED?

4 For simplicity's sake, we categorize a test in only one design in the Format Tabulations. Many tests, however, are applicable to more than one design. For example, a test to measure the difference between two samples (i.e. two sets of scores) may be used with a ONE GROUP DESIGN if there are repeated measures on the one group, or with a TWO GROUP DESIGN if the groups are matched.

The underlying rationale of all statistical tests rests upon SAMPLES (either related or unrelated) rather than GROUPS.

Think carefully, therefore, before selecting a particular test.

5 Where, as is often the case, *several* tests may be appropriate to your research problem, look at the examples given in the text to decide which one in particular meets your specific needs.

11.2 Format tabulation

RESEARCH DESIGN			TYPE OF TEST	NATURE OF DATA
Chapter	Design No.	ONE GROUP		
12	1	Single observation on one variable (categorized)		
13	2	One observation on each of two or more variables		
14	3	Repeated observations on same subjects under two conditions or, Before/After		
15	4	Multi-treatment	Parametric	Correlated
			Non-parametric	Correlated
				Uncorrelated

LEVELS OF MEASUREMENT					
Nominal		**Ordinal**		**Interval/Ratio**	
Design No.		Design No.		Design No.	
2	Tetrachoric correlation r, p 153			2	Pearson product moment correlation, p 138 eta, p 84
2	Point biserial correlation, p 151			2	Simple linear regression, p 141
				2	Partial correlation, p 154
				2	Multiple correlation, p 157
				2	multiple regression, p 159
				3	t-test (correlated means), p 170
				4	One way ANOVA for correlated means, p 181
3	McNemar test, p 176	1	One sample runs test, p 134		
4	Cochran Q test, p 193	2	Spearman rank order correlation coefficient, p 143		
3	Gart test, p 227				
		2	Kendall rank order correlation coefficient, p 148		
		2	Kendall partial rank correlation, p 156		
		3	Sign test, p 178		
		3	Wilcoxon matched pairs signed ranks test, p 172		
		4	Friedman two-way analysis of variance by ranks, p 185		
		4	Kendall's coefficient of concordance, p 189		
		4	Page's L trend test, p 195		
1	Chi square one sample test, p 126				
1	G test, p 127				
1	Binomial test, p 129				
1	Kolmogorov—Smirnov one sample test, p 132				
2	Percentage difference/ ratio, p 163				
2	Phi coefficient, p 164				
2	Yule's Q, p 165				
2	Contingency coefficient C, p 166				
2	Cramer's V, p 168				

RESEARCH DESIGN			TYPE OF TEST	NATURE OF DATA
Chapter	Design No.	TWO GROUP		
16	5	Static comparison on one or more variables	Parametric	Correlated
				Uncorrelated
			Non-parametric	Correlated
				Uncorrelated
		MULTI GROUP		
17	6	More than two groups, one single variable	Parametric	Correlated
				Uncorrelated
			Non-parametric	Correlated
				Uncorrelated
		FACTORIAL		
18	7	The effect of two independent variables	Parametric	Correlated
19	8	The effect of two independent variables (Repeated measures on ONE factor)		
20	9	The effect of two independent variables (Repeated measures on BOTH factors)		
				Uncorrelated

LEVELS OF MEASUREMENT					
Nominal		**Ordinal**		**Interval/Ratio**	
Design No.		**Design No.**		**Design No.**	
				5	t-test for correlated means (matched groups), p 170
				5	t-test for independent samples (pooled variance), p 197
				5	t-test for independent samples (separate variance), p 201
		5	Wilcoxon matched pairs signed ranks test (matched groups), p 172	5	Walsh test (matched groups), p 213
5	Fisher exact test, p 218	5	Mann–Whitney test, p 202		
5	Kolmogorov–Smirnov two sample test, p 211	5	Wald–Wolfwitz runs test, p 215		
5	Chi square $2 \times k$, p 208				
5	G test, p 210				
5	A median test (Conover), p 229				
				6	One way ANOVA for correlated means (matched groups), p 181
				6	One way ANOVA for independent samples, p 241
6	Chi square $k \times n$, p 257	6	Kruskal Wallis analysis of variance, p 248		
6	Somer's d, p 264				
6	Goodman and Kruskal gamma, p 264	6	A slippage test (Conover), p 252		
6	Kendall's tau b, p 264	6	Jonckheere Trend Test, p 254		
6	Guttman's lambda, p 267				
				8	Two-way ANOVA (repeated measures on ONE factor), p 283
				9	Two way ANOVA (repeated measures on BOTH factors), p 289
				7	Two way ANOVA (independent samples), p 277

CHAPTER 12
DESIGN 1

One group design: single observations on one variable

A   Ordinal, interval or ratio scores

Group 1	Observations$_{(X)}$
Subjects A	X_A
B	X_B
C	X_C
D	X_D

or

B   Nominal data

	Number of Subjects in Each Category			
	Category 1	Category 2	Category 3	Category k
Group 1. Subjects A, B, C, ..., N				
Observed frequency	$X_{1(o)}$	$X_{2(o)}$	$X_{3(o)}$	$X_{k(o)}$
Expected frequency	$X_{1(e)}$	$X_{2(e)}$	$X_{3(e)}$	$X_{k(e)}$

12.1 Using the CHI SQUARE ONE-SAMPLE TEST

Table 43 shows the findings of a town planning department looking into the means of transport used by a representative sample of factory workers in a new town.

Table 43 Workers' choice of transportation

	On foot	By bus	By train	By car	By pedal cycle	By motor cycle
$N =$	50	19	17	30	9	7

The planners wish to know whether any one form of transport is favoured significantly more than any other, the null hypothesis (H_0) in this case being that there is no significant difference in workers' choice of type of transportation.

The chi square (χ^2) one-sample test is appropriate to the problem. Chi square is probably the most used of all non-parametric tests. It is applicable when data are nominal and grouped into categories or boxes. It allows us to test the difference between *observed* frequencies and *expected* or theoretical frequencies.

PROCEDURE FOR COMPUTING CHI SQUARE

1 Estimate the expected frequency E for each category based upon the null hypothesis.

The null hypothesis (H_0) states that there is no significant difference in workers' choice of type of transportation. It follows therefore that each category is expected to attract the same number of choices, i.e. 22 (the number of subjects divided by the number of transport choices).

2 Tabulate the expected frequencies (E) and the observed frequencies (O) as shown below.

Table 44 Observed and expected frequencies

	On foot	By bus	By train	By car	By pedal cycle	By motor cycle
Observed frequency (O)	50	19	17	30	9	7
Expected frequency (E)	22	22	22	22	22	22

3 Calculate chi square (X^2) from the formula

$$\chi^2 = \Sigma \frac{(O-E)^2}{E}$$

where

O = the observed frequency in each category

E = the expected frequency in each category.

$$\chi^2 = \frac{(50-22)^2}{22} + \frac{(19-22)^2}{22} + \frac{(17-22)^2}{22} + \frac{(30-22)^2}{22} + \frac{(9-22)^2}{22}$$
$$+ \frac{(7-22)^2}{22}$$

$$= 35.6 + 0.4 + 1.1 + 2.9 + 7.7 + 10.2$$

$$= 57.9$$

4   Determine the degrees of freedom (d.f.). For a one sample chi square test degrees of freedom are given by the formula $k - 1$ where k = the number of categories or cells

$$\text{d.f.} = k - 1 = 6 - 1 = 5.$$

5   Enter the table in Appendix 4 at 5 degrees of freedom. The critical value at d.f. = 5 is 11.07 at the 5% level and 15.09 at the 1% level. The obtained value exceeds this. The town planning department should therefore reject the null hypothesis and conclude that there are significant differences in workers' choices of transportation.

Correction for chi square in 2 × 2 contingency tables

If the number of categories or cells in a chi square is only 2, giving just one degree of freedom then the chi square computed for such data is likely to be an overestimate and may lead to erroneous conclusions unless an adjustment is made to the formula.

Yates' correction for lack of continuity is frequently employed to effect such adjustment. It involves subtracting 0.5 from the numerator of each category or cell value of χ^2 as shown in the formula

$$\chi^2 = \sum \frac{(|O - E| - 0.5)^2}{E}$$

Recently it has been noted that Yates' correction *overcorrects*. Pirie and Hamden (1972) therefore propose a correction that does not change the value of chi square by as much as that of Yates' adjustment.

As a general rule, the chi square test is restricted to the use of randomly selected samples of an adequate size. The stability of the test is said* to be decreased if there are less than 5 expected frequencies in any one category or cell. One way around the problem of low expected frequencies is to combine categories or cells, reducing the total number of cells but increasing the frequencies within those remaining. (See 'Collapsing tables in contingency analyses', page 259.)

The power efficiency of a χ^2 test is not usually reported due to the fact that it is used when no suitable parametric test is available for the same data.

12.2   Using the *G*-TEST

The *G*-test is an alternative to the chi square test for analysing frequencies. The two methods are interchangeable: if a chi square test is appropriate then so too is a *G*-test and the assumptions are the same. Moreover, the outcome of the *G*-test is a test statistic, *G*, which is compared with the distribution of chi square in the same tables as the chi square test. Why then do we need a second test that serves exactly the same purpose?

---

*   However, an alternative view has been argued by Everitt (1977) page 40. See also W. R. Pirie and M. A. Hamden (1972). 'Some revised continuity corrections for discrete data.' *Biometrics*, **28** 693-701.

First, the *G*-test is easier to execute with a desk-top hand calculator, especially with contingency tables. (See for example, page 210.) Second, mathematicians believe that the *G*-test has theoretical advantages which are beyond the scope of this text. Despite these advantages, however, it seems that 'old habits die hard'; chi square remains the most widely used method of analysing frequencies in social science journals.

Applying the G-test in a one-way classification of frequencies

The *G*-test may be used for testing *homogeneity* and for *goodness of fit* of frequencies arranged in a one-way classification. The procedure in each is the same, remembering that expected frequencies in a goodness of fit test are generated according to a mathematical model and the rule for determining the degrees of freedom is different. A correction factor (*Williams' correction*) is applied in the *G*-test irrespective of the number of degrees of freedom. (But see G-test in $2 \times k$ contingency table, page 210). The adjusted value of *G* is a symbolized $G_{adj.}$.

To emphasize the similarity between the *G*-test and the chi square test, we use the data employed in Table 44. Observed and expected frequencies are:

$$O \quad 50 \quad 19 \quad 17 \quad 30 \quad 9 \quad 7$$

$$E \quad 22 \quad 22 \quad 22 \quad 22 \quad 22 \quad 22$$

The formula for *G* is:

$$G = 2 \times \sum^{k} O \ln \frac{O}{E}$$

where O and E are *observed* and *expected* frequencies, respectively; $\sum$ means the sum of the products $O\ln(O/E)$ for all k categories or frequency classes; and 'ln' means natural logarithm. The stepwise procedure is as follows:

Step 1   For each category or frequency class multiply O by $\ln(O/E)$ and add up the total. Thus:

$$50 \times \ln(\tfrac{50}{22}) + 19 \times \ln(\tfrac{19}{22}) + 17 \times \ln(\tfrac{17}{22}) + 30 \times \ln(\tfrac{30}{22}) + 9 \times \ln(\tfrac{9}{22}) + 7 \times \ln(\tfrac{7}{22})$$

$$= 41.049 + (-2.785) + (-4.383) + 9.304 + (-8.044) + (-8.015)$$

$$= 27.126$$

Step 2   Double this number:

$$27.126 \times 2 = 54.252$$

Step 3   Divide *G* by a correction factor which is applied irrespective of the number of degrees of freedom.

$$\text{Correction factor} = 1 + (k^2 - 1)/6nv$$

where k is the number of categories or frequency classes, n is the total number of observed frequencies and v is the degrees of freedom $(k - 1)$. Thus:

$$\text{Correction factor} = 1 + (36 - 1)/6 \times 132 \times 5$$

$$= 1 + \frac{35}{3960} = 1.0088$$

$$G_{adj.} = G/\text{Correction factor} = \frac{54.252}{1.0088} = 53.778$$

Step 4 Compare the value of G_{adj} against the chi-square distribution in Appendix 4 for $(k - 1) = 5$ d.f.

The calculated value of G_{adj} is very similar to the value of 57.9 obtained in the chi square test.

There is no difference in the execution of the G-test when there is only one degree of freedom. In using the G-test for goodness of fit, the rules for determining the degrees of freedom are the same as in the chi square test (see Section 10.11).

12.3 Using the BINOMIAL TEST

The binomial distribution is the sampling distribution of the proportions that we might observe if random samples are drawn from a *two-class population* such as male/female, pass/fail, married/single, etc. We illustrated the binomial distribution in our discussion of coin-tossing on pages 29 to 31. Now, our task is to outline the statistical testing procedure (the *binomial test*) which is appropriate whenever sampling is done on a dichotomous population that can be likened to a population of coin tosses. Initially this involves us in some more words about coin-tossing.

Suppose we toss 10 coins at once, assuming of course that all ten are perfectly symmetrical so that none is more likely to fall 'heads' than 'tails', there are 11 possible outcomes in terms of the *number of heads*.

10, 9, 8, 7, 6, 5, 4, 3, 2, 1, and 0 heads.

To compute the probability of each of these eleven outcomes we apply the multiplication rule for independent events. Take the case of '10 heads' for example:

Each outcome can be thought of as an event set composed of 10 independent trials. '10 heads', is made up of 'heads' on coin 1, 'heads' on coin 2, etc. for all 10 coins. Assuming perfectly symmetrical coins, the probability of a 'head' on any one throw is 0.5. Applying the multiplication rule, we compute the probability for '10 heads' as:

$$\text{probability of '10' heads} = \tfrac{1}{2} \cdot \tfrac{1}{2} \cdot \tfrac{1}{2} \cdot \tfrac{1}{2} \cdot \tfrac{1}{2} \cdot \tfrac{1}{2} \cdot \tfrac{1}{2} \cdot \tfrac{1}{2} \cdot \tfrac{1}{2} \cdot \tfrac{1}{2}$$
$$= (\tfrac{1}{2})^{10} = 0.001$$

Consider now the case of the result, '9 heads'. Whereas the only way we can get '*10 heads*' is that all coins must land 'heads', we can see that this is not the case with 9 coins. For example, coin 1 might land 'tails' and all the rest 'heads' or coin 3 might land 'tails' and all the rest 'heads', and so on.

The number of possible arrangements that could occur when we are not concerned with the order of heads and tails with respect to one another is given by the formula:

$$\frac{N!*}{x! \, (N - x)!}$$

---

* $N!$ means N factorial, that is to say, $N(N - 1)(N - 2)(N - 3) \ldots (2)(1)$. Thus, $5! = (5)(4)(3)(2)(1) = 120$.

$$\frac{N!}{x!(N - x)!} = \binom{N}{x}$$

The table in Appendix 13(b) gives binomial coefficients (i.e. N/x) for values up to $N = 20$.

where

$$N = \text{the number of } trials\text{—n our present example, } N = 10$$

$$x = \text{the number of 'heads'—in our present example, } x = 9$$

and

$$N - x = \text{the number of 'tails'—in our present example, } N - x = 1$$

Thus the number of arrangements that could occur in the result

$$\begin{Bmatrix} 9 \text{ heads} \\ 1 \text{ tail} \end{Bmatrix} = \frac{N!}{x!\,(N-x)!} = \frac{10!}{9!\,(1!)}$$

From the factorial table in Appendix 2 we see that

$$\frac{10!}{(9!)\,(1!)} = \frac{3,628,800}{(362,880)(1)} = 10$$

What exactly does this mean? Simply that we can get 9 heads out of 10 tosses in *10 different ways*. What this does *not* tell us, however, is the *probability of getting 9 heads*. This we can easily compute from:

$$(\tfrac{1}{2})^x (\tfrac{1}{2})^{N-x}$$

Substituting our values into the formula:

$$(\tfrac{1}{2})^9 (\tfrac{1}{2})^1 = (\tfrac{1}{2})^{10} = .001$$

We now have the following information:

The *probability* of *any one arrangement* of 9 'heads' and 1 'tail' = .001
The *number* of such *possible arrangements* of 9 'heads' and 1 'tail' = 10

What we must now do is to apply the *addition rule* to compute the probability of getting *either* one *or* another of the ten possible arrangements. Thus:

$$.001 + .001 + .001 + .001 + .001 + .001 + .001 + .001 + .001 + .001 = .010$$

There is a quicker way of working out the probability of x 'heads' in one category and $N - x$ 'tails' in the other category and that is by applying the general formula:

$$\frac{N!}{x!\,(N-x)!}\,(P)^x (Q)^{N-x}$$

where $P = \text{the probability of a 'head' on a single trial}$
$O = \text{the probability of a 'tail' on a single trial}$

and $N!x!$ and $(N - x)!$ are as before.

Under our null hypothesis (H_o) and our assumption of perfectly symmetrical coins, $P = .5$ for all trials. In a binomial population Q must always equal $1 - P$. Thus $Q = .5$. Let's now look at the probabilities of a number of other results from tossing 10 coins.

$$\begin{Bmatrix} 8 \text{ heads} \\ 2 \text{ tails} \end{Bmatrix} = \frac{N!}{x!\,(N-x)!}\,(P)^x (Q)^{N-x} = \frac{10!}{8!\,2!}\left(\frac{1}{2}\right)^8 \left(\frac{1}{2}\right)^2$$

We have already seen (footnote, page 129) that $\dfrac{N\,!}{x!\,(N-x)!}$ can be simplified to $\binom{N}{x}$—the binomial coefficient which can be obtained from the table in Appendix 13b. Moreover the expression

$$(\tfrac{1}{2})^8(\tfrac{1}{2})^2 = (\tfrac{1}{2})^{10} = .001$$

Therefore the probability of $\left\{\begin{array}{l}8\text{ heads}\\2\text{ tails}\end{array}\right\} = 45(.001) = .045$. Similarly, the probability of $\left\{\begin{array}{l}4\text{ heads}\\6\text{ tails}\end{array}\right\}$ is given by:

$$\frac{N!}{x!\,(N-x)!}(P)^x(Q)^{N-x} = \frac{10!}{4!\,6!}\left(\frac{1}{2}\right)^4\left(\frac{1}{2}\right)^6$$

Again, there is no need to compute $\dfrac{N!}{x!\,(N-x)!}$ as $\dfrac{3{,}628{,}800}{(24)(720)}$. Simply substitute $\binom{N}{x}$ by reference to the table in Appendix 13(b). Thus, the probability of $\left\{\begin{array}{l}4\text{ heads}\\6\text{ tails}\end{array}\right\}$ is given by:

$$\binom{10}{4} \times .001 = 210(.001) = .210$$

Clearly it is tedious and time-consuming to obtain probabilities of sampling distributions of tossed coins, males/females, successes/failures, Conservatives/Labourites, etc. in this manner.

Tables are available for the application of the binomial test, and it is to the use of such a table that we now turn.

Using the binomial test with small samples ($N \le 25$)

Frequently the researcher is concerned with cases where $P = Q = \frac{1}{2}$, that is to say, the proportion of cases expected in one category of a two-category class is equal to the proportion expected in the other.

Take for example the data we use to illustrate the use of the Fisher Exact Probability Test in which the following decision about strike action is observed (page 218).

	Decision to strike
Male	11
Female	3
Total	14

Another way in which we can explore the voting behaviour of the male and female union delegates is to propose the null hypothesis (H_0), that there is no difference between the probability of males and the probability of females voting for strike action. Our obtained values show that 11 cases fall into one category and 3 cases in the other. Hence $N = 14$ and $x = 3$. We can use the table in Appendix 13a to test the one-tailed probability of the occurrence under H_0 of $x = 3$ or fewer when $N = 14$.

Entering the table in Appendix 13a at $x = 3$, $N = 14$ we see that the *one-tailed* probability is $p = .029$. Since our prediction was simply that the two frequencies would differ, a *two-tailed* test is appropriate, and therefore:

$$p = .029 \times 2 = .058$$

Thus, for $N = 14$ and $x = 3$ the two-tailed probability associated with the occurrence under H_0 of such an extreme value of x is $p = .058$.

Using the binomial test with larger samples (N>25)

When $N > 25$ the table in Appendix 13(a) cannot be used. Instead, use the formula:

$$Z = \frac{(x \pm \frac{1}{2}) - \frac{1}{2}N}{\frac{1}{2}\sqrt{N}}$$

where N and x have the same meaning as before and where $\pm\frac{1}{2}$ is used as a correction factor for continuity. When x is less than $\frac{1}{2}N$, the correction factor is $+\frac{1}{2}$. When x is more than $\frac{1}{2}N$, the correction factor is $-\frac{1}{2}$.

Even when $N < 25$ (providing $P = \frac{1}{2}$) the above formula will give a good approximation as can be shown by using the same data that we employed in the Binomial Test for small samples. There $N = 14$ and $x = 3$. Substituting into the formula:

$$Z = \frac{(x \pm \frac{1}{2}) - \frac{1}{2}N}{\frac{1}{2}\sqrt{N}} = \frac{(3 + \frac{1}{2}) - 7}{\frac{1}{2}\sqrt{14}} = \frac{-3.5}{1.87} = -1.87$$

Reference to the table in Appendix 3(b) shows that the probability of $Z = -1.87$ is $2 \times .0307 = .061$, assuming again that a two-tailed test is appropriate. This is almost the same probability (.058) that we found by using the table of exact probabilities in Appendix 13a.

12.4   Using the KOLMOGOROV-SMIRNOV ONE-SAMPLE TEST

As part of their Social Work course, students are given the choice of one of five options in Statistical Methods, the options themselves differing only in the time taken to cover a fixed amount of material. We wish to find out whether or not student choice of a particular option is related to the period of time over which that option is scheduled. The data are set out in Table 45.

The null hypothesis (H_0) in this case is that students' option choice is unrelated to the duration of the option.

The Kolmogorov-Smirnov one-sample test is appropriate to the problem. This goodness of fit test enables us to test the degree of agreement between the distribution of an observed set of values with a specified theoretical distribution. The assumption governing the use of the Kolmogorov-Smirnov test is that the underlying distribution is continuous and that the data are at the nominal level of measurement.

Table 45   Students' choice of statistical methods option

Number of students choosing	Option 1 10 wks	Option 2 8 wks	Option 3 6 wks	Option 4 4 wks	Option 5 2 wks
$n = 30$	0	2	10	15	3

The formula employed in the Kolmogorov-Smirnov one-sample test is given by:

$$D = |(CP_b - CP_e)|\max$$

where

D = an obtained statistic which is compared with STATISTIC D in the table in Appendix 14

CP_o = an observed cumulative proportion

CP_e = an expected cumulative proportion

and

$|(CP_o - CP_e)|$ max = the greatest divergence between any two proportions.

PROCEDURE FOR COMPUTING THE KOLMOGOROV-SMIRNOV D STATISTIC

1 Convert the observed frequencies into cumulative frequencies.

2 Divide the cumulative frequencies by N to obtain cumulative proportions (CP_o).

3 Calculate a theoretical cumulative proportion (CP_e) on the basis that under the null hypothesis (H_o) in our example, each of the Statistics Methods Options would receive similar numbers of choices.

4 The observed cumulative proportion (CP_o) is then compared with the theoretical cumulative proportion (CP_e) to identify $|(CP_o - CP_e)|$ maximum, that is, the greatest divergence between any two proportions.

5 $|CP_o - CP_e|$ (max) = D, which is then compared with STATISTIC D in the table in Appendix 14 to find out whether the obtained D is statistically significant.

Let's see how D is computed in our example. We construct Table 46, which shows that

$$D = 10/30 = .33$$

Reference to the table in Appendix 14 shows that for $N = 30$ the obtained D value of .33 exceeds Statistic $D = .24$ at the 5% level and also exceeds Statistic $D = .27$ at the 2% level.

We therefore reject the null hypothesis and conclude that student choice of a particular option is significantly related to the length of time for which that option is scheduled.

The power efficiency of the Kolmogorov-Smirnov one-sample test is generally regarded as greater than the χ^2 test as individual observations are treated separately and very small samples can be used.

Table 46 Students' choice of statistical methods options: computation

Number of students choosing	Option 1 10 wks	Option 2 8 wks	Option 3 6 wks	Option 4 4 wks	Option 5 2 wks
CP_o	$\frac{0}{30}$	$\frac{2}{30}$	$\frac{12}{30}$	$\frac{27}{30}$	$\frac{30}{30}$
CP_e	$\frac{6}{30}$	$\frac{12}{30}$	$\frac{18}{30}$	$\frac{24}{30}$	$\frac{30}{30}$
$(CP_o - CP_e)$ max	$\frac{6}{30}$	$\frac{10}{30}$	$\frac{6}{30}$	$\frac{3}{30}$	$\frac{0}{30}$

12.5 Using the ONE-SAMPLE RUNS TEST

Recall that in an earlier discussion (page 29) of coin tossing we asked what possible relevance the distributions of tossed pennies have to everyday matters. It is time now to see.

Suppose that a coin is tossed 20 times and the following sequence of heads (H) and tails (T) is recorded:

H T H T H T H T H T H T H T H T H T H T

What might a casual observer make of this? Might he suspect, perhaps, that everything is not quite 'above board'?

Suppose the coin is tossed another 20 times and the following sequence is recorded:

H H H H H H H H H H T T T T T T T T T T

By now, our casual observer has grave doubts about the coin, the coin tosser or both! What bothers him is the non-randomness of the two distributions; in the first sequence there are too many regular fluctuations, in the second, there are too few. Instead of the word *fluctuations,* let's use the term *runs.*

A run is defined as a series of identical signs that are preceded or are followed by a different sign or no sign at all. Thus, in our second example of coin-tossing there are only 2 runs.

$$\underline{\text{H H H H H H H H H H}} \quad \underline{\text{T T T T T T T T T T}}$$
$$\qquad\qquad 1 \qquad\qquad\qquad\qquad\qquad 2$$

In our first example of coin-tossing there are 20 runs.

$$\underline{\text{H}}\ \underline{\text{T}}\ \underline{\text{H}}\ \underline{\text{T}}\ \underline{\text{H}}\ \underline{\text{T}}\ \underline{\text{H}}\ \underline{\text{T}}\ \underline{\text{H}}\ \underline{\text{T}}\ \underline{\text{H}}\ \underline{\text{T}}\ \underline{\text{H}}\ \underline{\text{T}}\ \underline{\text{H}}\ \underline{\text{T}}\ \underline{\text{H}}\ \underline{\text{T}}\ \underline{\text{H}}\ \underline{\text{T}}$$
$$1\ 2\ 3\ 4\ 5\ 6\ 7\ 8\ 9\ 10\ 11\ 12\ 13\ 14\ 15\ 16\ 17\ 18\ 19\ 20$$

When a researcher wishes to test the hypothesis that a sample is drawn at random from a population then the one-sample runs test is appropriate. Suppose, for example, that in the course of a morning's observation a sociologist records the following sequence of unemployed ethnic majority group males (1) and ethnic minority group males (0) entering an Employment Bureau in a racially mixed community.

$$\underline{\text{0 0 0 0 0}}\ \underline{\text{1}}\ \underline{\text{0 0 0 0 0}}\ \underline{\text{1 1 1 1 1}}\ \underline{\text{0 0 0 0 0}}$$
$$\qquad 1 \qquad\ \ 2 \qquad 3 \qquad\quad 4 \qquad\quad 5$$

The researcher suspects that the order of majority and minority group members is not random. He can explore his hunch by stating the null hypothesis, (H_0), that the order of ethnic majority group males and ethnic minority group males entering the Employment Bureau is random, and testing that hypothesis using the one-sample runs test (small samples).

PROCEDURE FOR COMPUTING THE ONE-SAMPLE RUNS TEST WITH SMALL SAMPLES ($n_1 \le 20$; $n_2 \le 20$)

1 Determine n_1, the number of majority group males and n_2, the number of minority group males.

2 Compute the R, the number of runs.

3 Determine the critical value of R in the table in Appendix 16.

Substituting the values from our example:

$$n_1 = 6 \quad n_2 = 15 \quad R = 5$$

Entering the table in Appendix 16 at $n_1 = 6$ and $n_2 = 15$ we see that the critical value of R at the 5% level of significance (two-tailed test) is $R \leq 5$. The obtained R is equal to that critical value. The researcher therefore rejects the null hypothesis and concludes that the order of unemployed ethnic majority and minority group males entering the Employment Bureau is not random.

PROCEDURE FOR COMPUTING THE ONE-SAMPLE RUNS TEST WITH LARGE SAMPLES ($n_1 > 20$; $n_2 > 20$)

When either n_1 or n_2 is larger than 20, the table in Appendix 16 cannot be used.

For larger samples, the sampling distribution of R approximates the normal curve and the null hypothesis (H_0) may be tested using the formula:

$$Z = \frac{R - \left(\dfrac{2n_1 n_2}{n_1 + n_2} + 1\right)}{\sqrt{\dfrac{2n_1 n_2 (2n_1 n_2 - n_1 - n_2)}{(n_1 + n_2)^2 (n_1 + n_2 - 1)}}}$$

Suppose now that the sociologist observes unemployed ethnic majority and minority group males entering the Employment Bureau over the course of a whole day with the following results:

$$\underset{1}{\underline{0\ 0\ 0\ 0\ 0}}\ \underset{2}{\underline{1}}\ \underset{3}{\underline{0\ 0\ 0\ 0\ 0}}\ \underset{4}{\underline{1\ 1\ 1\ 1}}\ \underset{5}{\underline{0\ 0\ 0\ 0\ 0}}\ \underset{6}{\underline{1}}$$

$$\underset{7}{\underline{0\ 0\ 0}}\ \underset{8}{\underline{1\ 1}}\ \underset{9}{\underline{0\ 0}}\ \underset{10}{\underline{1\ 1}}\ \underset{11}{\underline{0\ 0\ 0\ 0}}\ \underset{12}{\underline{1}}\ \underset{13}{\underline{0}}\ \underset{14}{\underline{1}}\ \underset{15}{\underline{0}}\ \underset{16}{\underline{1\ 1\ 1}}$$

$$\underset{17}{\underline{0\ 0\ 0\ 0\ 0\ 0}}\ \underset{18}{\underline{1}}\ \underset{19}{\underline{0\ 0\ 0}}\ \underset{20}{\underline{1}}\ \underset{21}{\underline{0}}\ \underset{22}{\underline{1}}\ \underset{23}{\underline{0}}\ \underset{24}{\underline{1}}\ \underset{25}{\underline{0}}\ \underset{26}{\underline{1}}\ \underset{27}{\underline{0\ 0}}$$

That is to say, $n_1 = 21$, $n_2 = 40$, $R = 27$. Substituting these values in the formula:

$$Z = \frac{27 - \left(\dfrac{2(21)(40)}{21 + 40} + 1\right)}{\sqrt{\dfrac{2(21)(40)[2(21)(40) - 21 - 40]}{(21 + 40)^2 (21 + 40 - 1)}}} = \frac{-1.54}{3.56} = -0.43$$

The significance of Z can be tested by reference to the table in Appendix 3(b).

A critical value of $Z = \pm 1.96$ is significant at the 5% level. The obtained value is less than this. The researcher therefore accepts the null hypothesis and concludes that the order of unemployed ethnic majority and minority group males entering the Employment Bureau during the course of the day is random.

12.6 A probability test for use with Likert-type scales

Cooper (1976) has outlined a test for use with certain types of Likert scales that often feature in social and educational research. Frequently in psychological and educational studies, subjects' attitudes towards various concepts are rated on 3-, 5- or 7-point scales. Thus, pupils might rate an activity in which they have engaged in the following form:

Enjoyment of classroom activity

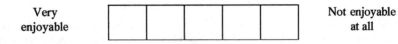

| Very enjoyable | | | | | | Not enjoyable at all |

The boxes in the Likert-type scale above represent categories which are equally spaced between the poles, subjects being invited to place an X in the box that best represents their view. Let us suppose that children have rated a particular classroom activity. How may the teacher determine whether that event is, or is not, seen as enjoyable in general?

The method of analysis that follows is based on the normal probability distribution. It may be applied when the categories are uniformly spread between the poles of a scale, when all categories have equal probabilities of being selected by each respondent and when, typically, the boxes are weighted 1, 2, 3 etc., greater weights being allotted to larger degrees of the quality of concern, in our example *enjoyment*.

When the technique is used with very small samples, an exact probability test is employed. (See Cooper (1976) for an outline of procedures and tables of probability.) Since, however, with samples as small as eight subjects rating categories on a 3-point scale, the test statistic S approaches normality, we now outline the 'large' sample approximation that may be used in most of the cases encountered in social and educational research.

Let us suppose that 25 pupils have rated their enjoyment of a classroom activity on a 5-point scale and that we have their responses as set out below.

Enjoyment of classroom activity

Very enjoyable	6	10	8	1	0	Not enjoyable at all
(Weightings)	(5)	(4)	(3)	(2)	(1)	

S is the total score obtained by summation across all the weighted categories.

$$S = (6 \times 5) + (10 \times 4) + (8 \times 3) + (1 \times 2) + (0 \times 1) = 96$$

The expectation of S is:

$$E(S) = \frac{N(r + 1)}{2}$$

where N = the number of respondents and r = the number of equally spaced categories.

$$E(S) = \tfrac{1}{2} \text{ of } 25(5 + 1) = 75$$

Variance is given by:

$$\text{Var}(S) = \frac{N(r^2 - 1)}{12} = \frac{25(25 - 1)}{12} = 50$$

We may now compute the Z score as follows:

$$Z = \frac{S - \frac{N(r+1)}{2}}{\sqrt{\frac{N(r^2-1)}{12}}} = \frac{(96-75) + \frac{1}{2}}{\sqrt{50}} \quad (\frac{1}{2} \text{ being the correction for continuity})$$

$Z = 3.04$

Entering the table in Appendix 3(b) we see that a Z of 3.04 is significant at the .01 level. We conclude that the value of $S = 96$ is significatly greater than the expected value of $E(S) = 75$. The children rate the classroom activity as *enjoyable*.

CHAPTER 13
DESIGN 2

One group—one observation per subject on each of two or more variables

Group 1	Observations on Variable 1 (X)		Observations on Variable 2 (Y)	
Subjects A	X	X_A	Y	Y_A
B		X_B		Y_B
C		X_C		Y_C
D		X_D		Y_D

13.1 Using the PEARSON PRODUCT MOMENT CORRELATION COEFFICIENT

A sample of 10 patients with anorexia nervosa is drawn at random and their anxiety and depression scores are obtained using the Crown-Crisp Experimental Index (CCEI) with the results shown in Table 47.

We wish to find out whether there is a relationship between anxiety and depression. The Pearson product moment correlation coefficient is a suitable measure of relationship when samples are randomly selected from normally distributed populations. The assumptions underlying the product moment correlation coefficient when it is used for inferential purposes, are (i) *homoscedasticity* that is to say, the variances in the Y values are comparable to variances in the X values, and (ii) the data are normally distributed.

The null hypothesis (H_0) in our example is that there is no relationship between anxiety and depression.

The Pearson product moment correlation coefficient is given by the formula:

$$r = \frac{n \sum XY - (\sum X)(\sum Y)}{\sqrt{[n \sum X^2 - (\sum X)^2][n \sum Y^2 - (\sum Y)^2]}}$$

Table 47 Patient anxiety and depression scores

Patient	Anxiety score	Depression score
A	8.2	6.4
B	7.9	5.8
C	6.3	4.9
D	9.1	7.2
E	5.4	3.9
F	10.3	7.9
G	4.8	5.0
H	6.5	4.2
I	8.3	7.1
J	7.5	5.3

where

r = product moment correlation
n = number of pairs of scores
X = scores on variable X
Y = scores on variable Y
Σ = 'sum of'.

PROCEDURE FOR COMPUTING THE PEARSON PRODUCT MOMENT CORRELATION COEFFICIENT

1 Total the scores on anxiety (ΣX) and the scores on depression (ΣY).

2 Square each patient's scores on anxiety (X^2) and depression (Y^2).

3 Sum the X^2 values giving ΣX^2 , and sum the Y^2 values giving ΣY^2.

5 Multiply each patient's score for anxiety (X) by her score on depression (Y) to give her XY value.

6 Sum the XY values (ΣXY).

Substituting our computed data from Table 48,

$$r = \frac{10\,(447.1) - (74.3)(57.7)}{\sqrt{[10(577.8) - (74.3)^2][10(349.2) - (57.7)^2]}}$$

$$= 0.90$$

Consult the table in Appendix 8 for the significance of a correlation when $n = 10$. Degrees of freedom are given by $n - 2 = 8$.

Our value of 0.90 exceeds the values in the table at the 5% level ($r = 0.63$) and at the 1% level ($r = 0.77$). d.f. = 8.

We therefore reject the null hypothesis and conclude that there is a significant relationship between anxiety and depression.

Without recourse to a table showing significance levels of Pearson product moment correlation values, the significance of an obtained value can be tested by calculating t from the following formula:

$$t = r\sqrt{\frac{n-2}{1-r^2}}$$

Table 48 Computing the correlation between anxiety and depression

Patient	Anxiety (X)	Depression (Y)	X^2	Y^2	XY
A	8.2	6.4	67.24	40.96	52.48
B	7.9	5.8	62.41	33.64	45.82
C	6.3	4.9	39.69	24.01	30.87
D	9.1	7.2	82.81	51.84	65.52
E	5.4	3.9	29.16	15.21	21.06
F	10.3	7.9	106.09	62.41	81.37
G	4.8	5.0	23.04	25.00	24.00
H	6.5	4.2	42.25	17.64	27.30
I	8.3	7.1	68.89	50.41	58.93
J	7.5	5.3	56.25	28.09	39.75
	$\Sigma X = 74.3$	$\Sigma Y = 57.7$	$\Sigma X^2 = 577.8$	$\Sigma Y^2 = 349.2$	$\Sigma XY = 447.1$

Substituting our data where $r = .90$ and $n = 10$

$$t = .90 \sqrt{\frac{8}{.19}} = 5.84$$

Entering the table in Appendix 5 at d.f. $= n - 2$, i.e. 8 we find that a t value of 3.50 is significant at the 1 % level. Our obtained value exceeds this. We therefore reject the null hypothesis and conclude that there is a significant relationship between anxiety and depression.

Testing the significance of the difference between two correlation coefficients for uncorrelated data

A researcher obtains the following results in her analysis of the relationship between *achievement motivation* (x) and *algebra grades* (y) in two groups consisting of male and female students respectively. She wishes to test the hypothesis that the two population r's do not differ.

Male students(1)	*Female students*(2)
$r_{xy(1)} = .65$	$r_{xy(2)} = .81$
$N_1 = 111$	$N_2 = 127$

Procedures

(a) Using the table in Appendix 3(c), transpose the r values to Z

$$r_{xy(1)} = .65 ; Z_1 = .775$$
$$r_{xy(2)} = .81 ; Z_2 = 1.127$$

(b) Compute the standard error of the difference between the two Z's as given by:

$$\text{S.E.}_z = \sqrt{\frac{1}{N_1 - 3} + \frac{1}{N_2 - 3}}$$

$$\text{S.E.}_z = \sqrt{\frac{1}{111 - 3} + \frac{1}{127 - 3}} = \sqrt{.00925 + .00806} = .132$$

(c) Z is given by:

$$Z = \frac{Z_1 - Z_2}{S.E._z}$$

$$Z = \frac{.775 - 1.127}{.132} = -2.66$$

(d) Reference to the table in Appendix 3(b) shows that a $Z = 2.66$ is significant at $p = .0078$ for a two-tailed test.

13.2 Using SIMPLE LINEAR REGRESSION

In Chapter 9 we describe the technique of regression analysis by means of which we are able to predict values of one variable (dependent variable) when we know or assume values of the other(s) (independent variable(s)).

Let's now consider the use of such an analysis in a practical example where we wish to examine the relationship between GESTATION TIMES and BIRTH WEIGHTS of infants and more specifically, to find an equation which will allow us to predict birth weights from given gestation times.

Data collected from a sample of ten infants are presented in Table 49.

Given that a linear relationship exists between the two variables the technique of simple linear regression is appropriate for our analysis.

1 Expand the data in Table 49 by computing XY, X^2, Y^2, ΣXY, ΣX^2, ΣY^2 and include these in Table 50.

Table 49 Birth weights and gestation times

Infants	Birth Weight (kg)	Gestation Time (days)
1	3.15	286
2	4.28	290
3	4.07	259
4	2.94	239
5	2.49	251
6	4.46	284
7	4.73	315
8	3.60	275
9	3.15	253
10	4.01	272

Table 50   Computing the regression equation

Infants	Birth Weight Dependent variable (Y)	Gestation Times Independent variable (X)	XY	Y^2	X^2
1	3.15	286	900.90	9.92	81,796
2	4.28	290	1241.20	18.32	84,100
3	4.07	259	1054.13	16.56	67,081
4	2.94	239	702.66	8.64	57,121
5	2.49	251	624.99	6.20	63,001
6	4.46	284	1266.64	19.89	80,656
7	4.73	315	1489.95	22.37	99,225
8	3.60	275	990.00	12.96	75,625
9	3.15	253	796.95	9.92	64,009
10	4.01	272	1090.72	16.08	73,984
Σ	36.88	2724	10,158.14	140.86	746,598

The regression of Y on X is given by:

$$Y = a + bX$$

where

$$b = \frac{\sum XY - \frac{(\sum X)(\sum Y)}{n}}{\sum X^2 - \frac{(\sum X)^2}{n}} \quad (1)$$

and

$$a = \frac{\sum Y}{n} - b\frac{(\sum X)}{n} \quad (2)$$

2   Calculate (b) using formula (1).

$$b = \frac{10{,}158.14 - \frac{(2724)(36.88)}{10}}{746{,}598 - \frac{(2724)^2}{10}}$$

$$= \frac{10{,}158.14 - 10{,}046.11}{746{,}598 - 742{,}017.6} = \frac{112.03}{4580.4}$$

$$= 0.02446$$

3   Calculate (a) using formula (2).

$$a = \frac{36.88}{10} - \frac{(0.02446)(2724)}{10}$$

$$= 3.688 - 6.662$$

$$= -2.974$$

The regression equation for Y (birth weight) on X (gestation time) can be defined as:

$$Y = 0.02446(X) - 2.974$$

Using this equation we are now able to predict birth weight from gestation time. For example, if we wish to estimate the probable birth weight of an infant whose gestation time was 240 days we substitute this value into the equation and solve for Y.

$$Y = 0.02446(240) - 2.974$$

$$= 2.90$$

This tells us that the probable infant birth weight after a gestation time of 240 days is 2.90 kg.

It is at this stage of our analysis, however, that caution must be exercised.

From our earlier discussion on inferential statistics in Section 10.9 we know that there is always inaccuracy when predicting population parameters from sample statistics due to chance errors of sampling. It is therefore unlikely that the equation arrived at in this analysis will give us a totally accurate prediction of the population from which the sample was drawn. To overcome this problem we must take into account the sampling error or Standard Error of the Estimate. This can be computed from:

$$S.E_{(Y.X)} = \sqrt{\frac{\sum y^2 - \frac{(\sum xy)^2}{\sum x^2}}{n - 2}}$$

where

$$\sum y^2 = \sum Y^2 - \frac{(\sum Y)^2}{n} = 140.86 - \frac{(36.88)^2}{10} = 4.85$$

$$\sum x^2 = \sum X^2 - \frac{(\sum X)^2}{n} = 746,598 - \frac{(2724)^2}{10} = 4580.4$$

$$\sum xy = \sum XY - \frac{(\sum X)(\sum Y)}{n} = 10,158.14 - \frac{(36.88)(2724)}{10}$$

$$= 112.03$$

Therefore

$$S.E_{(Y.X)} = \sqrt{\frac{4.85 - \frac{(112.03)^2}{4580.4}}{10 - 2}} = \sqrt{0.26} = 0.51$$

When this standard error is applied to our predicted score of $Y = 2.90$ it can be interpreted (using the rationale outlined in Section 10.9) that we are 68.28% confident that the infant's actual birth weight will be 2.90 ± 0.51 kg ($2.90 \pm 1 \times$ S.E.). Similarly we can be 95% confident that the birth weight will be 2.90 ± 1 kg ($2.90 \pm 1.96 \times$ S.E.) and 99% confident that the birth weight will be 2.90 ± 1.3 kg ($2.90 \pm 2.54 \times$ S.E.).

13.3 Using SPEARMAN'S RANK ORDER CORRELATION COEFFICIENT (rho)

In a study of the politicization of university students a researcher rates ten students on a liberalism/conservatism scale and on the degree to which they are actively involved in campus politics. He then puts the students into rank order on both measures. The data are set out in Table 51 below.

Table 51 Rank ordering on liberalism/conservatism and on political involvement: data for 10 subjects

Student	Variable (liberalism/conservatism)	X Variable (political involvement)
A	4	6
B	5	3
C	6	5
D	7	9
E	1	1
F	10	8
G	9	10
H	2	2
I	8	7
J	3	4

We wish to find out whether there is a significant relationship between political attitudes and involvement in student affairs in this small group. The null hypothesis (H_0) in this case is that political attitudes and activity in campus politics are not related.

Spearman's rank order correlation coefficient (r_s) is appropriate to our task. The assumptions for using r_s are that the sample is randomly selected and that the data are at the ordinal level of measurement.

Spearman's r_s is given by:

$$r_s = 1 - \frac{6 \sum d^2}{n(n-1)(n+1)}$$

where

d = the difference in rank between the items in a pair

n = the number of items

Σ = the sum of

Table 52 Rank ordering on liberalism/conservatism and on political involvement: computation

Student	Variable X (liberalism/conservatism) rank X	Variable Y (political involvement) rank Y	d	d^2
A	4	6	−2	4
B	5	3	2	4
C	6	5	1	1
D	7	9	−2	4
E	1	1	0	0
F	10	8	2	4
G	9	10	−1	1
H	2	2	0	0
I	8	7	1	1
J	3	4	−1	1
			Σd^2	= 20

Substituting from Table 52 into the formula:

$$r_s = 1 - \frac{6 \sum d^2}{n(n-1)(n+1)} = 1 - \frac{6 \times 20}{10(9)(11)} = 1 - \frac{120}{990}$$

$$r_s = 0.88 \cdot$$

Recalling our rough-and-ready guide to the strength of a correlation coefficient (see page 83) there is a very high relationship between liberalism and active involvement in student affairs. The significance of the relationship can be tested by reference to the table in Appendix 9.

Interpolating in that table at $n = 10$ we see that an r_s value of 0.56 is necessary for significance at the 5% level and that an r_s value of 0.75 must be reached for significance at the 1% level. Our obtained value exceeds both of these values. We therefore reject the null hypothesis and conclude that there is a significant correlation between the liberalism/conservatism and the political involvement rankings.

With larger samples (where $n = 10$ or more) an alternative way of testing the significance of r_s is suggested by Kendall (1948) using the t test.

The formula is given by:

$$t = r_s \sqrt{\frac{n-2}{1-r_s^2}}$$

In our example above

$$t = 0.88 \sqrt{\frac{10-2}{1-0.773}}$$

$$= 0.88\sqrt{35.4} = 5.24$$

Degrees of freedom are determined by d.f. $= n - 2$ where n is the number of ranks being compared. Reference to the table in Appendix 5 shows that for d.f. $= 8$, a t value of 2.31 is significant at the 5% level and a t value of 3.36 is significant at the 1% level. Our obtained value of $t = 5.24$ exceeds both of these values. We therefore reject the null hypothesis and conclude that there is a significant relationship between the rank ordering of our two variables, liberalism and involvement in campus politics.

The power efficiency of the Spearman rank order correlation test relative to the Pearson product moment test is reported to be 91.2% (Gibbons 1976).

Correcting for ties

When a small number of ties occur their effect on the rank order correlation coefficient is so negligible that it can be ignored. When the proportion of ties is large, however, a correction factor is necessary.

T, the correction factor, is given by:

$$T = \frac{t^3 - t}{12}$$

where t = the number of observations tied at a given rank. Siegel (1956) recommends the use of a different* form of the $r_s = 1 - (6 \sum d^2 / n (n - 1)(n + 1))$ formula in correcting for ties.

---

* In fact, the recommended formula for use with ties is the formula from which $r_s = 1 - (6 \sum d^2/n (n - 1)(n + 1))$ is derived. See Siegel (1956) pages 202-204 for a fuller discussion.

The recommended formula for r_s when correcting for extensive ties is:

$$r_s = \frac{\sum x^2 + \sum y^2 - \sum d^2}{2\sqrt{\sum x^2 \sum y^2}} \quad \text{where } \sum x^2 = \frac{n^3 - n}{12} \quad \text{and} \quad \sum y^2 = \frac{n^3 - n}{12}$$

$\sum x^2$ is corrected as follows:

$$\sum x^2 = \frac{n^3 - n}{12} - \sum T_x$$

where T_x is the correction factor for variable X observations tied at a given rank or ranks. Similarly, $\sum y^2$ is corrected as:

$$\sum y^2 = \frac{n^3 - n}{12} - \sum T_y$$

where $\sum T_y$ is the correction factor for variable Y observations tied at a given rank or ranks.

Let's see how the correction for ties is computed in a concrete example.

Suppose that the researcher into the politicization of university students referred to in Table 51 obtains similar data in respect of 12 more subjects. This time, however, their actual scores on liberalism/conservatism and political involvement are set out in Table 53.

Table 53 Scores on liberalism/conservatism and on political involvement: data for 12 subjects

Student	Variable X (liberalism/conservatism)	Variable Y (political involvement)
A	30	70
B	30	70
C	25	68
D	27	63
E	23	52
F	21	50
G	27	68
H	23	59
I	23	52
J	30	70
K	28	70
L	25	64

The scores from Table 53 have been rank ordered in Table 54. Observe that a large number of ties occur in respect of the scores on both variables X and Y.

In correcting for ties, the values of $\sum x^2$ and $\sum y^2$ must first be determined.

The correction factor $\sum T_x$ for the ties on the x variable is given by:

$$T_x = \frac{t^3 - t}{12}$$

where t = the number of observations tied at a given rank. The correction factor $\sum T_y$ for the ties on the Y variable is given by:

$$T_y = \frac{t^3 - t}{12}$$

where t = the number of observations tied at a given rank.

146

Notice that in Table 54, observations tied at a given rank are assigned the average of the ranks that would have been given had no ties occurred. (Lowest scores are assigned lowest ranks.)

Table 54 Rank ordering on liberalism/conservatism and on political involvement: data for 12 subjects, computation

Student	Rank order on variable X	Rank order on variable Y	d	d^2
A	11	10.5	0.5	.25
B	11	10.5	0.5	.25
C	5.5	7.5	−2.0	4.00
D	7.5	5	2.0	4.00
E	3	2.5	0.5	.25
F	1	1	0	0
G	7.5	7.5	0	0
H	3	4	−1.0	1.0
I	3	2.5	0.5	.25
J	11	10.5	0.5	.25
K	9	10.5	−1.5	2.25
L	5.5	6	−0.5	.25
				$\Sigma d^2 = 12.75$

For variable X

$$\Sigma T_x = \frac{t^3 - t}{12} = \frac{(3^3 - 3)}{12} + \frac{(2^3 - 2)}{12} + \frac{(2^3 - 2)}{12} + \frac{(3^3 - 3)}{12} = 5.0$$

For variable Y

$$\Sigma T_y = \frac{t^3 - t}{12} = \frac{(2^3 - 2)}{12} + \frac{(2^3 - 2)}{12} + \frac{(4^3 - 4)}{12} = 6.0$$

$$\Sigma x_2 \text{ corrected for ties} = \frac{n^3 - n}{12} - \Sigma T_x = \frac{12^3 - 12}{12} - 5.0 = 138$$

$$\Sigma y_2 \text{ corrected for ties} = \frac{n^3 - n}{12} - \Sigma T_y = \frac{12^3 - 12}{12} - 6.0 = 137$$

From Table 54 we see that $\Sigma d^2 = 12.75$. Substituting these values into the formula, we have:

$$r_s = \frac{\Sigma x^2 + \Sigma y^2 - \Sigma d^2}{2\sqrt{\Sigma x^2 \Sigma y^2}} = \frac{138 + 137 - 12.75}{2\sqrt{(138)(137)}} = .955$$

Suppose that we had computed r_s from the formula:

$$r_s = 1 - \frac{6 \Sigma d^2}{n(n-1)(n+1)}$$

that is without correcting for ties. Then

$$r_s = 1 - \frac{76.50}{12 (11)(13)} = .95542$$

We can see that in our example the correction for ties makes hardly any difference to the value of r_s.

147

13.4 Using KENDALL'S RANK ORDER CORRELATION COEFFICIENT (tau)

Wherever Spearman's rank order correlation coefficient is applicable, so too is Kendall's tau. Although it is a little more difficult to compute than Spearman's rho, when there are no ties the solution is short and simple.

The data in Table 55 are taken from our example of the correlation between student liberalism and active involvement in campus politics (Table 51). Again, we wish to find out whether there is a significant relationship between these two variables, the null hypothesis (H_o) being that liberalism and active involvement in campus politics are not related.

Kendall's rank order correlation coefficient (τ) is appropriate to our task. The assumptions for using τ are that the sample is randomly selected and that the data are at the ordinal level of measurement.

Kendall's τ is given by:

$$\tau = \frac{S}{\frac{1}{2}N(N-1)}$$

where

$$S = P - Q$$

P = the sum of the number of ranks that are larger.

Q = the sum of the number of ranks that are smaller.

PROCEDURE FOR COMPUTING THE KENDALL RANK ORDER CORRELATION COEFFICIENT

1 Rank the subjects on variable X from 1 to N, assigning 1 to be the highest ranking and so on.

2 Rank the subjects on variable Y from 1 to N, assigning 1 to be the highest ranking and so on.

3 Arrange the list of N subjects so that the X ranks are in their natural order, that is 1, 2, 3,... N.

4 Taking the data in Table 55 as our example, in the column for variable Y, for each individual, compute the number of ranks *below* him that are *larger* and enter in column P. Compute the number of ranks *below* him that are *smaller* and enter in column Q. Thus, in the case of subject B, there are 5 ranks below him that are larger (ranks 5, 9, 7, 10, and 8) and 0 ranks below him that are smaller.

5 Compute P by summing the values in column P.

6 Compute Q by summing the values in column Q.

7 Compute S, that is: $P - Q$.

8 Where two or more observations on either the X or the Y variable are tied, each subject is assigned the average of the ranks that would have been given had no ties occurred. The effect of ties is to change the denominator in the formula as follows:

$$\tau = \frac{S}{\sqrt{\frac{1}{2}N(N-1) - T_X}\sqrt{\frac{1}{2}N(N-1) - T_Y}}$$

where

$T_x = \frac{1}{2}\Sigma \, t(t - 1)$, t being the number of tied observations in each group of ties on variable X.

and

$T_Y = \frac{1}{2}\Sigma \, t(t - 1)$, t being the number of tied observations in each group of ties on variable Y.

The effect of correcting for ties is relatively small (see Siegel, 1956, pages 217-219 for a fuller discussion).

Table 55 Computing Kendall's rank order correlation coefficient—no ties

Student	Variable X (liberalism/conservatism) rank X	Variable Y (political involvement) rank Y	Number of ranks larger (P)	Number of ranks smaller (Q)
E	1	1	9	0
H	2	2	8	0
J	3	4	6	1
A	4	6	4	2
B	5	3	5	0
C	6	5	4	0
D	7	9	1	2
I	8	7	2	0
G	9	10	0	1
F	10	8	0	0
			P = 39	Q = 6

Substituting from Table 55 into the formula:

$$\tau = \frac{S}{\frac{1}{2}N(N-1)} = \frac{33}{5(9)} = 0.73$$

Recalling our rough and ready guide to the strength of a correlation coefficient (see page 83) there is a high relationship between liberalism and active involvement in student affairs. The significance of the relationship can be tested by reference to the table in Appendix 10.

Interpolating in that table at $N = 10$ and $S = 33$ we see that the probability associated with the occurrence (one-tailed) under H_0 of any value as extreme as our observed S is .001. We therefore reject the null hypothesis and conclude that there is a significant correlation between the liberalism/ conservatism and the political involvement rankings.

Where N is 10 or less, use the table in Appendix 10 to determine the probability associated with the occurrence under H_0. When N is larger than 10, τ may be considered to be normally distributed and the probability associated with the occurrence under H_0 of any value as extreme as the observed τ may be determined by computing Z and then finding the significance of Z by reference to Appendix 3(b).

Notice that although r_s (rho) and τ (tau) are computed from the same data in our example above, tau is the smaller. For a fuller discussion of the comparison of r_s and τ, see Siegel (1956) page 219.

The power efficiency of the Kendall rank order correlation test relative to the Pearson product moment test is reported to be 91% (Gibbons, 1976).

Correcting for ties: an example

Suppose that on a management training course ten potential executives have been rated on a personality measure (introversion/extroversion) and on an attitude-to-change scale. Their scores have been ranked and ordered in Table 56 and P and Q have been calculated in the usual way.

Table 56 Rank order of 10 management trainees on introversion/extroversion and attitude-to-change measures: computation with correction for ties

Subject	Variable x (Introversion/extroversion)	Variable y (Attitude-to-change)	P	Q
E	1	3.5	6	2
G	2	1.5	7	0
C	3	3.5	6	1
A	4	1.5	6	0
H	5	5.5	4	0
I	6	9.5	0	3
B	7	5.5	3	0
D	8	7	2	0
F	9	9.5	0	1
J	10	8	0	0
			$P = 34$	$Q = 7$

Kendall's τ with correction for ties is given by:

$$\tau = \frac{S}{\sqrt{\frac{1}{2}N(N-1) - T_x}\sqrt{\frac{1}{2}N(N-1) - T_y}}$$

For T_x, there are no ties among the scores of introversion-extroversion. Hence $T_x = 0$.

For T_y, there are four sets of tied ranks. That is to say, two subjects are tied at rank 1.5, two are tied at rank 3.5, two are tied at rank 5.5 and two are tied at rank 9.5.

T_y is computed as follows:

$$T_y = \tfrac{1}{2}\Sigma\, t(t-1)$$

$$= \tfrac{1}{2}[2(2-1) + 2(2-1) + 2(2-1) + 2(2-1)]$$

$$= 4$$

Thus $T_x = 0$, $T_y = 4$, $S = P - Q = 27$, and $N = 10$.

Substituting into the formula:

$$\tau = \frac{27}{\sqrt{\frac{1}{2}(10)(9) - 0}\sqrt{\frac{1}{2}(10)(9) - 4}}$$

$$= .628$$

Suppose we had not corrected for ties and had used the formula:

$$\tau = \frac{S}{\frac{1}{2}N(N-1)}$$

Our value for τ would have been .60. It can be seen that the effect of correcting for ties is very small.

13.5 Using THE POINT BISERIAL CORRELATION COEFFICIENT (r_{pb})

Sometimes a researcher wishes to assess the relationship between two variables but finds that the levels of measurement he has employed in gathering his data do not permit the use of the Pearson product moment correlation coefficient.

Suppose, for example, that a social work tutor wishes to examine the relationship between the scores that students obtain on a 'Theories of Counselling' examination paper and their pattern of residence over the duration of the counselling course, i.e. on-campus or off-campus living (see Table 57). This latter variable is truly dichotomous; that is to say, it can take only two values.

The point biserial correlation coefficient is used in computing the relationship between a *continuous** variable and one that is truly dichotomous.

The null hypothesis (H_o) in our example is that student performance in the examination is not related to their pattern of residence.

r_{pb} is given by the formula:

$$r_{pb} = \frac{M_p - M_q}{SD_X}\sqrt{pq}$$

where

M_p = the mean score of on-campus students

M_q = the mean score of off-campus students

SD_x = the standard deviation of the exam scores

p = the proportion of on-campus students

q = the proportion of off-campus students

PROCEDURE FOR COMPUTING THE POINT BISERIAL
CORRELATION COEFFICIENT

1 Compute the mean score of on-campus students (M_p)

2 Compute the mean score of off-campus students (M_q)

3 Compute the standard deviation of the exam scores (SD_X)

4 Compute the proportion of on-campus students (p)

5 Compute the proportion of off-campus students (q)

6 Compute the square root of pq ($\sqrt{pq}$)

7 Substitute the values M_p, M_q, SD_X, $\sqrt{pq}$ into the formula.

---

* Most variables in research can be represented by many points on a scale of measurement. The examination scores in the Theories of Counselling paper, for example, fall into this category. Such variables are referred to as *continuous variables.*

Table 57 Counselling examination scores of on-campus and off-campus social work students

Student	Examination score	Pattern of residence
A	85	on-campus
B	77	on-campus
C	63	off-campus
D	78	on-campus
E	71	off-campus
F	85	on-campus
G	62	off-campus
H	79	on-campus
I	64	off-campus
J	90	on-campus

$$M_p = \frac{85+77+78+85+79+90}{6} = \frac{494}{6} = 82.333$$

$$M_q = \frac{63+71+62+64}{4} = \frac{260}{4} = 65.0$$

$SD_x = 10.01$ (see page 38 for computation of SD)

$p = \frac{6}{10} = .60$

$q = \frac{4}{10} = .40$

Substituting these values into the formula:

$$r_{pb} = \frac{82.333 - 65.0}{10.01}\sqrt{(.60)(.40)}$$

$$= .848$$

From our rough-and-ready guide to the meaning of a correlation (page 83) we see that there is a high correlation between success in examination and on-campus residence.

Incidentally, r_{pb} does not behave exactly like r or rho in that it is not always possible to obtain values of +1.0 or −1.0. These values can only occur when p and q each equal 0.50. (Schmidt, 1979).

The significance of our result can be tested using the following formula:

$$t = \frac{r_{pb}\sqrt{N-2}}{\sqrt{1-r_{pb}^2}}$$

$$t = \frac{.848\sqrt{8}}{\sqrt{1-.848^2}}$$

$$= 4.526$$

Entering the table in Appendix 5 at $N - 2 = 8$ degrees of freedom we see that a t value of 3.36 is significant at the 1% level. Since our obtained value exceeds this we therefore reject the null hypothesis and conclude that students' performance in the examination is related to their pattern of residence.

13.6 Using the CORRELATION COEFFICIENT TETRACHORIC r

In our discussion of the phi coefficient (ϕ) (page 164) we use as our example the association between sex and driving test success, both variables being *truly dichotomous,* that is to say, they can take only two values. Suppose however that we wish to correlate two *dichotomized* variables, variables that is, that are recorded in only two categories but which, in reality, are continuous and normally distributed? Why bother, the reader may ask? Why not simply use the Pearson product moment correlation coefficient r? Regrettably it's often the case that the requisite data necessary to compute Pearson's r are not at hand. If our data are available only as dichotomies we have no alternative no matter how continuous and normal the underlying distributions may be! In such cases, tetrachoric r (r_t for short) is appropriate to our problem.

Because the computation of r_t is laborious, an approximation is generally used.* All we need to know is the value AD/BC from our fourfold table. We then read off the table in Appendix 12 the estimation of the tetrachoric coefficient r_t.

Suppose for example, that we have classified the achievement of 100 university students on a basic computing course as satisfactory or unsatisfactory on the basis of their high level or low level mathematics ability on entry to the course. The data are set out in Table 58.

$$\text{Substituting for } \frac{AD}{BC} \text{ we get } \frac{(40)(30)}{(10)(20)} = 6.00$$

Entering the table in Appendix 12 we see that for AD/BC = 5.81–6.03, the estimated r_t is .61.

Suppose we decide to compute the value of r_{cos} from the formula given at the bottom of this page rather than interpolating from the table in Appendix 12.

Table 58 Satisfactory/unsatisfactory ratings on a basic computing course

		Satisfactory	Unsatisfactory	
High level maths on entry	A	40	B 10	
Low level maths on entry	C	20	D 30	= 100

$$r_{cos} = \cos \frac{1}{1 + \sqrt{\dfrac{AD}{BC}}} 180°$$

$$r_{cos} = \cos \frac{180°}{1 + \sqrt{6.00}} = \frac{180°}{3.449} = 52° \ 19'$$

---

* This is cosine pi correlation coefficient, so called because its value is a function (the cosine) of an angle whose size is expressed as a fraction of 180°. The formula for cosine pi correlation coefficient is:

$$r_{cos} = \cos \frac{1}{1 + \sqrt{\dfrac{AD}{BC}}} 180° \quad \text{or} \quad r_{cos} = \cos \frac{\pi}{1 + \sqrt{\dfrac{AD}{BC}}}$$

Entering the table of natural cosines in Appendix 28 we see that cos 52° 19' = 0.6117, the same value as that had we used the table in Appendix 12.

Our result is subject to exactly the same type of interpretation that applies to Pearson's r since it is an estimate of that correlation coefficient. However, the standard error of r_t is considerably larger than for r. Moreover, the approximation obtained from using the table in Appendix 12 works best when both variables have been dichotomized on the basis of a 50–50 split.

Finally, if AD is found to be less than BC, simply use the ratio BC/AD in entering the table in Appendix 12. Remember, the larger of the two products is always placed in the numerator.

13.7   Using PARTIAL CORRELATION

When a researcher needs to compute the relationship between two variables, which may be wholly or partly due to the effect of a third variable, he can use the technique of partial correlation to control for the effects of that third variable.

Take for example the correlations found to exist between such variables as height, weight, general knowledge and vocabulary. Since all of these attributes increase between birth and early adulthood, the high positive correlation that generally obtains between any two of them is probably due to the common factor of age that is highly correlated with both variables. Eliminate the effects due to maturity and the correlation may well drop to zero. The partial correlation coefficient does precisely this. It eliminates or 'partials out' the effects of the third variable.

The general formula for the partial product moment correlation coefficient is given by:

$$r_{12.3} = \frac{r_{12} - r_{13}r_{23}}{\sqrt{(1 - r_{13}^2)(1 - r_{23}^2)}}$$

where

$r_{12.3}$ = the correlation between variables 1 and 2 with variable 3 held constant or partialed out

r_{12} = the correlation between variables 1 and 2

r_{13} = the correlation between variables 1 and 3

r_{23} = the correlation between variables 2 and 3

Suppose that we have the following three variables:

1. chronological age

2. weight

3. scores on a general knowledge test

and that we have computed correlations among the variables on a sample of 100 primary school pupils with the following results:

$$r_{12} = .80$$

$$r_{13} = .70$$

$$r_{23} = .60$$

Consulting the table in Appendix 8 we find that the correlation between weight and general knowledge score is significant. Suppose now that we compute the partial correlation between weight and general knowledge scores, controlling for the effect of age.

$$r_{23.1} = \frac{r_{23} - r_{12}r_{13}}{\sqrt{(1 - r_{12}^2)(1 - r_{13}^2)}}$$

$$= \frac{.60 - (.80)(.70)}{\sqrt{(1 - .80^2)(1 - .70^2)}}$$

$$= \frac{.04}{.42}$$

$$= .09$$

With the effects of chronological age removed, there is no significant relationship between weight and general knowledge scores. The significance of the partial r is interpreted in the same way as r, the Pearson product moment correlation coefficient.

In our example $r_{23.1} = .09$ and $n = 100$. Entering the table in Appendix 8 we see that for d.f. $= 98$ (i.e. $n - 2$) an r value of .20 is significant at the 5% level. Our obtained value fails to reach this. We therefore accept the null hypothesis and conclude that weight and general knowledge scores are unrelated.

Combining independent significance tests of partial relations

Sometimes researchers seek to discover whether a partial relation exists between two variables after controlling for the possible interaction effect of another independent variable by drawing several samples from the same population and then exploring whether this set of samples displays the same relation.

Let us suppose that a researcher has done this and has obtained data at the nominal scale of measurement from three samples. Controlling for *level of social deprivation* (Variable X_2), the investigator wishes to examine the relationship between the subjects' *ethnicity* (Variable X_1) and their *academic achievement* (Variable Y). The data are set out in Table 58.1.

Table 58.1 Partial relation between two nominal scale variables (X_1: ethnicity; Y: academic achievement) with a third nominal scale variable (X_2: level of social deprivation) held constant

Academic achievement (Y)	X_2 Level of social deprivation					
	Sample 1 Ethnicity (X_1)		Sample 2 Ethnicity (X_1)		Sample 3 Ethnicity (X_1)	
	A	B	A	B	A	B
HIGH	5	11	11	21	19	29
LOW	22	16	33	26	49	41
	$\chi_1^2 = 2.22$ ns. $\alpha = .14$		$\chi_2^2 = 3.05$ ns. $\alpha = .07$		$\chi_3^2 = 2.20$ ns. $\alpha = .14$	

Fisher (1941, pages 97–98) shows that the product of several independent αs may be transformed into a function having a χ^2 distribution by the application of the following formula based on *natural logarithms* and readily computed on a scientific calculator.

$$\chi^2 = -2 \log_e (\alpha_1) (\alpha_2) (\alpha_3) \ldots (\alpha_k)$$

Note: it follows that because all αs are less than 1.00, so too will be their product. Moreover, the logarithm of a number less than 1.00 is negative. Multiplying by -2 therefore gives a positive product, the significance of which can be determined by reference to the chi-square table in Appendix 4 at $2 \times k$ degrees of freedom.

$$\chi^2 = -2 \log_e (.14) (.07) (.14)$$

$$= -2 \log_e (.001372)$$

$$= -2 (-6.591)$$

$$= 13.18 \text{ significant at } p \le .05$$

Should tables of *common logarithms* be all that is available, an alternative form of the formula is given by:

$$\chi^2 = -4.60517 \log_{10} (\alpha_1) (\alpha_2) (\alpha_3) \ldots (\alpha_k)$$

$$= -4.60517 \log_{10} (.14) (.07) (.14)$$

$$= -4.60517 (-2.8626)$$

$$= 13.18 \text{ significant at } p \le .05$$

The product of the three αs has resulted in a χ^2 of 13.18 which at 6 d.f. (i.e. twice the number of the independent tests combined) is significant beyond the 5% level.

13.8 Using KENDALL'S PARTIAL RANK CORRELATION COEFFICIENT ($\tau_{12.3}$)*

There is always the possibility that the relationship between two variables may be influenced by a third variable. Suppose, for example, that we have rank ordered the scores of a group of secondary school pupils on measures of their (1) racism, (2) neuroticism, and (3) self-esteem and have computed the following rank correlation coefficients

$$\tau_{12} = .50$$

$$\tau_{13} = .80$$

$$\tau_{23} = .60$$

Both racism and neuroticism are strongly related to self-esteem and they appear to be related to each other. But is this last relationship a true one or is it simply the effect of the two variables being related to the common one, self-esteem? The technique of partial correlation allows us to control (or to partial out) the effect of the third variable in order to answer our question.

The partial τ formula is given by:

$$\tau_{12.3} = \frac{\tau_{12} - \tau_{13}\tau_{23}}{\sqrt{(1 - \tau_{13}^2)(1 - \tau_{23}^2)}}$$

---

* Read as follows: 'the correlation between variables 1 and 2 with variable 3 held constant or partialed out.' Notice that the formula for Kendall's $\tau_{12.3}$ is directly comparable to the general formula for the parametric partial product moment correlation (Siegel, 1956 footnote p. 226), in our example on page 154.

where

τ_{12} = the correlation between variables 1 and 2.

τ_{13} = the correlation between variables 1 and 3.

τ_{23} = the correlation between variables 2 and 3.

Applying the formula to our data:

$$\tau_{12.3} = \frac{.50 - (.80)(.60)}{\sqrt{(1 - .80^2)(1 - .60^2)}}$$

$$= \frac{.02}{.48}$$

$$= .04$$

It can be seen that with the effects of self-esteem partialed out, there is a very low correlation between pupils' scores on racism and neuroticism.

Comparable equations to that for $\tau_{12.3}$ can be written for $\tau_{13.2}$ and $\tau_{23.1}$

$$\tau_{13.2} = \frac{\tau_{13} - \tau_{12}\tau_{23}}{\sqrt{(1 - \tau_{12}^2)(1 - \tau_{23}^2)}}$$

and

$$\tau_{23.1} = \frac{\tau_{23} - \tau_{12}\tau_{13}}{\sqrt{(1 - \tau_{12}^2)(1 - \tau_{13}^2)}}$$

13.9   Using the MULTIPLE CORRELATION COEFFICIENT R

When a researcher wishes to know the relationship between one variable and two or more other variables considered simultaneously he can employ the technique of multiple correlation. The general formula for the multiple correlation coefficient is given by:

$$R_{1.23} = \sqrt{\frac{r_{12}^2 + r_{13}^2 - 2r_{12}r_{13}r_{23}}{1 - r_{23}^2}}$$

Where $R_{1.23}$ is the multiple correlation between variable 1 and a combination of variables 2 and 3; r_{12} is the product-moment correlation coefficient between variables 1 and 2; r_{13} is the product-moment correlation coefficient between variables 1 and 3; r_{23} is the product-moment correlation coefficient between variables 2 and 3.

Suppose, for example that a university admissions tutor is dissatisfied with selection procedures for undergraduate entry which rely solely on the quality of candidates' A-level performance and wishes to find out whether the addition of a measure of student personality will enable a better prediction to be made of first-year examination results. He obtains the following data in respect of 100 students:

1. First-year examination results

2. Weighted A-level scores

3. Neuroticism.

He then computes correlations among the three variables as follows:

$$r_{12} = .50$$
$$r_{13} = -.60$$
$$r_{23} = -.20$$

Substituting these values in the multiple correlation formula,

$$R_{1.23} = \sqrt{\frac{(.50)^2 + (-.60)^2 - 2(.50)(-.60)(-.20)}{1 - (-.20)^2}}$$

$$R_{1.23} = \sqrt{\frac{.61 - .12}{.96}} = .71$$

What exactly does this result mean in practical terms to the admissions tutor? Simply this. Whereas he is able to attribute 25% (i.e. $(.50)^2$) of the variance in first-year examination results to variance in the A-level scores on their own and 36% (i.e. $(-.60)^2$) of the variance in the examination results to student anxiety scores on their own, by the combination of A-level and anxiety measures, 50.4% (i.e. $(.71)^2$) of the variance in first-year examination performance can now be accounted for.

Testing the significance of the multiple correlation coefficient

Analysis of variance may be used to test the significance of R.

(i)   The square of R gives the variance due to the three measures (first-year exams, A-level scores, and neuroticism level).

(ii) $1 - R^2$ is the residual variance.

(iii)   The total number of degrees of freedom is given by $(N - 1)$, where N = the number of students tested.

Testing the significance of multiple R by the F ratio

Source of variance	Sum of Squares	d.f.	Mean Square	F
All three measures	.504	3	.168	$F = \dfrac{.168}{.005}$
Residual	.496	96	.005	$= 33.6$
Total	1.000	99		

Interpolating in the table in Appendix 6 at $v_1 = 3$, $v_2 = 96$ we see that an F value of 4.03 is significant at the 1% level. Our obtained value exceeds this indicating that the multiple R is significant.

13.10   Using MULTIPLE REGRESSION ANALYSIS

Multiple regression analysis, as we explained in Section 9.3, is used to estimate values of one variable (dependent variable) from a knowledge of two or more variables (independent variables). The assumptions for this technique are the same as those for the product moment correlation and simple linear regression, namely: homoscedasticity, that is to say, the variances in Y values are comparable to variances in X values; and the data are normally distributed when the analysis is used for inferential purposes.

By way of further illustration let's consider the following hypothetical project.

As part of an enquiry into the effectiveness of the National Health Service it was decided to examine the extent to which the average length of stay in hospital for surgical and medical conditions was affected by available and occupied beds. Data gathered from a sample of Local Health Authorities are displayed in Table 59.

Table 59   NHS hospitals: number of beds and patient flow

Sample authority	Average length of stay (days)	Available beds (Thousands)	Occupied beds (Thousands)
1	9.0	26	21
2	9.3	28	24
3	8.2	14	11
4	9.8	30	25
5	10.7	31	26

While the sample size selected for this example is far too small to satisfy the requisite assumptions for the analysis technique it does allow us to illustrate the technique without unduly increasing the computational burden.

1   Expand the data in Table 59 by computing X_1Y, X_2Y, X_1X_2, Y^2, X_1^2, X_2^2.

2   Compute the totals (Σ) for each column of the table.

Table 60   Computing the multiple regression equation

Sample	Length of stay (Y)	Available beds (X₁)	Occupied beds (X₂)	X_1Y	X_2Y	X_1X_2	Y^2	X_1^2	X_2^2
1	9.0	26	21	234.0	189.0	546	81	676	441
2	9.3	28	24	260.4	223.2	672	86.49	784	576
3	8.2	14	11	114.8	90.2	154	67.24	196	121
4	9.8	30	25	294.0	245.0	750	96.04	900	625
5	10.7	31	26	331.7	278.2	806	114.49	961	676
	47	129	107	1234.9	1025.6	2928	445.26	3517	2439

The regression equation for this example is given by:

$$Y = a + b_1X_1 + b_2X_2$$

where b_1 and b_2 are the partial regression coefficients.

3   Substitute the appropriate values from Table 60 into the following equations.

$$\sum Y = Na + b_1 \sum X_1 + b_2 \sum X_2 \tag{1}$$

$$\sum X_1 Y = a \sum X_1 + b_1 \sum X_1^2 + b_2 \sum X_1 X_2 \tag{2}$$

$$\sum X_2 Y = a \sum X_2 + b_1 \sum X_1 X_2 + b_2 \sum X_2^2 \tag{3}$$

where n = sample size. Therefore

$$47 = 5a + b_1 129 + b_2 107 \tag{1}$$

$$1234.9 = a\,129 + b_1 3517 + b_2 2928 \tag{2}$$

$$1025.6 = a\,107 + b_1 2928 + b_2 2349 \tag{3}$$

4   Solve the three linear equations simultaneously. A step-by-step explanation of the process by which equations are solved simultaneously is shown in Section 9.3 pages 95 to 98.
Using that method on our data in this example we see that

$$b_1 = 0.08629$$

$$b_2 = 0.03589$$

$$a = 6.40568$$

5   Complete the computed multiple regression equation.

$$Y_c = 6.40568 + 0.08629X_1 + 0.03589X_2$$

Given this equation we are now in a position to predict the value of Y_c (length of stay) for particular values of X_1 (available beds) and X_2 (occupied beds). For example, if in a selected Health Authority there were 24 thousand available beds and 20 thousand occupied beds, how long might we expect the average stay in hospital to be?

$$Y_c = 6.40568 + 0.08629(24) + 0.03589(20)$$

$$= 6.40568 + 2.07096 + 0.71780$$

$$= 9.19444$$

The predicted length of stay in hospital is approximately 9.2 days.

As in the case of simple linear regression the accuracy of the above prediction will depend upon the degree of sampling error. In order to complete our multiple regression analysis this error must be calculated.
To compute the STANDARD ERROR OF THE ESTIMATE for multiple regression the following formula is employed:

$$SE_{(Y \cdot X_1 X_2)} = SD_Y \sqrt{1 - R^2_{(Y \cdot X_1 X_2)}}$$

where

R^2 = coefficient of multiple correlation squared (coefficient of multiple determination)
SD_Y = standard deviation of the dependent variable Y

1   To solve the equation for standard error we first have to calculate the coefficient of multiple correlation using the formula on page 157,* namely;

$$R = \sqrt{\frac{(r_{Y \cdot X_1})^2 + (r_{Y \cdot X_2})^2 - 2(r_{Y \cdot X_1})(r_{Y \cdot X_2})(r_{X_1 X_2})}{1 - (r_{X_1 X_2})^2}}$$

where $r_{Y \cdot X_1}, r_{Y \cdot X_2}, r_{X_1 X_2}$ are the separate product moment correlations.

(a)   Correlation 1.

For length of stay (Y) and available beds (X_2)

$$r_{Y \cdot X_1} = \frac{n \sum X_1 Y - (\sum X_1)(\sum Y)}{\sqrt{[n \sum X_1^2 - (\sum X_1)^2][n \sum Y^2 - (\sum Y)^2]}}$$

$$= \frac{5(1234.9) - (129)(47)}{\sqrt{[5(3517) - (129)^2][5(445.26) - (47)^2]}}$$

$$= \frac{6174.5 - 6063}{\sqrt{[17585 - 16641][2226.3 - 2209]}}$$

$$= 0.8725$$

(b)   Correlation 2.

For length of stay (Y) and occupied beds (X_2)

$$r_{Y \cdot X_2} = \frac{n \sum X_2 Y - (\sum X_2)(\sum Y)}{\sqrt{[n \sum X_2^2 - (\sum X_2)^2][n \sum Y^2 - (\sum Y)^2]}}$$

$$= \frac{5(1025.6) - (107)(47)}{\sqrt{[5(2439) - (107)^2][5(445.26) - (47)^2]}}$$

$$= \frac{5128 - 5029}{\sqrt{(12195 - 11449)(2226.3 - 2209)}}$$

$$= 0.8715$$

(c)   Correlation 3.

For available beds (X_1) and occupied beds (X_2)

$$r_{X_1 X_2} = \frac{n \sum X_1 X_2 - (\sum X_1)(\sum X_2)}{\sqrt{[n \sum X_1^2 - (\sum X_1)^2][n \sum X_2^2 - (\sum X_2)^2]}}$$

$$= \frac{5(2928) - (129)(107)}{\sqrt{[5(3517) - (129)^2][5(2439) - (107)^2]}}$$

$$= \frac{14640 - 13803}{\sqrt{[17585 - 16641][12195 - 11449]}}$$

$$= 0.997$$

---

*   If there are more than two independent variables calculate R^2 by using β weights, the procedure for which is explained in Section 9.5 pages 98 to 100.

2   Calculate the standard deviation of Y (length of stay).

$$SD_Y = \sqrt{\frac{\sum Y^2 - \frac{(\sum Y)^2}{N}}{N-1}}$$

$$= \sqrt{\frac{445.26 - 441.8}{4}}$$

$$= 0.93$$

3   Compute the coefficient of multiple determination.

$$R^2_{Y \cdot x_1 x_2} = \frac{(0.8725)^2 + (0.8715)^2 - 2(0.8725)(0.8715)(0.997)}{1-(0.997)^2}$$

$$= \frac{0.00462}{0.00599}$$

$$= 0.77129$$

4   Correct for degrees of freedom
In regression analysis, if we wish to make inferences about population parameters from sample measures it is advisable, particularly when sample sizes are small, to adjust the coefficient of multiple determination for degrees of freedom.* This is done by using

$$R^2_c = 1-(1-R^2) \cdot \left(\frac{N-1}{N-k}\right)$$

where

R^2_c = Adjusted or corrected coefficient of multiple determination.
R = Sample coefficient of determination.
N = Sample size.
k = Number of constants in regression equation.

(If the sample sizes are large and number of independent variables small we see that $((N-1)/(N-k))$ is nearly 1 and therefore R approximates R^2_c.) For this example

$$R^2_c = 1-(1-0.77129)\left(\frac{5-1}{5-3}\right)$$

$$= 0.5426$$

---

*   A detailed discussion of the rationale for this adjustment is beyond the scope of this text

5 Compute the standard error of the estimate.

$$SE_{(Y \cdot X_1 X_2)} = SD_Y \sqrt{1 - R^2_{c(Y \cdot X_1 X_2)}}$$
$$= 0.93\sqrt{1 - 0.5426}$$
$$= 0.6290$$

When this standard error is applied to our predicted value of $Y = 9.1944$ it can be interpreted that we are 68.28% confident that the actual length of stay in hospital will be 9.1944 ± 0.6290 ($Y \pm 1(SE)$). Similarly we can be 95% confident that the length of stay will be 9.1944 ± 1.2328 ($Y \pm 1.96(SE)$), and 99% confident that the length of stay will be 9.1944 ± 1.5977 ($Y \pm 2.54(SE)$).

13.11   Using the PERCENTAGE DIFFERENCE

One of the simplest asymmetric* measures of association, the *percentage difference* consists of comparing percentages in different columns of the same row, or conversely, in different rows of the same column. Look, for example, at the data in Table 61.

Table 61   Percentage of Public Library members by social class status

Public library membership	Social class status	
	Middle class	Working class
Member	87	38
Non-member	13	62
Total	100	100

What information can we glean from the table?

By comparing percentages in different columns of the same row we can see that 49% more middle class persons are members of libraries than are working class persons.

By comparing percentages in different rows of the same column we can see that 74% more middle class persons are members rather than non-members.

The data suggest, do they not, an association between the social class status of individuals and their membership of Public Libraries.

A second way in which we can make use of the data in Table 61 is to compute *percentage ratio* (%R). Take the information in the second row of the table for instance. By dividing 62 by 13 (%R = 4.8) we can say that almost five times as many working class persons are not members of Public Libraries as are middle class persons.

The *percentage difference* ranges from 0% when there is complete independence to 100% when there is complete association in the direction being examined. It is easy to calculate and simple to comprehend. Notice however that the percentage difference as we have defined it can only be employed when there are only two categories in the variable along which we percentage and only two categories in the variable in which we compare.

---

*   An *asymmetric* measure is a measure of *one-way association*. That is to say, it measures the extent to which one phenomenon implies the other but not vice versa. Measures which are concerned with the extent to which two phenomena imply each other are referred to as *symmetric* measures.

13.12   Using the PHI COEFFICIENT φ

Suppose that we have the 'first-time' driving test results of a sample of 200 individuals and that we have classified the data by sex and by success/failure as shown in Table 62.

We wish to explore the association between the two variables, the null hypothesis in this case being that there is no relationship between sex and success/failure in driving test results.

Table 62   Sex and success or failure in first-time driving test results

Sex	Success	Failure	
Male	70	28	(98)
Female	50	52	(102)
	(120)	(80)	(200)

When each of the variables is *truly dichotomous,* that is to say, can take only two values (male/female; pass/fail; right/wrong) then the phi coefficient is an appropriate test of association.

Phi is given by the formula:

$$\phi = \frac{ad - bc}{\sqrt{klmn}}$$

where the cells in a 2 × 2 contingency table and the marginal totals are lettered as follows:

a	b	(k)
c	d	(l)
(m)	(n)	(N)

PROCEDURE FOR COMPUTING THE PHI COEFFICIENT

1 Cast the data in the form of a 2 × 2 contingency table. Compute the marginal totals $k, l, m, n,$ and the grand total N.

2 Compute ad and subtract bc.

3 Divide ad − bc by the square root of the *klmn*.

Substituting our data from Table 62 above:

$$\phi = \frac{ad - bc}{\sqrt{klmn}} = \frac{3640 - 1400}{\sqrt{(98)(102)(120)(80)}}$$

$$= \frac{2240}{\sqrt{95,961,600}} = 0.229$$

The significance of the result can be tested using the following formula:

$$\chi^2 = N\phi^2$$

where degrees of freedom are given by:

$$(r - 1)(c - 1), r = \text{rows}, c = \text{columns in the contingency table.}$$

In our example $(r - 1)(c - 1) = (2 - 1)(2 - 1) = 1$ df. Substituting our obtained phi value in the formula:

$$\chi^2 = 200(.0524)$$
$$= 10.48$$

Entering the table in Appendix 4 we see that a χ^2 value of 6.63 is significant at the 1% level. Our obtained value exceeds this. We therefore reject the null hypothesis and conclude that there is a significant relationship between the sex of the driver and first-time driving test success.

13.13   Using YULE'S Q

Yule's Q is a measure developed especially for 2×2 tables where both measures are dichotomous. Q belongs to a group of measures of association based on a pair-by-pair comparison of all entries with all other entries (see pages 69–74 for a fuller discussion).

Q is given by the formula:

$$Q = \frac{ad - bc}{ad + bc}$$

where a, b, c, and d represent the cells of a 2×2 contingency table as set out in the figure below.

	A Present	A Absent
B Present	a	b
B Absent	c	d

The extent to which A is associated with B is measured by examining every possible pair. Thus:

A *positive association* between A and B is recorded for all pairs in which one member is from the a cell and one member is from the d cell.

A *negative association* between A and B is recorded for all pairs in which one member is from the b cell and one member is from the c cell.

Pairs formed from other combinations of cells are alike in possession of one or both attributes and are thus irrelevant, that is to say, if both possess attribute A or attribute B, then the pair can throw no light on whether the attributes are associated.

PROCEDURE FOR COMPUTING YULE'S Q

1   Multiply a times d and b times c to give ad and bc.

2   Subtract bc from ad.

3   Add bc and ad.

4   Divide the result of Step 2 by the result of Step 3 to give Q.

In the example in Table 63, we are interested in the degree to which inoculation prevents the outbreak of whooping cough in young babies, but not in the converse. Clearly a one-way measure of association is required.

Table 63 Incidence of whooping cough in babies who have been and have not been inoculated

Contracted whooping cough	Inoculated		Total
	Yes	No	
Yes	2	40	42
No	498	460	958
Total	500	500	1000

$$Q = \frac{ad - bc}{ad + bc} = \frac{(2 \times 460) - (40 \times 498)}{(2 \times 460) + (40 \times 498)} = -0.91$$

What exactly does our result of $Q = -0.91$ mean? Simply that there is a strong negative association between inoculation and contracting whooping cough. Put somewhat more technically, Q is a measure of the proportionate preponderance of relevant instances supporting the idea that inoculation and freedom from whooping cough are associated over relevant instances supporting the idea that an association exists between inoculation and the contracting of whooping cough.

Some characteristics of Q

Where there is complete lack of association, the coefficient equals zero. Q varies between +1.0 and – 1. 0, the direction of the association being indicated by the sign. Unlike some measures of association (gamma for example, page 265) Q is margin-free, that is to say, it does not vary with the size of the marginal proportions.

13.14 Using the CONTINGENCY COEFFICIENT C

A sociologist interested in the relationship between social class and educational achievement obtains data on comprehensive school pupils' placement in different strata of the school's curriculum and on their social class backgrounds. The data are set out in Table 64.

Table 64 School curriculum placement and social class

Curriculum placement	Social Class				
	I and II	III	IV	V	
G.C.S.E.	22	41	22	13	(98)
N.V.Q.	11	36	40	19	(106)
No examinations	1	12	48	62	(123)
Total	(34)	(89)	(110)	(94)	(327)

He wishes to determine the association between curriculum placement and social class, his null hypothesis (H_o) being that social class and school curriculum placement are unrelated.

The contingency coefficient C is a measure of the association between two sets of attributes. It is particularly useful when one or both of those attributes are at the nominal level of measurement as in the present example.

C is given by the formula:

$$C = \sqrt{\frac{x^2}{N + x^2}}$$

That is to say, the contingency coefficient C is based on the chi-square statistic and in order to compute C we must first work out the value of x^2.

PROCEDURES FOR COMPUTING THE CONTINGENCY COEFFICIENT C

1 Compute chi square (x^2) from the formula:

$$x^2 = \sum \frac{(O - E)^2}{E}$$

where O represents the observed frequency for any given cell and E represents the expected frequency in that cell, E being calculated by multiplying the column and row totals for that cell and dividing by N.

2 Compute C by inserting the chi-square value into the formula:

$$C = \sqrt{\frac{x^2}{N + x^2}}$$

The expected frequencies in our example of social class and school curriculum placement are given in Table 65 below.

Table 65 School curriculum placement and social class: expected frequencies

	Social Class			
Curriculum placement	I and II	III	IV	V
G.C.S.E.	10.19	26.67	32.96	28.17
N.V.Q.	11.02	28.85	35.66	30.47
No examination	12.79	33.48	41.38	35.36

$$x^2 = \sum \frac{(O - E)^2}{E}$$

$$= 13.69 + 7.70 + 3.64 + 8.16 + 0.00 + 1.77 + 0.53$$
$$+ 4.32 + 10.87 + 13.78 + 1.06 + 20.07$$

$$= 85.59$$

Substituting into our formula for C:

$$C = \sqrt{\frac{\chi^2}{N+\chi^2}} = \sqrt{\frac{85.59}{327+85.59}} = 0.45$$

Testing the significance of the contingency coefficient

The significance of C is determined from the value of χ_2 for the data in hand. Degrees of freedom are given by d.f. = $(k-1)(r-1)$ where k = columns and r = rows.

In our example d.f. = $(4-1)(3-1) = 6$.

Entering the table in Appendix 4 at d.f. = 6 we see that a critical value of 16.81 is significant at the 1% level. The obtained value exceeds this. The sociologist should reject the null hypothesis and conclude that there is a significant association between social class and placement in the school curriculum.

Limitations of the contingency coefficient C

Siegel (1956) identifies the following limitations in connection with the use of C.

(i)   While the contingency coefficient equals zero when there is no association between two variables, its upper limit is something less than 1 although as the number of columns and rows increase, C approaches 1.

(ii)   The second limitation of C is a direct consequence of (i) above, namely that since the upper limit of C depends upon the sizes of k and r, two contingency coefficients are not comparable unless they are derived from contingency tables of the same size.

(iii)   A third limitation of C is that the data must be amenable to the computation of χ_2 before C can be used.

(iv)   A fourth limitation of C is that it is not directly comparable with other measures of correlation such as Pearson's r, Spearman's rho or Kendall's tau.

Despite these deficiencies, Siegel observes that the contingency coefficient is an extremely useful measure of association because of its wide applicability. Moreover its freedom from assumptions and requirements permits its use as an indicator of relationship when many other measures of association are inapplicable.

13.15   Using Cramer's V

Overcoming the size limitation of the contingency coefficient C (above) Cramer's V is a commonly used measure of association ranging from 0 to 1 regardless of table size. Like the phi coefficient (f) and the contingency coefficient, Cramer's V is based on the chi-square statistic.

Cramer's V is given by:

$$V = \sqrt{\frac{\chi^2}{N \min. (r-1)(c-1)}}$$

where min. $(r-1)$, $(c-1)$ refers to either $r-1$ or $c-1$ whichever is the *smaller* and N is the sample size.

Thus, computing V for the data in Table 65 we have:

$$V = \sqrt{\frac{85.59}{327 \times 2}} = 0.362$$

It isn't possible to say, precisely, what is a strong or weak value of Cramer's V and this is why it needs to be read in conjunction with the chi-square test of significance.

13.16 Choosing a measure of association

1   When we wish to compare the extent to which one nominal level variable implies another nominal level variable *but not vice-versa* (i.e. when we require an asymmetric measure of association, see page 69), and our data are in the form of percentages, then the *percentage difference* and *percentage ratio* are appropriate.

2   When we wish to compare the association between two variables each at the nominal level and each *truly dichotomous,** then the phi coefficient (ϕ) is appropriate.

3   When we wish to compare the association between two variables each at the nominal level and *each dichotomous,* then Yule's Q is appropriate.

4   When we wish to compare the association between two sets of variables and one or both sets are at the nominal level then the contingency coefficient C is appropriate. See however the limitations of C outlined on page 168.

5   When we wish to compare the association between two variables each of which is at the ordinal level then Spearman's rho or Kendall's tau is appropriate, Spearman's rho being the easier to compute.

6   When we wish to compare the association between two variables when a third variable is held constant, all three variables being at the ordinal level, then Kendall's partial rank order correlation coefficient is appropriate.

7   When we wish to compare the association between two variables each of which is at the interval/ratio level and where samples have been randomly selected from normally distributed populations, then the Pearson product moment correlation coefficient is appropriate.

---

*   That is to say, they can take only two values.

CHAPTER 14
DESIGN 3

One group: repeated observations on the same subjects under two conditions

Group 1	Condition 1	Condition 2
Subjects A	X_{A_1}	X_{A_2}
B	X_{B_1}	X_{B_2}
C	X_{C_1}	X_{C_2}
D	X_{D_1}	X_{D_2}

or

Before and after treatment

Group 1	Before Treatment Observations (b)	After Treatment Observations (a)
Subjects A	$X_{A_{(b)}}$	$X_{A_{(a)}}$
B	$X_{B_{(b)}}$	$X_{B_{(a)}}$
C	$X_{C_{(b)}}$	$X_{C_{(a)}}$
D	$X_{D_{(b)}}$	$X_{D_{(a)}}$

14.1 Using the t TEST FOR CORRELATED DATA

A researcher investigating the effectiveness of different forms of advertising randomly selects ten subjects to take part in an experiment to determine if reaction time to a visible stimulus is different from reaction time to an audible stimulus. The null hypothesis (H_o) in this case is that there is no difference between reaction time to visual and auditory stimuli. The data are set out in Table 66.

Table 66   Reaction times to visual and auditory stimuli

Subjects	Reaction time to visual stimulus (m. secs)	Reaction time to auditory stimulus (m. secs)
A	259	201
B	275	198
C	304	245
D	285	287
E	288	190
F	314	250
G	291	285
H	304	295
I	285	231
J	246	201

When data are at the interval level of measurement and taken from one randomly selected sample on two occasions or from two matched samples on one occasion, the difference of the two sets of data can be estimated using the *t* test for correlated means.

t is given by the formula:

$$t = \frac{\sum D}{\sqrt{\dfrac{N \sum D^2 - (\sum D)^2}{N-1}}}$$

where *D* is the difference between each subject's two scores and *N* is the number of subjects.

PROCEDURE FOR COMPUTING *t* FOR CORRELATED DATA

1 Compute the difference (*D*) between each subject's two scores.

2 Square the *D*'s for each subject.

3 Sum the *D* values to give $\sum D$.

4 Sum the D^2 values to give $\sum D^2$.

Table 67   Reaction times to visual and auditory stimuli: computation

Subjects	Visual stimulus	Auditory stimulus	D	D²
A	259	201	58	3364
B	275	198	77	5929
C	304	245	59	3481
D	285	287	−2	4
E	288	190	98	9604
F	314	250	64	4096
G	291	285	6	36
H	304	295	9	81
I	285	231	54	2916
J	246	201	45	2025

$\sum D = 468$ $\sum D^2 = 31536$

Substituting our data from Table 67

$$t = \frac{468}{\sqrt{\dfrac{10(31536) - (468)^2}{9}}} = 4.52$$

Degrees of freedom are given by the formula:

$$\text{d.f} = N - 1 = 9$$

Enter the table in Appendix 5 at d.f. = 9. If the obtained value is larger than the values in the table then it is significant at those levels of confidence. The obtained value of $t = 4.52$ exceeds the critical value at the 5% level ($t = 2.26$) and the critical value at the 1% level (3.25). The researcher should therefore reject the null hypothesis and conclude that there is a significant difference between reaction times to visual and audible stimuli.

Example   Using Sandler's A Statistic

A simple alternative to t for correlated data is suggested by Sandler (1955).* Sandler's A is given by the formula:

$$A = \frac{\text{the sum of the squares of the differences } (\sum D^2)}{\text{the square of the sum of the differences } (\sum D)^2}$$

Substituting the data from the example of visual and audible stimuli (Table 67 above)

$$A = \frac{(\sum D^2)}{(\sum D)^2} = \frac{31536}{(468)^2} = 0.144$$

Degrees of freedom are given by $N - 1 = 9$.

Enter the Table in Appendix 27 at $N - 1 = 9$. If our obtained value is *equal* to or *less than* the values in the table then it is significant at those levels of confidence.

Our obtained value of $A = 0.144$ is less than the critical value at the 1% level ($A = 0.213$) and at the 0.5 % level ($A = 0.185$). We therefore reject the null hypothesis, and conclude that there is a significant difference between reaction times to visual and audible stimuli. Our conclusion is precisely the same as the researcher would arrive at using the t test for correlated data. This is bound to be so since the A Statistic and the t test are mathematically equivalent.

14.2   Using the WILCOXON MATCHED-PAIRS SIGNED-RANKS TEST

A psychiatric social worker interested in changes in maternal behaviour from pre-sibling to post-sibling birth in respect of mothers' interactions with their first-born children obtains the data in Table 68 on 15 mothers' verbal prohibitioning during extended clinic sessions.

The social worker wishes to know whether the observed frequencies represent a significant change in mothers' behaviour towards their first-born. The null hypothesis (H_0) in this case is that there is no difference in the rate of verbal prohibitioning in the pre-sib and post-sib conditions.

The Wilcoxon matched-pairs signed-ranks test is appropriate to the problem. It can be used to assess the significance of difference between two samples consisting of matched pairs of subjects.

---

* J. Sandler (1955) A test of the significance of the difference between the means of correlated measures based on a simplification of Student's t. Brit. J. of Psychology, 46, 225-226.

Table 68 Verbal prohibitions as % of all verbal interactions (1000 10-sec units)

Subjects	Pre-sib-birth	Post-sib-birth
A	15.1	21.2
B	14.5	17.8
C	16.5	22.0
D	19.2	18.8
E	16.9	19.2
F	14.3	17.1
G	16.2	16.4
H	22.5	27.0
I	20.4	22.8
J	19.5	18.5
K	14.2	21.8
L	16.9	16.1
M	20.5	21.7
N	26.2	37.5
O	12.2	14.5

Matched pairs of subjects would, of course, include two measures taken on the same subject as in the present example. The Wilcoxon test is the non-parametric counterpart of the t test for correlated data that we illustrated earlier (see page 170).

PROCEDURE FOR COMPUTING THE WILCOXON MATCHED-PAIRS
SIGNED-RANKS TEST

1   For each subject or pair determine the difference in scores (d).

2   Rank these differences ignoring the plus or minus signs and differences of 0. When ranks are tied assign the average of the tied ranks.

3   Assign each rank the + or – sign of the difference it represents.

4   Identify T, the smaller sum of the like-signed ranks. That is to say, compute the total ranks with + signs and the total ranks with – signs. Select the *smaller* of the two.

5   In the case of small samples ($N \leq 25$), consult the table in Appendix 17 to determine the significance of T for various sizes of N. When the observed value of T is *equal to or less than* the critical value given in the table, then the null hypothesis (H_o) may be rejected at that particular level of significance.

6   When $N > 25$ the table in Appendix 17 is not applicable and the T value has to be transformed to a Z score using the formula:

$$Z = \frac{T - \dfrac{N(N+1)}{4}}{\sqrt{\dfrac{N(N+1)(2N+1)}{24}}}$$

When extensive ties occur a correction factor $(\Sigma u^3 - \Sigma u)/48$ is introduced as follows:

$$Z = \frac{T - \dfrac{N(N+1)}{4}}{\sqrt{\dfrac{N(N+1)(2N+1)}{24} - \dfrac{\Sigma u^3 - \Sigma u}{48}}}$$

Example where N<25:

Table 69 Verbal prohibitions in pre-sibling and post-sibling conditions: computation

Subjects	Pre-sibling conditions	Post-sibling conditions	d	Rank of d	Smaller like-signed ranks
A	15.1	21.2	−6.1	−13	
B	14.5	17.8	−3.3	−10	
C	16.5	22.0	−5.5	−12	
D	19.2	18.8	0.4	2	2
E	16.9	19.2	−2.3	−6.5	
F	14.3	17.1	−2.8	−9	
G	16.2	16.4	−0.2	−1	
H	22.5	27.0	−4.5	−11	
I	20.4	22.8	−2.4	−8	
J	19.5	18.5	1.0	4	4
K	14.2	21.8	−7.6	−14	
L	16.9	16.1	0.8	3	3
M	20.5	21.7	−1.2	−5	
N	26.2	37.5	−11.3	−15	
O	12.2	14.5	−2.3	−6.5	

The total (T) = the smaller sum of the like-signed ranks.

$$T+ = 2 + 4 + 3 = 9$$

Entering the table in Appendix 17 we find that a critical value of 19 is significant at the 2% level for $N = 15$. When an obtained value of T is less than that in the table there is a significant difference between the sets of scores at that level. The obtained value of $T = 9$ is less than the critical value. The psychiatric social worker should reject the null hypothesis and conclude that there is a significant increase in the verbal prohibitioning of first-born children in the sample of mothers observed.

Ties

To illustrate the computation of Z in large samples when extensive ties occur, the data in Table 70 relate to pre- and post-test scores of management science students after exposure to an intensive short accountancy course. The researcher wishes to determine whether attitudes to accountancy as a profession have changed as a consequence of the course, the null hypothesis (H_0) being that no change has occurred.

Table 70 Attitudes to accountancy as a profession

Subject	Pre-intensive course	Post-intensive course	d	Rank of d	Smaller like-signed ranks
1	12	15	−3	−11	
2	8	10	−2	−5.5	
3	9	6	3	11	11
4	7	15	−8	−25	
5	10	13	−3	−11	
6	11	12	−1	−2	
7	14	10	4	16.5	16.5
8	15	19	−4	−16.5	
9	8	12	−4	−16.5	
10	6	10	−4	−16.5	
11	11	18	−7	−22.5	
12	11	16	−5	−19.5	
13	20	18	2	5.5	5.5
14	14	21	−7	−22.5	
15	8	7	1	2	2
16	12	19	−7	−22.5	
17	15	18	−3	−11	
18	14	19	−5	−19.5	
19	19	26	−7	−22.5	
20	14	13	1	2	2
21	9	7	2	5.5	5.5
22	12	14	−2	−5.5	
23	16	19	−3	−11	
24	14	11	3	11	11
25	12	9	3	11	11

$$\sum T = 64.5$$

In Table 70 we have assigned the average of tied ranks in the column headed 'Rank of d'. $\sum T$, the sum of the smaller like-signed ranks = 64.5. The correction factor $(\sum u^3 - \sum u)/48$ for tied ranks is applied as follows:

u = the number of absolute differences that are tied for a given rank and $\sum u$ is the sum over all sets of u tied ranks.

In Table 70 there are 6 sets of tied ranks

the 1s make a set of 3 tied ranks, or $u_1 = 3$

the 2s make a set of 4 tied ranks, or $u_2 = 4$

the 3s make a set of 7 tied ranks, or $u_3 = 7$

the 4s make a set of 4 tied ranks, or $u_4 = 4$

the 5s make a set of 2 tied ranks, or $u_5 = 2$

and the 7s make a set of 4 tied ranks, or $u_6 = 4$

We sum these values of u over the 6 sets to get the correction factor:

$$\frac{\sum u^3 - \sum u}{48} = \frac{(3^3 + 4^3 + 7^3 + 4^3 + 2^3 + 4^3) - (3 + 4 + 7 + 4 + 2 + 4)}{48} = 11.375$$

We may now compute Z.

$$Z = \frac{T - \dfrac{N(N+1)}{4}}{\sqrt{\dfrac{N(N+1)(2N+1)}{24} - \dfrac{\sum u^3 - \sum u}{48}}}$$

$$= \frac{64.5 - \dfrac{25\,(26)}{4}}{\sqrt{\dfrac{25\,(26)\,(51)}{24} - 11.375}} = -2.65$$

The significance of the obtained Z value may be determined by reference to the table in Appendix 3(b). There we see that the two-tailed p of $Z = -2.65$ is .008. The researcher should reject the null hypothesis and conclude that there is a significant change of attitudes as a consequence of the accountancy course.

The power efficiency of this test relative to the t test for correlated means is reported as 95.5% (Gibbons, 1976).

14.3 Using the McNEMAR TEST FOR THE SIGNIFICANCE OF CHANGE

Often when we are interested in the effects of a particular activity or intervention upon the behaviour of individuals we are able to organize our data into a before-and-after design in which each person acts as his/her own control.

For example, listed below are the ratings of 25 children who were judged on their level of sociability on referral to a Child Guidance Clinic and after an intensive course of group therapy with professional workers there. The sociability ratings range from (5) high level of sociability to (1) low level of sociability.

We wish to know whether what appears to be an overall improvement in the sociability of the children is, in fact, a significant change. The null hypothesis (H_o) in this case is that the observed improvement is no different from what could be expected to occur by chance. The data are set out in Table 71.

The McNemar test for the significance of change is appropriate to our task. The assumptions for using the McNemar test are that the pairs of scores or observations are randomly drawn from a population and that the data are either at nominal or at ordinal levels of measurement.

Table 71 Direction of change in levels of sociability following group therapy sessions ($n = 25$)

Child	Sociability rating on referral	Sociability rating after therapy	Direction of change
A	1	2	+
B	3	4	+
C	4	2	−
D	2	3	+
E	2	3	+
F	2	2	0
G	1	1	0
H	2	4	+
I	2	5	+
J	1	3	+
K	2	1	−
L	2	2	0
M	1	3	+
N	2	4	+
O	2	4	+
P	1	4	+
Q	1	1	0
R	2	3	+
S	3	2	−
T	2	4	+
U	1	3	+
V	3	5	+
W	1	3	+
X	1	1	0
Y	2	4	+

The McNemar test for the significance of change is given by the formula:

$$\chi^2 = \frac{(|A - D| - 1)^2}{A + D}$$

where the data have been cast into a fourfold table such as the one below

After

		−	+
Before	+	A	B
	−	C	D

and where the −1 in the formula is a correction factor for continuity (see Siegel, 1956, page 64 for a fuller discussion).

PROCEDURE FOR COMPUTING THE McNEMAR TEST

1 For each pair of scores or observations, record the direction of change. Sum to obtain the values of A and D.

2 If $\frac{1}{2}$ of $(A + D)$ is less than 5 then the McNemar test cannot be employed. Use the Sign Test instead (see next Section and Appendix 13).

3 Compute χ^2 and determine significance by reference to the table in Appendix 4.

The data in Table 71 are cast in a fourfold table. If a child's sociability rating remains unchanged enter his tally either in Box C or Box B. It doesn't matter which box is used since we are only interested in Boxes A and D, that is, those Boxes that indicate that changes have occurred.

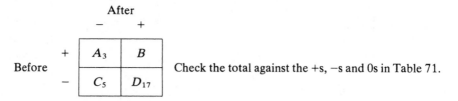

Check the total against the +s, −s and 0s in Table 71.

The McNemar test for the significance of change is given by:

$$\chi^2 = \frac{(|A - D| - 1)^2}{A + D} = \frac{(|3 - 17| - 1)^2}{20} = 8.45$$

From the table in Appendix 4 we see that at d.f. = 1, a critical value of 3.84 is significant at the 5% level and a critical value of 6.13 is significant at the 1% level. Our obtained value exceeds these. We therefore reject the null hypothesis and conclude that the change in sociability ratings for the group of children as a whole is significant.

14.4   Using the SIGN TEST

As part of a Social Psychology course, students are required to participate in an experiment called 'Knowledge of Results'. The experiment aims to find out whether or not knowledge of results aids performance in judging (by sense of touch) the length of a wooden rod.

Briefly, each participant is blindfolded and allowed 50 attempts (in a series of 5 × 10 tries) to estimate the length of a 9 cm wooden rod and to draw a line of the same length on a piece of paper.

In the 'before treatment' condition participants are given NO FEEDBACK at all about their 50 attempts to match the length of the wooden rod in their drawings.

In the 'treatment' condition, each time a line is drawn that is within ±5 mm of the correct length of 9 cm, subjects are informed, 'That is correct'.

In Table 72 below are the results of 16 subjects. We wish to know whether feedback significantly affects performance in the experimental task. The null hypothesis (H_0) in this case is that task performance is not related to feedback.

The Sign test is appropriate to our problem. As the name implies, the Sign test uses plus and minus signs rather than quantitative data. The test makes no assumptions about the form of the distribution of differences between the 'before treatment' and 'treatment' conditions, nor does it require that our subjects are randomly selected or even drawn from the same population. The only assumption made in using the Sign test is that the variable under investigation has a continuous distribution.

PROCEDURE FOR COMPUTING THE SIGN TEST WITH SMALL SAMPLES $(N < 25)$

1   Record the sign of the difference between the scores for each subject as shown in the last column of our example in Table 73 below.

Table 72 Knowledge of results data

Group 1	'Before treatment' NO FEEDBACK No. of correct responses in each of 10 tries					'Treatment' FEEDBACK No. of correct responses in each of 10 tries				
Subject	10	10	10	10	10	10	10	10	10	10
A	0	1	1	2	0	2	4	5	5	6
B	0	0	0	0	0	0	2	1	2	3
C	4	0	4	4	4	5	5	6	7	8
D	0	0	0	0	1	0	0	0	0	0
E	0	1	0	0	0	0	0	0	2	2
F	0	0	0	0	0	0	0	0	0	0
G	0	1	1	2	2	1	3	2	3	4
H	0	0	1	1	1	1	1	2	2	1
I	3	2	0	1	3	4	4	5	6	6
J	0	0	1	0	0	0	0	0	0	0
K	2	0	2	2	3	3	2	4	4	1
L	2	3	2	4	2	4	4	4	6	4
M	1	0	1	1	1	2	0	2	2	1
N	2	1	2	1	1	2	3	3	3	4
O	0.	1	1	2	2	0	2	2	1	4
P	0	1	0	1	0	2	2	2	4	2

2 Drop from the analysis any subject for whom a difference of 0 is recorded.

3 Count up the number of *fewer* signs. Let these be x.

4 Count up the total number of paired observations against which either a + or a − has been recorded. Let these be N.

5 For $N \leq 25$ enter the table in Appendix 13(a) down the left-hand column for N and across the top row for x. At their intersection is the *one-tailed* probability associated with a value as small as x. For a *two-tailed* test multiply p by 2.

From Table 73 we find that $N = 15$ (notice that we drop subject F from the analysis), and $x = 2$.

Entering the table in Appendix 13(a) at $N = 15$ and $x = 2$, the one-tailed probability of our obtained value occurring by chance is 0.004. Since we made no prediction about the direction in which differences in the 'before treatment' and 'treatment' comparisons would occur, we apply a two-tailed test to our findings:

$$2 \times 0.004 = 0.008$$

We therefore reject the null hypothesis and conclude that since only 8 in 1000 times could our result have occurred by chance, feedback has a significant effect upon the performance of our subjects.

COMPUTING THE SIGN TEST WITH LARGER SAMPLES ($N > 25$)

Where N is greater than 25, use the formula:

$$Z = \frac{(x \pm \frac{1}{2}) - \frac{1}{2}N}{\frac{1}{2}\sqrt{N}}$$

179

Table 73 Computation of knowledge of results data

Group 1 Subject	'Before treatment' NO FEEDBACK No. of correct responses in each of 10 tries					TOTAL	'Treatment' FEEDBACK No. of correct responses in each of 10 tries					TOTAL	Sign of difference
	10	10	10	10	10		10	10	10	10	10		
A	0	1	1	2	0	4	2	4	5	5	6	22	+
B	0	0	0	0	0	0	0	2	1	2	3	8	+
C	4	0	4	4	4	16	5	5	6	7	8	31	+
D	0	0	0	0	1	1	0	0	0	0	0	0	−
E	0	1	0	0	0	1	0	0	0	2	2	4	+
F	0	0	0	0	0	0	0	0	0	0	0	0	0
G	0	1	1	2	2	6	1	3	2	3	4	13	+
H	0	0	1	1	1	3	1	1	2	2	1	7	+
I	3	2	0	1	3	9	4	4	5	6	6	25	+
J	0	0	1	0	0	1	0	0	0	0	0	0	−
K	2	0	2	2	3	9	3	2	4	4	1	14	+
L	2	3	2	4	2	13	4	4	4	6	4	22	+
M	1	0	1	1	1	4	2	0	2	2	1	7	+
N	2	1	2	1	1	7	2	3	3	3	4	15	+
O	0	1	1	2	2	6	0	2	2	1	4	9	+
P	0	1	0	1	0	2	2	2	2	2	4	12	+

$x = 2$
$N = 15$

where N and x have the same meaning as before and where $\pm\frac{1}{2}$ is used as a correction factor. When x is less than $\frac{1}{2}N$ (i.e. $x < \frac{1}{2}N$) the correction factor is $+\frac{1}{2}$. When x is greater than $\frac{1}{2}N$ (i.e. $x > \frac{1}{2}N$) the correction factor is $-\frac{1}{2}$.

Suppose, for example, that a researcher looking into the effects of Health Visitor advice in a large sample of one-parent families has obtained the following data:

$$N=100$$

$$x = 30$$

In this event, the correction factor will be $+\frac{1}{2}$ since 30 is less than $\frac{1}{2}N$ (i.e. 50). Substituting in the formula:

$$Z = \frac{(x + \frac{1}{2}) - \frac{1}{2}N}{\frac{1}{2}\sqrt{N}} = \frac{30.5 - 50}{\frac{1}{2}\sqrt{100}} = \frac{-19.5}{5} = -3.9$$

Reference to the table in Appendix 3(b) shows that the probability of a $Z = -3.9$ is $2 \times (0.00005) = 0.0001$, assuming again that a two-tailed test is appropriate. In light of this result, the researcher would reject the null hypothesis and conclude from the data that his result is significant.

The power efficiency of the Sign Test relative to the t test is reported as 63.7% for large samples (Gibbons 1976).

CHAPTER 15
DESIGN 4

One group—multi-treatment (trials): treatments as independent variable

Group	After treatment 1	After treatment 2	After treatment 3	After treatment 4
Subjects A	X_{A_1}	X_{A_2}	X_{A_3}	X_{A_4}
B	X_{B_1}	X_{B_2}	X_{B_3}	X_{B_4}
C	X_{C_1}	X_{C_2}	X_{C_3}	X_{C_4}
D	X_{D_1}	X_{D_2}	X_{D_3}	X_{D_4}

15.1 Using ONE-WAY ANALYSIS OF VARIANCE FOR CORRELATED MEANS (with repeated measures on the same sample or separate measures on matched samples)

Analysis of variance techniques are designed to test differences between the means of several (>2) groups of scores and are based upon the analysis of factors known or unknown which account for the variability of scores.

Recall that in Section 8.12 (page 86) we explained how the *total* variability or variance of a set of scores can be divided up or partitioned into *systematic variance* and *error variance*. We said that systematic variance is that part of the total variance caused by factors which lean scores in one direction or the other and that error variance is that part of the total variance caused by chance fluctuations in measures.

Analysis of variance involves the further partitioning and subsequent analysis of those systematic and error variances. As we shall see, the methods of partitioning differ depending upon the particular experimental design that we employ.

Differences between the means of groups of scores are tested by calculating the statistic F which compares the variability between group measures (means) with the variability between individual scores within the group.

Use of analysis of variance demands that the following assumptions be met:

(a)   Scores are measured on at least an interval scale.

(b)   Samples are taken at random.

(c)   Samples are taken from normally distributed populations.

(d)   The variances of the sample populations are equal.

Example   Suppose that as part of an investigation into the changes in atmospheric pollution the smoke concentrations at various sites throughout a city were measured at various times over one week at three different times in the year. We wish to determine whether average concentrations of smoke levels change over the three occasions. Our data are set out in Table 74.

Table 74   Atmospheric pollution: average smoke concentrations at urban sites (microgrammes per cubic metre)

	Smoke concentrations		
Sites	October (X_1)	February (X_2)	May (X_3)
A	26	25	25
B	24	25	22
C	23	24	25
D	27	27	26
E	27	25	24
F	27	23	23
G	23	24	23
H	24	26	27

The null hypothesis in this case is that there is no change in average smoke levels over the given time period.

The *total* variance in this repeated measures design can be partitioned into:

SYSTEMATIC EFFECTS

1   BETWEEN TRIALS VARIANCE is the variance between the means of each group of scores caused by the independent variable, that is, the time period between testing.

2   BETWEEN SITE VARIANCE is the variance due to the same sites being measured on three different occasions, thus giving correlated data. Some sites, for example, score consistently higher than others on all three occasions.

If measures had been taken on people this variance would be called 'between subject' variance.

ERROR EFFECTS

3   INTERACTION or RESIDUAL VARIANCE is caused by individual variation or site variation on the three different testing occasions. It measures factors due not only to the individual effects of sites, or time of testing, but the *joint* effect of these variables; hence the term *interaction*.

$$F_{\text{BETWEEN TRIALS}} = \frac{\text{Between trials variance}}{\text{Interaction or residual variance}}$$

$$F_{\text{BETWEEN SITES}} = \frac{\text{Between sites variance}}{\text{Interaction or residual variance}}$$

$$\text{where Variance} = \frac{\text{Sum of squares}}{\text{Degrees of freedom}}$$

PROCEDURE FOR COMPUTING ONE-WAY ANALYSIS OF VARIANCE
(CORRELATED MEANS)

1 Square all the scores, sum the columns and rows and include in Table 75.

Table 75 Atmospheric pollution: computational procedures

| | Smoke Concentrations | | | | | | |
| | October | | February | | May | | Totals |
Sites	X_1	X_1^2	X_2	X_2^2	X_3	X_3^2	$(X_1+X_2+X_3)$
A	26	676	25	625	25	625	76
B	24	576	25	625	22	484	71
C	23	529	24	576	25	625	72
D	27	729	27	729	26	676	80
E	27	729	25	625	24	576	76
F	27	729	23	529	23	529	73
G	23	529	24	576	23	529	70
H	24	576	26	676	27	729	77
	$\sum X_1 = 201$		$\sum X_2 = 199$		$\sum X_3 = 195$		$\sum X = 595$
	$\sum X_1^2 = 5073$		$\sum X_2^2 = 4961$		$\sum X_3^2 = 4773$		$\sum X^2 = 14{,}807$

2 Calculate the grand total GT (correction factor)

$$GT = \frac{(\sum X)^2}{N} = \frac{(\sum X_1 + \sum X_2 + \sum X_3)^2}{N}$$

where N = Total number of scores (24).

$$GT = \frac{(595)^2}{24} = \frac{354{,}025}{24}$$

$$= 14{,}751.04$$

3 Compute the TOTAL sum of squares (SS_{TOTAL})

$$SS_{\text{TOTAL}} = \sum X^2 - GT$$

$$= 14{,}807 - 14{,}751.04$$

$$= 55.96$$

183

4 Compute BETWEEN TRIALS sum of squares (SS_{TRIALS})

$$SS_{TRIALS} = \frac{(\sum X_1)^2}{N_1} + \frac{(\sum X_2)^2}{N_2} + \frac{(\sum X_3)^3}{N_3} - GT$$

$$= \frac{(201)^2}{8} + \frac{(199)^2}{8} + \frac{(195)^2}{8} - 14{,}751.04$$

$$= 5050.13 + 4950.13 + 4753.13 - 14{,}751.04$$

$$= 2.335$$

5 Compute BETWEEN SITES sum of squares (SS_{SITES})

$$SS_{SITES} = \frac{(\sum X_A)^2}{N_T} + \frac{(\sum X_B)^2}{N_T} + \cdots + \frac{(\sum X_H)^2}{N_T} - GT$$

where $\sum X_A$ = Site A's total scores on all three occasions (row total).
N_r = Number of trials.

$$= \frac{(76)^2}{3} + \frac{(71)^2}{3} + \frac{(72)^2}{3} + \frac{(80)^2}{3} + \frac{(76)^2}{3} + \frac{(73)^2}{3} + \frac{(70)^2}{3} + \frac{(77)^2}{3} - 14{,}751.04$$

$$= 1925.33 + 1680.33 + 1728 + 2133.33 + 1925.33 + 1776.33 + 1633.33 + 1976.33 - 14{,}751.04$$

$$= 27.291$$

6 Compute INTERACTION or RESIDUAL sum of squares (SS_{INT})

$$SS_{INT} = SS_{TOTAL} - (SS_{TRIALS} + SS_{SITES})$$

$$= 55.96 - (2.335 + 27.291)$$

$$= 26.334$$

7 Compute the degrees of freedom for each sum of squares.

$$\text{d.f. for } SS_{TOTAL} = (N - 1) = (24 - 1) = 23$$

where N = total number of scores.

$$\text{d.f. for } SS_{TRIALS} = (k - 1) = (3 - 1) = 2$$

where k = number of trials or columns.

$$\text{d.f. for } SS_{SITES} = (r - 1) = (8 - 1) = 7$$

where r = number of sites or rows.

$$\text{d.f. for } SS_{INT} = (r - 1)(k - 1) = (8 - 1)(3 - 1)\ 14$$

8 Estimate the variances.

$$\text{Variance} = \frac{\text{Sum of squares}}{\text{d.f.}}$$

$$\text{Var}_{TRIALS} = \frac{SS_{TRIALS}}{\text{d.f.}_{TRIALS}} = \frac{2.335}{2} = 1.168$$

$$\text{Var}_{SITES} = \frac{SS_{SITES}}{\text{d.f.}_{SITES}} = \frac{27.291}{7} = 3.899$$

$$\text{Var}_{\text{INT}} = \frac{SS_{\text{INT}}}{\text{d.f.}_{\text{INT}}} = \frac{26.334}{14} = 1.881$$

9   Compute F values for between trials and between sites.

$$F_{\text{TRIALS}} = \frac{\text{Var}_{\text{TRIALS}}}{\text{Var}_{\text{INT}}} = \frac{1.168}{1.881} = 0.621$$

$$F_{\text{SITES}} = \frac{\text{Var}_{\text{SITES}}}{\text{Var}_{\text{INT}}} = \frac{3.899}{1.881} = 2.073$$

10   Enter results in an Analysis of Variance table.

Table 76   Analysis of variance table

Source of variation	Sum of squares	d.f.	Variance	F
Between trials	2.335	2	1.168	0.621
Between sites	27.291	7	3.899	2.073
Interaction	26.334	14	1.881	
Total	55.96	23		

11   Refer to the table in Appendix 6 to determine the significance of each F value. Our F value for trials (0.621) is less than the value in the table (3.72) for 2 and 14 degrees of freedom at the 0.05 level. We therefore accept the null hypothesis and conclude that there is no difference in mean smoke concentrations over the specified period.

Our F value for sites (2.073) is less than the value in the table (2.78) for degrees of freedom 7 and 14 at the 0.05 level. We therefore accept the null hypothesis and conclude that there is no significant difference between sites when considering them without regard for trials.

If significant differences had been found between trials a further test called a Tukey Test would have to have been applied. This test is illustrated on pages 246 to 247.

15.2   Using the FRIEDMAN TWO-WAY ANALYSIS OF VARIANCE BY RANKS (χ_r^2)

At the end of an interview each of six management consultants is independently asked to place five candidates in order of suitability for appointment to a top executive post. The selection committee chairman wishes to find out whether or not there is a significant bias towards one particular candidate on the part of the selectors. The data are set out in Table 77.

Table 77 Selectors' preferences for candidates

Management consultants	Candidates				
	L	M	N	O	P
A	2	4	5	1	3
B	1	5	3	2	4
C	3	2	1	4	5
D	2	4	3	1	5
E	3	5	1	2	4
F	2	3	4	1	5

The null hypothesis (H_0) in this case is that consultants' choices are not directed towards one particular candidate. Put differently, do the overall rankings of the five candidates differ significantly?

The Friedman two-way analysis of variance by ranks is appropriate to the problem. In our example we are looking at the same group of subjects (the six management consultants) under each of five conditions (their choices of five candidates). The Friedman test assumes ordinal level data and is given by the formula:

$$\chi_r^2 = \frac{12H}{NK(K+1)} - 3N(K+1)$$

where

N = the number of rows

K = the number of columns

and

R^2 = the square of the rank total

and

H = the sum of the squares of the rank totals, i.e. ΣR^2

PROCEDURE FOR COMPUTING THE FRIEDMAN TEST (χ_r^2)

1 Calculate the necessary elements for the formula. Thus:

Table 78 Selectors' preferences for candidates: computation

Management consultants	Candidates				
	L	M	N	O	P
A	2	4	5	1	3
B	1	5	3	2	4
C	3	2	1	4	5
D	2	4	3	1	5
E	3	5	1	2	4
F	2	3	4	1	5
	$R_1 = 13$	$R_2 = 23$	$R_3 = 17$	$R_4 = 11$	$R_5 = 26$

$$N = 6$$

$$K = 5$$

$$(R_1)^2 = 169; \quad (R_2)^2 = 529; \quad (R_3)^2 = 289$$

$$(R_4)^2 = 121; \quad (R_5)^2 = 676$$

$$H = (R_1)^2 + (R_2)^2 + (R_3)^2 + (R_4)^2 + (R_5)^2 = 1784$$

Substituting into the formula:

$$\chi_r^2 = \frac{12H}{NK(K+1)} - 3N(K+1) = \frac{12(1784)}{(6)(5)(5+1)} - 18(5+1) = 10.9$$

Testing the significance of χ_r^2

The significance of χ_r^2 may be tested by entering the table in Appendix 18 at the appropriate column (i.e. $N = 6$, $K = 5$). The critical value of χ_r^2 is 9.07 at the 5% level. The obtained value exceeds this. The committee chairman may therefore reject the null hypothesis and conclude that the management consultants' choices are directed towards one particular candidate, candidate O.

The power efficiency of the Friedman test relative to the parametric F test for correlated means is reported as $95.5k/(k + 1)\%$ where k is the number of columns.

Correcting for ties

When observations are tied at a given rank they are assigned the average of the ranks that they would have been given had no ties occurred. Suppose that in our example of management consultants' choices of candidates (Table 77) each of the six judges has been unable to distinguish between the merits of certain candidates and that the final preferences are as set out in Table 79 below. A correction for ties, $1 - T/(N(K^3 - K))$, is introduced into the formula as follows:

$$\chi_r^2 = \frac{\dfrac{12H}{NK(K+1)} - 3N(K+1)}{1 - \dfrac{T}{N(K^3 - K)}}$$

where T = the sum of the corrections for ties, that $\Sigma\ (t^3 - t)$.

Table 79 Selectors' preferences for candidates showing tied rankings

Management consultants	Candidates				
	L	M	N	O	P
A	1.5	4	5	1.5	3
B	1	5	3.5	2	3.5
C	3	2	1	4.5	4.5
D	2.5	4	2.5	1	5
E	3	5	1.5	1.5	4
F	2	3.5	3.5	1	5
	$R_1 = 13$	$R_2 = 23.5$	$R_3 = 17.0$	$R_4 = 11.5$	$R_5 = 25.0$

To illustrate the computation of T, the data from Table 79 are again set out below. Look at Judge A. His assessment shows 2 ranks at 1.5, one at 3, one at 4, and one at 5. In this case the corrections will be

$$\Sigma\ (t^3 - t) = (2^3 - 2)\ + (1^3 - 1) + (1^3 - 1) + (1^3 - 1) = 6.$$

Table 80   Preferences for candidates showing tied rankings calculation

| Judges | \multicolumn{5}{c}{Candidates} | |
	L	M	N	O	P	$\Sigma\ (t^3 - t)$
A	1.5	4	5	1.5	3	$(2^3-2)+(1^3-1)+(1^3-1)+(1^3-1)=6$
B	1	5	3.5	2	3.5	$(1^3-1)+(1^3-1)+(2^3-2)+(1^3-1)=6$
C	3	2	1	4.5	4.5	$(1^3-1)+(1^3-1)+(1^3-1)+(2^3-2)=6$
D	2.5	4	2.5	1	5	$(2^3-2)+(1^3-1)+(1^3-1)+(1^3-1)=6$
E	3	5	1.5	1.5	4	$(1^3-1)+(1^3-1)+(2^3-2)+(1^3-1)=6$
F	2	3.5	3.5	1	5	$(1^3-1)+(2^3-2)+(1^3-1)+(1^3-1)=6$

Down the right-hand side of Table 80 all the corrections have been calculated.

$$T = \Sigma\ (t^3 - t)\ = 36.$$

The computation of χ_r^2 is exactly the same as before:

Step 1. The data are cast in a two-way table with K columns and N rows.

Step 2. The scores in each row are ranked from 1 to K, average ranks being assigned when ties occur.

Step 3. R, the sum of the ranks in each column is determined and then $H = \Sigma R^2$.

Step 4. χ_r^2 is then computed.

For the data in Table 80.

$$N = 6$$
$$K = 5$$
$$(R_1)^2 = (13)^2 = 169$$
$$(R_2)^2 = (23.5)^2 = 552.25$$
$$(R_3)^2 = (17)^2 = 289$$
$$(R_4)^2 = (11.5)^2 = 132.25$$
$$(R_5)^2 = (25)^2 = 625$$
$$H = (R_1)^2 + (R_2)^2 + (R_3)^2 + (R_4)^2 + (R_5)^2 = 1767.5$$

Substituting into the formula:

$$\chi_r^2 = \frac{\dfrac{12H}{NK(K+1)} - 3N(K+1)}{1\dfrac{T}{N(K^3-K)}}$$

$$\chi_r^2 = \frac{\dfrac{12(1767.5)}{(6)(5)(5+1)} - 18(5+1)}{1 - \dfrac{36}{6(125-5)}}$$

$$= \frac{9.83}{0.95} = 10.35$$

Testing the significance of χ_r^2

The significance of $\chi_r^2 = 10.35$ is tested by entering the table in Appendix 4 at the appropriate column (i.e. $N = 6$, $K = 5$). The critical value χ_r^2 is 9.07 at the 5% level.

The obtained value is more than this. In this case, the committee chairman may reject the null hypothesis and conclude that the management consultants' choices are directed towards a particular candidate, candidate O.

15.3 Using KENDALL'S COEFFICIENT OF CONCORDANCE (W)

Suppose that researchers in a Department of Town and Country Planning wish to identify factors influencing the relocation of commercial enterprises within a large metropolitan conurbation. They draw a sample of five recently relocated businesses, each one in a different area of the region and ask managing directors to rank-order reasons for the movement and relocation of their offices. The researchers wish to discover whether certain specific factors are influential in the decision to relocate, the null hypothesis (H_0) in this case being that relocation site and reason for movement are unrelated. The data are set out in Table 81.

Table 81 Factors influencing office movement and location choice

Location in metropolitan area	(a) Expansion	(b) Economy	(c) End of lease	(d) Demolition	(e) Staff hiring problems	(f) Traffic congestion	(g) Company reorganization	(h) Superior accommodation	(i) Access to rail links	(j) Access to airport	(k) Favourable rents, rates	(l) Access to city centre	(m) Proximity to hotels	(n) Ease of access to clients
Central city	1	2	3	4	5	6	7	8	9	10	11	12	13	14
Northern suburbs	6.5	6.5	3	13	13	3	1	6.5	13	10	6.5	10	10	3
Southern suburbs	3.5	8	3.5	3.5	3.5	11	11	8	3.5	3.5	8	13	11	14
Eastern suburbs	5	4	2	1	6	11	9	10	3	7	8	14	12	13
Western suburbs	1	2	4	4	4	6.5	11.5	6.5	8.5	11.5	11.5	14	8.5	11.5

When we wish to determine the relationship among three or more sets of ranks, Kendall's coefficient of concordance (W) is appropriate.

Kendall's W is given by the formula:

$$W = \frac{S}{1/12k^2(N^3 - N)}$$

where S is the sum of the squares of the observed deviations from the mean of R_j and is given by the formula

$$S = \sum \left(R_j - \frac{R_j}{N} \right)^2$$

and R_j is obtained by summing the ranks in each column of a $k \times N$ table.

In our example $N = 14$, that is, the number of reasons for the relocation of businesses; k = 5, the number of businesses sampled.

PROCEDURE FOR COMPUTING KENDALL'S COEFFICIENT OF CONCORDANCE

1 Sum the ranks (R_j) in each column of the table ($\sum R_j$).

2 Compute the mean value of R_j by dividing R_j by N.

3 With R_j and the mean value

$$M = \frac{\sum R_j}{N}$$

S can be computed.

4 S is the sum of the squares of the observed deviations from the mean of R_j and is given by:

$$S = \sum \left(R_j - \frac{\sum R_j}{N} \right)^2$$

5 $1/12k^2$ ($N^3 - N$) is then computed to give the maximum possible sum of the squared deviations. That is to say, the sum S which would occur given perfect agreement among the k rankings.

6 The Kendall coefficient of concordance is given by the formula:

$$W = \frac{S}{1/12k^2(N^3 - N)}$$

In our example there are a number of tied rankings. The effect of tied rankings is to depress the value of W as computed by the formula given above (Siegel, 1956, p. 234). A correction factor is therefore introduced as follows:

$$T = \frac{\sum (t^3 - t)}{12}$$

where t = the number of observations in any group that are tied for a given rank; and $k \sum_T T$ = all the values of T for all k rankings are to be summed.

Kendall's W with the correction factor for tied ranks is given by the formula:

$$W = \frac{S}{1/12k^2(N^3 - N) - k \sum_T T}$$

Table 82 Factors influencing office movement and location choice: computation

Location	Reasons													
	a	b	c	d	e	f	g	h	i	j	k	l	m	n
Central city	1	2	3	4	5	6	7	8	9	10	11	12	13	14
Northern suburbs	6.5	6.5	3	13	13	3	1	6.5	13	10	6.5	10	10	3
Southern suburbs	3.5	8	3.5	3.5	3.5	11	11	8	3.5	3.5	8	13	11	14
Eastern suburbs	5	4	2	1	6	11	9	10	3	7	8	14	12	13
Western suburbs	1	2	4	4	4	6.5	11.5	6.5	8.5	11.5	11.5	14	8.5	11.5
$R_i =$	17	22.5	15.5	25.5	31.5	37.5	39.5	39	37	42	45	63	54.5	55.5

$$\sum R_i = 525 \qquad M = \frac{\sum R_i}{N} = \frac{525}{14} = 37.5$$

The correction for ties is computed first. From Table 82:

$$K_1 \text{ Central city } T = 0$$

$$K_2 \text{ Northern suburbs } T = \frac{(3^3 - 3) + (4^3 - 4) + (3^3 - 3) + (3^3 - 3)}{12} = 11$$

$$K_3 \text{ Southern suburbs } T = \frac{(6^3 - 6) + (3^3 - 3) + (3^3 - 3)}{12} = 22$$

$$K_4 \text{ Eastern suburbs } T = 0$$

$$K_5 \text{ Western suburbs } T = \frac{(3^3 - 3) + (2^3 - 2) + (2^3 - 2) + (4^3 - 4)}{12} = 8$$

$$k \sum_T T = 41$$

Next we compute S

$$S = \sum \left(R_i - \frac{\sum R_i}{N} \right)^2$$

$$= (17 - 37.5)^2 + (22.5 - 37.5)^2 + (15.5 - 37.5)^2 + (25.5 - 37.5)^2$$

$$+ (31.5 - 37.5)^2 + (37.5 - 37.5)^2 + (39.5 - 37.5)^2 + (39 - 37.5)^2$$

$$+ (37 - 37.5)^2 + (42 - 37.5)^2 + (45 - 37.5)^2 + (63 - 37.5)^2$$

$$+ (54.5 - 37.5)^2 + (55.5 - 37.5)^2$$

$$= 2655.5$$

We are now ready to compute W

$$W = \frac{2655.5}{1/12(5^2)[(14)^3 - 14] - 5(41)} = 0.484$$

Testing the significance of W in larger samples (N > 7)

Where $N > 7$ the significance of W may be determined by reference to the table in Appendix 4 applying the following formula:

$$\chi^2 = k\,(N-1)\,W \text{ where degrees of freedom are given by } N - 1$$

Substituting our data from above:

$$\chi^2 = 5(13).484$$

$$= \underline{31.46}$$

From the table in Appendix 4 we see that a χ^2 value of 27.69 (d.f. = 13) is significant at the 1% level. Our obtained value exceeds this. We therefore reject the null hypothesis and conclude that irrespective of area of the metropolitan region that is chosen, relocation of business is related to specific reasons among those listed.

Testing the significance of W when N > 7 by means of the F distribution

An alternative to using χ^2 in testing the significance of W in larger samples (where $N > 7$) is suggested by Maxwell (1971, p. 120).

It involves an adjustment (W_1) to the value of W in order to correct for continuity. W_1 is given by the formula:

$$W_1 = \frac{12(S-1)}{k^2(N^3-N)+24}$$

A variance ratio or F value is then obtained as follows:

$$F = \frac{(k-1)\,W_1}{1-W_1}$$

F is then determined from the table in Appendix 6, with v_1 and v_2 degrees of freedom for the horizontal and vertical sides of the table respectively where:

$$v_1 = (N-1) - \frac{2}{k}$$

and

$$v_2 = (k-1)\left[(N-1) - \frac{2}{k}\right]$$

Since v_1 and v_2 will not be whole numbers, interpolation of the F table will be necessary. Substituting our obtained values from above:

$$S = 2655.5$$

$$k = 5$$

$$N = 14$$

$$W_1 = \frac{12(S-1)}{k^2(N^3-N)+24} = \frac{12(2654.5)}{25(14^3-14)+24} = 0.46$$

$$F = \frac{(k-1)\,W_1}{1-W_1} = \frac{4(.46)}{.54} = 3.40$$

$$\nu_1 = (N-1) - \frac{2}{k} = 13 - \frac{2}{5} = 12.6$$

$$\nu_2 = (k-1)\left[(N-1) - \frac{2}{k}\right] = 4(12.6) = 50.4$$

Interpolating the table in Appendix 6 at ν_1 (horizontal) 12.6 and ν_2 (vertical) 50.4, we see that a value of 2.58 is significant at the 1% level. Our obtained value exceeds this. We therefore reject the null hypothesis and conclude that irrespective of the area of the metropolitan region that is chosen, relocation of business is related to specific reasons among those listed.

Testing the significance of W when N < 7

Where N is less than 7, the significance of W may be found by reference to the table in Appendix 25.

15.4   Using the COCHRAN Q TEST

A psychologist experimenting in response latency effects presents subjects with four problems in booklet form. He arranges the problems in all possible permutations of four and each subject is required to work out all four problems, the tasks themselves being such that no overall practice effects occur. The experimenter wishes to find out whether the problems differ in difficulty as shown by the proportion of subjects solving each task correctly. The data are set out in Table 83 below.

Table 83   Frequency of correct solutions to each of four problems by ten subjects

	Problem number			
Subject	1	2	3	4
A	1	0	1	1
B	1	1	1	0
C	1	1	1	1
D	1	0	0	1
E	0	1	0	1
F	1	1	0	1
G	1	0	1	1
H	0	1	1	1
I	0	1	1	1
J	0	1	1	1

The null hypothesis (H_0) in this case is that there is no difference in the difficulty of the four problems.

The Cochran Q test is appropriate to the problem. Cochran's Q is an extension of the McNemar test for two related samples that we illustrate on pages 176 to 178. It provides a way of testing whether three or more matched sets of frequencies or proportions differ significantly among themselves. Matching may be based on characteristics of the different subjects or, as in our example, on the fact that the same subjects are used under different conditions. Cochran's Q is particularly suitable when the data are at the nominal level or are dichotomized ordinal information.

Practical Statistics for Students

Cochran's Q is given by the formula:

$$Q = \frac{(k-1)(k \sum C^2 - T^2)}{kT - \sum R^2}$$

where

k = the number of groups

R = the total of the rows

C = the total of the columns

T = the grand total.

PROCEDURE FOR COMPUTING COCHRAN'S Q TEST

1 Compute the row totals (R).

2 Square the row totals (R^2).

3 Compute the column totals (C).

4 Square the column totals (C^2).

5 Sum the row totals ($\sum R$).

6 Sum the column totals ($\sum C$).

7 Sum the squares of the row totals ($\sum R^2$).

8 Sum the squares of the column totals ($\sum C^2$).

Table 84 Frequency of correct solutions to each of four problems by ten subjects: computation

Subject	Problem number				Totals	
	1	2	3	4	R	R^2
A	1	0	1	1	3	9
B	1	1	1	0	3	9
C	1	1	1	1	4	16
D	1	0	0	1	2	4
E	0	1	0	1	2	4
F	1	1	0	1	3	9
G	1	0	1	1	3	9
H	0	1	1	1	3	9
I	0	1	1	1	3	9
J	0	1	1	1	3	9
Totals C	6	7	7	9	$T = 29$	$\sum R^2 = 87$
C^2	36	49	49	81	$\sum C^2 = 215$	

Substituting our data from Table 84:

$$Q = \frac{(k-1)(k \sum C^2 - T^2)}{kT - \sum R^2} = \frac{(4-1)(4(215)-(29)^2)}{4(29)-87}$$

$$Q = \frac{57}{29} = 1.96$$

We refer to the table in Appendix 4 since Q approximates χ^2 when the numbers of rows are not too small. Degrees of freedom are given by:

$$\text{d.f.} = k - 1 = 4 - 1 = 3$$

We see that for d.f. = 3 a Q (or a chi-square) value of 7.81 must be obtained to be significant at the 5% level. The obtained value fails to reach this critical value. The psychologist should therefore accept the null hypothesis and conclude that there is no significant difference in the difficulty of the four problems.

15.5 Using PAGE'S *L* TREND TEST

The Page *L* trend test is an extension of the Friedman two-way analysis of variance by ranks (see page 185). It is appropriate when the researcher is looking at *trends* between three or more conditions, that is to say, when in light of previous experience or some theory, he can predict the *ordering* of results under the various experimental treatments.

Suppose for example that a child psychologist is interested in the cognitive consequences of play and exploration in economically disadvantaged preschool children and has assigned six sets of matched subjects (five in each set) to each of four conditions in which the amounts of sociodrama, play and exploration training are varied. The psychologist's hunch is that cognitive gains are a direct consequence of the degree to which the children are exposed to these stimulating activities. In other words, he is able to predict the *ordering of results* under his four experimental conditions. The data are set out in Table 85.

Table 85 Cognitive gain scores in predicted order under four learning conditions

Six sets of matched subjects	Training conditions			
	1	2	3	4
A	16	17	15	19
B	13	14	18	20
C	16	13	20	22
D	15	19	20	19
E	14	16	18	20
F	12	12	15	24
	$M = 17.2$	18.2	21.2	24.8

The null hypothesis (H_0) in this case is that cognition gains are not related to the predicted order of experimental treatments.

Page's L is given by:

$$L = \sum (t_c \times c)$$

where

t_c = the rank totals for each column

c = the numbers allotted to the conditions (i.e. the columns) from left to right.

PROCEDURE FOR COMPUTING PAGE'S L

1 Arrange the conditions in the order predicted.

2 Rank the scores across each row separately (see Table 86).

3 Sum the ranks for each *column*.

Table 86 Ranked cognitive gain scores in predicted order under four learning conditions: computation

Six sets of matched subjects	Training conditions 1 Ranks	2 Ranks	3 Ranks	4 Ranks
A	2	3	1	4
B	1	2	3	4
C	2	1	3	4
D	1	2.5	4	2.5
E	1	2	3	4
F	1.5	1.5	3	4
Rank totals	8.5	12	17	22.5

Substituting into the formula:

$$L = \Sigma \, (t_c \times c) = (8.5 \times 1) + (12 \times 2) + (17 \times 3) + (22.5 \times 4)$$
$$= 173.5$$

Testing the significance of Page's L

The significance of our obtained value may be tested by entering the table in Appendix 19 at the appropriate intersection of *number of subject sets* and *number of conditions* (i.e. 6 subject sets and 4 conditions). If our obtained value is equal to or larger than the critical value we are in order to reject the null hypothesis. The critical value, we find, is 172 at the 0.1% level. Our obtained value exceeds this. We therefore reject the null hypothesis and conclude that cognitive gains in our sample of deprived children are related to the predicted order of experimental treatments.

The critical values in the table in Appendix 19 apply only to a one-tailed hypothesis when, as in Page's L, a trend is predicted in one direction.

CHAPTER 16
DESIGN 5

Two group research designs: static comparison on one variable

	Group 1	Subjects A, B, C, D	Group 2	Subjects E, F, G, H
	X_1	X_a	X_2	X_e
		X_b		X_f
		X_c		X_g
		X_d		X_h

OR

Two group: static comparison on one or more variables

	Variable A	Variable B	Variable C	Variable D	Variable E
Group 1	N_{1A}	N_{1B}	N_{1C}	N_{1D}	N_{1E}
Group 2	N_{2A}	N_{2B}	N_{2C}	N_{2D}	N_{2E}

16.1 Using the *t* TEST FOR INDEPENDENT SAMPLES (POOLED VARIANCE)

A Social Work Department wishing to validate an empathy scale gives the measure to intending Social Workers and to a group of students matched by age and sex whose career choices are other than Social Work. Table 87 shows the scores of the two groups on the empathy measure.

Do the scores indicate that intending Social Workers have higher empathy scores than non-Social Workers? Our null hypothesis (H_o) in this case is that there is no significant difference between the two groups. Notice that in testing the null hypothesis here, we are looking at the direction in which an expected difference between Social Workers and non-Social Workers might lie. Thus, a one-tailed test is appropriate.

Table 87   Empathy scores of Social Work and non-Social Work students

Social workers (X_1)	Non-social workers (X_2)
80	68
79	71
78	58
69	62
68	52
78	67
75	63
74	70
73	59
81	61

The t test allows us to determine whether or not the means of the two samples differ so much that the samples are unlikely to have been drawn from the same population. The parametric t test is the one most commonly used by student researchers who wish to test the significance of the difference between the means of two independent samples.

There are various formulae for t, depending upon the particular circumstances governing the data. In this first example of the use of t, the following assumptions are made:

(i)   that the groups are independent and have been randomly sampled;

(ii)   that the population variances are equal;

(iii)   that the population distributions are normal.

The formula for t is given as:

$$t = \frac{\dfrac{\sum X_1}{N_1} - \dfrac{\sum X_2}{N_2}}{\sqrt{\left(\dfrac{\sum X_1^2 - [(\sum X_1)^2/N_1] + \sum X_2^2 - [(\sum X_2)^2/N_2]}{N_1 + N_2 - 2}\right)\left(\dfrac{N_1 + N_2}{N_1 N_2}\right)}}$$

where

$$\sum X_1 = \text{sum of scores, group 1}$$

$$\sum X_2 = \text{sum of scores, group 2}$$

$$\sum X_1^2 = \text{sum of squares, group 1}$$

$$\sum X_2^2 = \text{sum of squares, group 2}$$

$$N_1 = \text{numbers in group 1}$$

$$N_2 = \text{numbers in group 2}$$

PROCEDURE FOR COMPUTING *t* (POOLED VARIANCE)

1 Square the individual scores in both groups.

Table 88 Computing Social Worker and non-Social Worker scores

Social Workers		Non-Social Workers	
X_1	X_1^2	X_2	X_2^2
80	6400	68	4624
79	6241	71	5041
78	6084	58	3364
69	4761	62	3844
68	4624	52	2704
78	6084	67	4489
75	5625	63	3969
74	5476	70	4900
73	5329	59	3481
81	6561	61	3721
$\sum X_1 = 755$	$\sum X_1^2 = 57{,}185$	$\sum X_2 = 631$	$\sum X_2^2 = 40{,}137$

2 Sum the scores ($\sum X$) and the squares of scores ($\sum X^2$) for each group. Substituting our data from Table 88 above:

$$t = \frac{\dfrac{755}{10} - \dfrac{631}{10}}{\sqrt{\left(\dfrac{57185 - [(755)^2/10] + 40137 - [(631)^2/10]}{10 + 10 - 2}\right)\left(\dfrac{(10+10)}{(10)(10)}\right)}}$$

$$= \frac{12.4}{2.365}$$

$$= 5.24$$

Our value of 5.24 is larger than the values in the *t* table Appendix 5, for the 0.05 level ($t = 1.73$) and the 0.01 level ($t = 2.55$). We therefore reject the null hypothesis and conclude that intending Social Workers do have significantly higher empathy scores than non-Social Workers.

The formula used in our example above is called the *pooled variance formula*. When dealing with large groups of unequal standard deviations, it is more appropriate to employ the *separate variance formula*:

$$t = \frac{M_1 - M_2}{\sqrt{\dfrac{SD_1^2}{N_1} + \dfrac{SD_2^2}{N_2}}}$$

where

M_1 = mean of sample 1

M_2 = mean of sample 2

$$SD_1 = \text{standard deviation of sample 1}$$
$$SD_2 = \text{standard deviation of sample 2}$$
$$N_1 = \text{numbers in sample 1}$$
$$N_2 = \text{numbers in sample 2}$$

In consulting the table in Appendix 5 for the significance of t found by the separate variance formula, the t value that our estimated value is compared with is an *average* value if sample sizes are unequal.

For example, if we calculated a t value of 2.4, using groups of 10 and 15 respectively, and we wanted to determine if that value was significant at the 0.05 level, we would first have to look into the table for 9 d.f. ($N_1 - 1 = 10 - 1$, group 1) and find the t at the 0.05 level ($t = 2.26$). We would then consult the table for 14 d.f. ($N_2 - 1 = 15 - 1$, group 2) and find the t at the 0.05 level ($t = 2.15$).

The t value that our value is compared with is the *average of the two values.*

$$\text{Average } t \text{ at 0.05 level} = \frac{2.26 + 2.15}{2}$$

$$= 2.21$$

We therefore conclude that our value of 2.4 is significant at the 0.05 level, being larger than the average t value (2.21) in the table.

For a more detailed discussion about the use of pooled or separate variance formulae, see Popham and Sirotnik (1973) pages 139–142.

The difference between two proportions in two independent samples

A researcher compares the proportion of prejudiced individuals in two large, independent, random samples drawn from 18-year-olds without 'A' level qualifications and those with 'A' levels, and finds that the proportion of prejudiced persons (p_{s1}) in the sample without 'A' levels ($N_1 = 533$) is 59% and the proportion (p_{s2}) in the sample with 'A' levels ($N_2 = 587$) is 49%. Can she conclude that there is a significant difference between the two samples at the .01 level?

Blalock (1979) has shown that the difference between two proportions in independent random samples can be treated as a special case of a difference between two means. Interested readers are referred to a full exposition in Blalock (1979, pages 232–234) where it is shown that:

$$Z = \frac{(p_{s1} - p_{s2}) - 0}{\hat{\delta} p_{s1} \, p_{s2}}$$

where

$$p_{s1} = \text{the proportion of prejudiced persons in the } N_1 \text{ sample}$$
$$p_{s2} = \text{the proportion of prejudiced persons in the } N_2 \text{ sample}$$

and

$$\hat{\delta} p_{s1} - p_{s2} = \sqrt{\hat{p}_\mu \hat{q}_\mu} \; \sqrt{\frac{N_1 + N_2}{N_1 N_2}}$$

where

$\hat{\delta} p_{s1} - p_{s2}$ = the estimated standard deviation of the difference between the sample proportions

$\hat{p}_{\mu}$ = a pooled estimate of the sample error, obtained as a weighted average of the sample proportions where

$$\hat{p}_{\mu} = \frac{N_1 p_{s1} + N_2 p_{s2}}{N_1 + N_2}$$

$$\hat{q}_{\mu} = 1 - \hat{p}_{\mu}$$

In computing the test statistic Z when comparing the proportions of prejudiced persons in the example above, we must first obtain $\hat{p}_{\mu}$ and $\hat{\delta} p_{s1} - p_{s2}$.

$$\hat{p}_{\mu} = \frac{N_1 p_{s1} + N_2 p_{s2}}{N_1 + N_2} = \frac{533(.59) + 587(.49)}{533 + 587} = .538$$

therefore $\hat{q}_{\mu} = 1 - \hat{p}_{\mu} = .462$.

$$\hat{\delta} p_{s1} - p_{s2} = \sqrt{\hat{p}_{\mu}\hat{q}_{\mu}} \sqrt{\frac{N_1 + N_2}{N_1 N_2}} = \sqrt{(.538)(.462)} \sqrt{\frac{533 + 587}{(533)(587)}} = .0298$$

We may now proceed to compute the test statistic Z:

$$Z = \frac{(p_{s1} - p_{s2})}{\hat{\delta} p_{s1} - p_{s2}} - 0 = \frac{.59 - .49}{.0298} = 3.36$$

Because the direction of the difference between the two sample proportions was not predicted ahead of time, a two-tailed test is indicated. A Z score of 2.58 taken at each end of the normal curve cuts off 1% of the total area of the curve as shown in Figure 56 (page 115: see also Appendix 5). Our obtained value exceeds this. Our researcher should conclude that there is a significant difference beyond the 1% level with respect to the proportions of prejudiced persons in the two samples.

16.2 Using the *t* TEST FOR INDEPENDENT SAMPLES (SEPARATE VARIANCE)

The mean scores and standard deviations of two groups consisting of 71 nursery school children and 64 non-nursery school children on a social perceptiveness measure are set out in Table 89 below. We wish to find out whether the mean scores differ significantly. The null hypothesis (H_0) in this case is that nursery school and non-nursery school children's scores are no different from one another.

Table 89 Social perceptiveness scores of nursery school and non-nursery school children

Nursery school	Non-nursery school
$N_1 = 71$	$N_2 = 64$
$M_1 = 19.5$	$M_2 = 15.3$
$SD_1 = 3.4$	$SD_2 = 4.6$

The t test allows us to determine whether or not the means of the two samples differ so much that the samples are unlikely to have been drawn from the same population.

There are various formulae for t depending upon the particular circumstances governing the data. In the example above the following assumptions are made:

(i)   that the groups are independent and have been randomly sampled,

(ii) that the population variances are unequal or heterogeneous, and

(iii) that the population distributions are normal.

PROCEDURE FOR COMPUTING t

The formula for computing t is:

$$t = \frac{M_1 - M_2}{\sqrt{\dfrac{SD_1^2}{N_1} + \dfrac{SD_2^2}{N_2}}}$$

1   Substitute in the formula:

$$t = \frac{19.5 - 15.3}{\sqrt{\dfrac{(3.4)^2}{71} + \dfrac{(4.6)^2}{64}}} = \frac{4.2}{0.70} = 6.0$$

Before we can test the significance of our t value of 6.0 obtained by the separate variance formula we have to compare it with an *average* value as we illustrate in the example on page 200. Our degrees of freedom in the present example are

Nursery school children d.f. $= N_1 - 1 = 71 - 1 = 70$

Non-nursery school children d.f. $= N_2 - 1 = 64 - 1 = 63$

In point of fact, averaging with such large degrees of freedom makes very little difference to our calculation. In the present example our averaged t value at the 5% level is 2.00 and 2.65 at the 1% level. Our obtained value of 6.0 exceeds both of these. We therefore reject the null hypothesis and conclude that there is a significant difference between the mean social perceptiveness scores of the two groups of children.

16.3   Using the MANN–WHITNEY U TEST (For small samples $N < 8$)

In assessing the difference between two independent samples the Mann–Whitney U Test provides a useful non-parametric alternative to the t test for uncorrelated data when the assumptions of the t test are not met. It must be made clear however that the Mann–Whitney U test is used to evaluate the difference between population distributions, *not* the difference between population means. With the test therefore it is possible to obtain a significant difference between groups when the means are, in fact, identical. However, when the distributions of groups are similar as is frequently the case in most experiments at student level, the Mann–Whitney U test does compare the central tendencies of the groups.

Suppose for example that a researcher had coded a nurse's responses towards a small group of severely mentally-retarded patients over a period of time according to whether those responses are broadly non-maintaining in respect of the interaction between nurse and patient (Group A responses) or maintaining of the interaction (Group B responses). He now wishes to determine whether, according to his coding schedule, the nurse's actions are primarily non-maintaining or maintaining. The data are set out in Table 90 below. The H_0 in this case is that the two groups of scores for non-maintaining and maintaining behaviour do not differ.

Table 90 Coding of nurse's responses to severely mentally-handicapped patients

Group A		Group B	
Category	**Non-maintaining** *f*	**Category**	**Maintaining** *f*
ignoring	110	inviting interaction	60
redirecting	125	expanding verbalisation	52
negative verbalisation	89	positive verbalisation	33
initiation not maintained	90	continuous referral	10
negative non-verbal	48	smiling, encouraging	28

f = frequency $\Sigma f = 645$

PROCEDURES IN COMPUTING THE MANN–WHITNEY *U*

1 Arrange the scores in order of their size and identify the group from which they are drawn.

2 Determine *U* by counting how many scores from Group A precede (i.e. are lower than) *each* Group B score.

3 Determine *U* by counting how many scores from Group B precede (i.e. are lower than) *each* Group A score.

4 Using the smaller value of *U* determine its significance by reference to the table in Appendix 20.

Returning to our example:

Scores	10	28	33	48	52	60	89	90	110	125
Group	B	B	B	A	B	B	A	A	A	A

$$U = 0 + 0 + 0 + 1 + 1 = 2$$

That is to say: for B score 10; no Group A score precedes it
for B score 28; no Group A score precedes it
for B score 33; no Group A score precedes it
for B score 52; 1 Group A score precedes it
for B score 60; 1 Group A score precedes it

Similarly

$$U = 3 + 5 + 5 + 5 + 5 = 23$$

That is to say: for A score 48; 3 Group B scores precede it
for A score 89; 5 Group B scores precede it
for A score 90; 5 Group B scores precede it
for A score 110; 5 Group B scores precede it
for A score 125; 5 Group B scores precede it.

We enter the table in Appendix 20 at the part for equal sample sizes and find that for $n = 5$ the critical value of *U* at the 5% level is 2. The obtained value of *U* must be equal to or smaller than the value shown in the table to be significant. Since the obtained value is, in fact, 2, the researcher should reject the null hypothesis and conclude that the two groups of scores for non-maintaining and maintaining behaviour on the nurse's part do differ significantly.

16.4 Using the MANN–WHITNEY U TEST (For moderately large samples N_2 between 9 and 20)

Suppose that a sociologist interested in the relationships between spatial variations, social class, and educational attainment has obtained literacy scores for children of non-manual workers and semi-skilled/unskilled workers in selected areas in the Inner London Education Authority. Table 91 presents part of his data for the Lewisham district.

Table 91   Literacy scores in Lewisham by social group

Children of non-manual workers	Children of semi-skilled and unskilled workers
93.5	94.1
104.1	92.3
103.6	93.3
106.3	98.3
105.2	103.0
94.2	93.8
105.1	90.2
91.1	93.4
101.8	91.9
108.4	97.6
93.9	95.1
105.8	89.9
105.9	92.5
	93.4
	85.8
	89.6
	83.7

The sociologist wishes to know whether there is a significant difference between these two sets of scores, his null hypothesis (H_o) being that the scores of non-manual and semi-skilled/unskilled workers' children do not differ.

PROCEDURE FOR COMPUTING THE MANN–WHITNEY U

1   Rank all the scores as though they are in one group, giving rank 1 to the lowest score.

2   To those ranks that are tied assign the average of the tied ranks.

3   Sum the ranks for each group to obtain R_1 and R_2.

4   Compute U from the formulae:

$$U = N_1 N_2 + \frac{N_1(N_1+1)}{2} - R_1 \qquad \text{Formula A}$$

and

$$U = N_1 N_2 + \frac{N_2(N_2+1)}{2} - R_2 \qquad \text{Formula B}$$

where

$$R_1 = \text{the sum of ranks for group with } n_1 \text{ subjects}$$

$$R_2 = \text{the sum of ranks for group with } n_2 \text{ subjects.}$$

5 Rather than substitute into both Formula A and Formula B, find one value and substitute into:

$$U = N_1 N_2 - U'$$

6 Enter the table in Appendix 20 for the critical values of U. If the value of U in the table is *larger* than the smaller estimated value for the particular size of the samples, then there is a significant difference between the groups.

Table 92 Literacy scores in Lewisham by social group: computation

Children of non-manual workers		Children of semi-skilled and unskilled workers	
Score	Rank	Score	Rank
93.5	13	94.1	16
104.1	24	92.3	8
103.6	23	93.3	10
106.3	29	98.3	20
105.2	26	103.0	22
94.2	17	93.8	14
105.1	25	90.2	5
91.1	6	93.4	11.5
101.8	21	91.9	7
108.4	30	97.6	19
93.9	15	95.1	18
105.8	27	89.9	4
105.9	28	92.5	9
		93.4	11.5
		85.8	2
		89.6	3
		83.7	1
$R_1 = 284$		$R_2 = 181$	

For Formula A

$$U = (13)(17) + \frac{13(13+1)}{2} - 284 = 28$$

Substituting into the formula $U = n_1 n_2 - U'$

$$U = (13)(17) - 28$$

$$= 193$$

Entering the table in Appendix 20 for the critical value of U at $N_L = 17$ and $N_S = 13$, we find $U = 49$ at the 1% level. This exceeds the smaller estimated U value in our example. The sociologist should reject the null hypothesis and conclude that there is a significant difference between the two social groups in respect of their literacy scores.

205

16.5 Using the MANN–WHITNEY U TEST (For large samples $N_2 > 20$)

When one or both of the sample sizes are larger than 20, we convert our obtained U value into a Z score in order to interpret the significance. This is done using the formula:

$$Z = \frac{U - \dfrac{N_1 N_2}{2}}{\sqrt{\dfrac{N_1 N_2 (N_1 + N_2) + 1}{12}}}$$

Having obtained our Z value we then use the table in Appendix 3(b) to find the probability of its occurrence under the normal curve.

The power efficiency of the Mann–Whitney U Test relative to the parametric t test for independent samples is reported as 95.5% (Gibbons 1976).

Ties

When a variable which has underlying continuity can be measured extremely accurately (the literacy scores for example in Table 92) the probability of ties occurring is remote. Often, however, the relatively crude measures that are employed result in the occurrence of ties being more frequent.

When tied scores do occur, each of the tied observations is assigned the average of the ranks that they would have received if no ties had occurred.

$$T \text{ is given by: } T = \frac{t^3 - t}{12}$$

where t is the number of observations tied for a given rank. ΣT is computed by summing all the Ts over all groups of tied observations. With the correction for ties we compute Z by:

$$Z = \frac{U - \dfrac{n_1 n_2}{2}}{\sqrt{\left(\dfrac{n_1 n_2}{N(N-1)}\right)\left(\dfrac{N^3 - N}{12} - \Sigma T\right)}}$$

where $n_1 + n_2 = N$.

Look at the data in Table 93. They* relate to aggression scores of two randomly selected groups of children, 12 of whom have been exposed to a series of violent television programmes and 10 of whom have not witnessed these aggressive video presentations. The experimenter wishes to know whether there is a significant difference between the two sets of scores, his null hypothesis (H_0) being that the scores of the two groups of children do not differ.

In Table 94 we have ranked all scores as though they are in one group giving rank 1 the lowest score. We have then assigned the average of the tied ranks to those that are tied.

For Formula A

$$U = (12)(10) + \frac{12(12+1)}{2} - 162 = 36$$

---

* For ease of computation in illustrating the procedures to be used when ties occur, only 12 scores are included in n_1 and only 10 scores are included in n_2 despite our use of the Mann–Whitney formula for $n_2 > 20$.

Table 93 Aggression scores of two groups of children

Exposed to violent TV presentations		Not exposed to violent TV presentations	
10	10	10	11
14	14	8	15
16	13	6	13
18	15	12	10
20	10	13	16
15	12		

Substituting into the formula $U = n_1 n_2 - U'$

$$U = (12)(10) - 36 = 84$$

Table 94 Aggression scores of two groups of children: computation

Exposed to violent TV presentations		Not exposed to violent TV presentations	
Score	Rank	Score	Rank
10	5	10	5
10	5	11	8
14	14.5	8	2
14	14.5	15	17
16	19.5	6	1
13	12	13	12
18	21	12	9.5
15	17	10	5
20	22	13	12
10	5	16	19.5
15	17		
12	9.5		
$R_1 = 162$		$R_2 = 86$	

We now apply the correction for tied ranks. From Table 94, t, the number of observations tied for a given rank is found to be:

t for aggression score 10 = 5

t for aggression score 12 = 2

t for aggression score 13 = 3

t for aggression score 14 = 2

t for aggression score 15 = 3

t for aggression score 16 = 2

Practical Statistics for Students

To find $\Sigma\,T$, we sum the values of $t^3 - t/12$ for each of these tied groups. Thus:

$$\Sigma T = \frac{5^3 - 5}{12} + \frac{2^3 - 2}{12} + \frac{3^3 - 3}{12} + \frac{2^3 - 2}{12} + \frac{3^3 - 3}{12} + \frac{2^3 - 2}{12}$$

$$= 10 + 0.5 + 2 + 0.5 + 2 + 0.5$$

$$= 15.5$$

and $n_1 + n_2 = N = 12 + 10 = 22$.
    Substituting these in the formula:

$$Z = \frac{U - \dfrac{n_1 n_2}{2}}{\sqrt{\left(\dfrac{n_1 n_2}{N(N-1)}\right)\left(\dfrac{N^3 - N}{12} - \Sigma T\right)}}$$

$$= \frac{36 - \dfrac{(12)(10)}{2}}{\sqrt{\left(\dfrac{(12)(10)}{22(22-1)}\right)\left(\dfrac{22^3 - 22}{12} - 15.5\right)}}$$

$$= -1.596$$

From the table in Appendix 3(b) we see that a Z value of 1.96 is required for significance at the 5% level. The obtained value is less than this. The experimenter may therefore accept the null hypothesis (H_0) and conclude that there is no significant difference between the two sets of aggression scores.

16.6   Using χ^2, CHI SQUARE (2 × k)

In Table 95 below we show data to do with university students' preferences for various options in their degree course, the students being differentiated by sex.
    We wish to find out whether there is a significant difference in the range of option choices recorded by the two groups. The null hypothesis (H_0) in this case is that option choices and sex of student are not related.

Table 95 Students' option choices in Social Science degree course

			Options			
Students	Constitutional Law	Demography	Statistics	Criminology	Economics	
Men	10	11	12	14	13	(60)
Women	20	10	8	30	12	(80)

The χ^2 test is appropriate to our problem. Our data are in the form of frequencies that fall into discrete categories and are at the nominal level of measurement. The assumptions that are made in applying chi square to our data are that the two groups are independent, that the subjects in each group are randomly and independently selected and, as we noted earlier, that our observations fall into discrete categories. One further assumption in using chi square is to do with the value of expected frequencies, a term we explain below. Sample sizes have to be reasonably large in applying the χ^2 test such that no expected frequency is less than 5 when rows (r) or columns (c) exceed 2, or less than 10 when rows (r) or columns (c) equal 2.

The chi square technique tests whether an *observed* frequency distribution is sufficiently close to an *expected* frequency distribution to have occurred under the null hypothesis.

where

$$\chi^2 \text{ is given by the formula: } \chi^2 = \Sigma \frac{(O-E)^2}{E}$$

O = the observed frequencies

E = the expected frequencies

Σ = the sum of

PROCEDURE FOR COMPUTING CHI SQUARE

1 Set out the *observed* frequency distribution as shown below.

Observed Frequencies

	Constitutional Law	Demography	Statistics	Criminology	Economics	
Men	10	11	12	14	13	(60)
Women	20	10	8	30	12	(80)
	(30)	(21)	(20)	(44)	(25)	N = 140

2 Compute the *expected frequencies* (E) for each cell by multiplying the column and the row total for that cell and dividing by N. By way of example, take the cell at the intersection of MEN and CONSTITUTIONAL LAW. The *expected* frequency is given by:

$$\frac{30 \times 60}{140} = 12.8$$

Similarly, for the cell at the intersection of column DEMOGRAPHY and row WOMEN, the expected frequency is given by:

$$\frac{21 \times 80}{140} = 12.0$$

The expected frequencies for each cell are set out below.

Expected Frequencies

	Constitutional Law	Demography	Statistics	Criminology	Economics
Men	12.8	9.0	8.6	18.9	10.7
Women	17.2	12.0	11.4	25.1	14.3

3 For each cell compute $\dfrac{(O - E)^2}{E}$

	Constitutional Law	Demography	Statistics	Criminology	Economics
Men	.62	.44	1.34	1.27	.49
Women	.45	.33	1.01	.96	.37

$$\chi^2 = \sum \frac{(O-E)^2}{E} = .62 + .44 + 1.34 + 1.27 + .49 + .45 + .33 + 1.01 + .96 + .37$$

$$= 7.28$$

4 Before we can determine the significance of our obtained chi square value we must obtain the appropriate degrees of freedom. Degrees of freedom are given by:

$$\text{d.f.} = (r - 1)(c - 1)$$
$$= (2 - 1)(5 - 1)$$
$$= 4$$

Consulting the table in Appendix 4 we find that at d.f. = 4 a value of 9.49 is required for significance at the 0.05 level. Since our obtained value fails to reach this we accept the null hypothesis and conclude that there is no significant difference between men and women in respect of their selection of options in the degree course.

16.7 Using the G-test in a 2 × k contingency table

As we said earlier (page 127), the G-test is an alternative to the chi square test in analysing frequencies. The two methods are interchangeable. An advantage of the G-test is that it is easier to execute with a desk-top hand calculator, especially with contingency tables.

In using the G-test for analysing 2 × 2, 2 × k, and k × n contingency tables we do not have to distinguish between *observed* and *expected* frequencies. We symbolize the observed frequencies therefore simply as f. Observed frequencies are multiplied by the natural logarithm of themselves and the products summed ($\sum f . \ln f$). The operation is easily accomplished with a scientific calculator.

Example

Obtain $\sum f . \ln f$ for the frequencies 5 and 8.

Operation	Readout
5 × ln M+	8.04718956
8 × ln M+	16.63553233
MR	24.68272189

Thus, (5.ln5) + (8.ln8) = 24.683 to three decimal places. Note that the operation M+ executes both the multiplication of f by ln f and adds the product to the sum accumulated in the memory.

The procedure for using the G-test with a 2 × k contingency table is shown in connection with the data in Table 95.

Step 1 Display the observed frequencies in a contingency table showing row, column and grand totals.

Men	10	11	12	14	13	(60)
Women	20	10	8	30	12	(80)
	(30)	(21)	(20)	(44)	(25)	(140)

Step 2 Calculate $\Sigma f.\ln f$ for all observed frequencies:

$$(10.\ln10) + (11.\ln11) + (12.\ln12) + (14.\ln14) + (13.\ln13) + (20.n20) +$$
$$(10.\ln10) + (8.\ln8) + (30.\ln30) + (12.\ln12) = 380.943$$

Step 3 Calculate $f.\ln f$ for the grand total:

$$140.\ln140 = 691.829$$

Step 4 Calculate $\Sigma f.\ln f$ for all row and column totals:

$$(60.\ln60) + (80.\ln80) + (30.\ln30) + (21.\ln21) + (20.\ln20) + (44.\ln44) + (25.\ln25) = 1069.084$$

Step 5 Calculate Step 2 + Step 3 – Step 4:

$$380.943 + 691.829 - 1069.084 = 3.688$$

Step 6 Double this to give G:

$$3.688 \times 2 = 7.376$$

Step 7 No correction is applied. Compare G with the chi-square distribution in Appendix 4 at $(r - 1)(c - 1) = 1 \times 4 = 4$d.f. The calculated G is less than the critical value of 9.49 at the 0.05 level. We therefore conclude that there is no significant difference between men and women in their selection of options in their degree studies. The corresponding value of chi-square test statistic is 7.28, smaller than the calculated G.

16.8   Using the KOLMOGOROV–SMIRNOV TWO-SAMPLE TEST

The social class origins of in-patients attending an alcoholic unit are compared with a sample taken from census data. We wish to examine the possibility that the social class composition of patients attending this particular unit is different from that in the population at large. The data are set out in Table 96. The null hypothesis (H_0) in this case is that there is no difference in the social class composition of the two groups.

Table 96   Social class origins of in-patients at an alcoholic unit

SOCIAL CLASS

	I	II	III	IV	V
Patients	14	24	30	15	17
Census data	3	14	52	20	11

The Kolmogorov–Smirnov two-sample test is appropriate to our problem. The assumptions governing the Kolmogorov–Smirnov two-sample test are similar to those that we outlined in connection with the one-sample test (page 132). The two-sample test is used with nominal data grouped in continuous categories or arranged in rank order form. Comparisons are made between two cumulative frequency distributions as in the one-sample test. In the two-sample test however, both distributions are *observed*, rather than one *observed* and one *theoretical* or *expected*.

PROCEDURE FOR COMPUTING THE KOLMOGOROV–SMIRNOV TWO-SAMPLE TEST

1 Convert the frequencies (F) in Table 96 above into cumulative frequencies (CF) by serially adding.

Table 97 Cumulative frequency table

Patients CF_1	14	38	68	83	100
Census data CF_2	3	17	69	89	100

2 Estimate the cumulative frequency proportions (CP) by dividing by the sample size and determine the absolute difference (D) between the cumulative proportions within each interval. Ignore minus signs, see table 98.

3 Identify the largest of the differences D

$$D = |CP_1 - CP_2| \text{ MAX} = 0.21$$

4 Compute K

$$K = D\sqrt{\frac{n_1 n_2}{n_1 + n_2}}$$

Table 98 Cumulative frequency proportions

	(14/100) .14	(38/100) .38	(68/100) .68	(83/100) .83	(100/100) 1.00
Patients CP_1	(14/100) .14	(38/100) .38	(68/100) .68	(83/100) .83	(100/100) 1.00
Census data CP_2	(3/100) .03	(17/100) .17	(69/100) .69	(89/100) .89	(100/100) 1.00
$D = |CP_1 - CP_2|$	.11	.21	.01	.06	0

where

$$n_1 = \text{the number of patients}$$

and

$$n_2 = \text{the number of subjects from the Census data.}$$

$$K = .21\sqrt{\frac{(100)(100)}{100 + 100}} = 1.48$$

212

5 Consult the table in Appendix 15 for the significance of K. If the obtained K is larger than the value in the table it is significant at that level.

We see that our obtained value of $K = 1.48$ is greater than the critical value of K at the 2.5% level. We therefore reject the null hypothesis and conclude that there is a significant difference in the social class composition of in-patients at this particular Alcoholic Unit when compared with the population at large.

16.9 Using the WALSH TEST

As part of an airline's selection procedures for airframe apprenticeships, two matched groups of candidates are given a timed manual dexterity test, a practice run being permitted for one group. Table 99 shows the WITH and WITHOUT PRACTICE ERRORS MADE by the two groups of 11 candidates. We wish to find out whether practice significantly improves performance on the manual dexterity test.

Providing specific assumptions can be met, the Walsh test is appropriate to our problem. When the researcher can assume that the distributions in the two related samples are continuous and symmetrical, that the data are at the interval level of measurement and $N \leq 15$, then the Walsh test can be used. The test involves ranking difference (d_i) scores in order of size.

It's important that we fully grasp the meaning of the term, *symmetrical,* in connection with the Walsh test. Siegel (1956, page 84) comments as follows:

the assumption is *not* that the d_i's are from a normal population, (which incidentally is the assumption of the parametric t test), nor even that the d_i's have to be from the same population.

Table 99 With and without practice errors in a manual dexterity selection test

Candidate	Group 1 without practice	Group 2 with practice	d_i
A	11	6	5
B	4	2	2
C	5	4	1
D	9	3	6
E	5	5	0
F	13	7	6
G	5	6	−1
H	7	3	4
I	8	4	4
J	10	7	3
K	12	7	5

What the Walsh test *does* assume is that the populations are symmetrical, that is to say, the mean is an accurate representation of central tendency and is equal to the median.

Because the Walsh test assumes that the d_i's are from populations with symmetrical distributions, the null hypothesis (H_0) is that the average of the difference scores $= 0$. For a two-tailed test therefore, (H_1), the alternative hypothesis, is that the population mean is other than 0, and for a one-tailed test, the alternative hypothesis (H_1) is either that the population mean > 0 or is < 0.

213

PROCEDURE FOR COMPUTING THE WALSH TEST

1 Compute the signed difference (d_i) for each matched pair.

2 Compute N, the number of pairs.

3 Rank order the d_i's in increasing size from d_1 to d_N. Thus, d_1 is the largest negative d_i and d_N is the largest positive d_i.

4 Enter the table in Appendix 21 to determine whether the null hypothesis (H_0) may be rejected in favour of H_1 for the observed values $d_1, d_2, d_3, \ldots d_N$.

From the column headed d_i in Table 99 we see that the lowest value of $d_i = -1$. Therefore, $d_1 = -1$, $d_2 = 0$, $d_3 = 1$, $d_4 = 2$ and $d_5 = 3$. There are two values of $d_i = 4$, therefore $d_6 = 4$ and $d_7 = 4$. Proceeding in this fashion, $d_8 = 5$, $d_9 = 5$, $d_{10} = 6$ and $d_{11} = 6$.

Entering the table in Appendix 21 at $N = 11$ we find that the null hypothesis (H_0) is rejected at $p = .097$ (two-tailed test) level of significance if:

$$\text{MAX} \left[d_7, \tfrac{1}{2}(d_4 + d_{11}) \right] < 0$$

or

$$\text{MIN} \left[d_5, \tfrac{1}{2}(d_1 + d_8) \right] > 0$$

Let's see how the critical values in the table in Appendix 21 are applied to our actual data. Substituting our obtained values:

$$d_7 = 4, \text{ and } \tfrac{1}{2}(d_4 + d_{11}) = 4 \text{ (i.e. neither is less than 0)}.$$

On this evidence we cannot reject the null hypothesis.

What of the 'minimum' term?—that is $d_5 > 0$ and $\tfrac{1}{2}(d_1 + d_8) > 0$?

$$d_5 = 3 \text{ and } \tfrac{1}{2}(d_1 + d_8) = 2 \text{ (i. e. both are greater than 0)}.$$

On the evidence of the 'minimum' term, we may reject the null hypothesis at $p = .097$ level of significance.

Notice that there are four sets of 'tests' within $N = 11$ in the table in Appendix 21. Let's apply the most stringent of these tests to our data where the two-tailed significance level is $p = .011$, to see whether our null hypothesis (H_0) may be rejected at this level.

When $p = .011$ (two-tailed test) we can reject the null hypothesis if:

$$\text{MAX} \left[d_9, \tfrac{1}{2}(d_7 + d_{11}) \right] < 0$$

or

$$\text{MIN} \left[d_3, \tfrac{1}{2}(d_1 + d_5) \right] > 0$$

Substituting our obtained values:

$$d_9 = 5, \text{ and } \tfrac{1}{2}(d_7 + d_{11}) = 5 \text{ (i.e. neither is less than 0)}$$

$$d_3 = 1, \text{ and } \tfrac{1}{2}(d_1 + d_5) = 1 \text{ (i.e. both are greater than 0)}.$$

Again, we are able to reject the null hypothesis as a result of the 'minimum' term equations and conclude that the average of the difference scores following the practice run on the manual dexterity test is significantly different from 0. Put simply, practice does significantly improve performance.

16.10   Using the WALD–WOLFOWITZ RUNS TEST

When the researcher wishes to test the null hypothesis that two independent samples have been drawn from the same population against alternative hypotheses that the two groups differ in respect of, say, their medians, variability or skewness, then the Wald–Wolfowitz runs test is appropriate. Because the test can be addressed to any sort of difference it is particularly useful to the researcher. The assumptions governing its use are that the variable under consideration has a distribution that is continuous and that the measurement of that variable is at least at the ordinal level.

Suppose, for example, that in a study of the effects of television violence, a researcher has obtained the aggression scores of a group of nursery school children who have watched a cartoon showing a considerable degree of aggressive behaviour (Condition A) and the scores of another group of nursery school children who have watched a neutral film containing no violence (Condition B). The data set out in Table 100 are the combined scores of Conditions A and B arranged in descending order.

The null hypothesis (H_0) in this case is that the aggression scores of both groups of children are selected from a common population.

PROCEDURE FOR COMPUTING THE WALD–WOLFOWITZ RUNS TEST

1   Arrange group 1 (n_1) and group 2 (n_2) scores in a single ordered series from highest to lowest.

2   Compute R, the number of runs.

3   Where n_1 and n_2 are < 20, use the table in Appendix 22 to determine the critical value of R.

Table 100   Aggression scores in 20 nursery school children following violent and neutral film cartoons

Score	Condition	Score	Condition
60	A	38	B
58	A	32	A
57	A	28	B
55	A	27	B
54	B	22	B
53	B	14	A
50	A	8	B
48	A	6	B
46	A	4	B
42	A	2	B

4   For larger samples $(n_1$ and $n_2 > 20)$ the sampling distribution of R approximates the normal curve and the null hypothesis (H_0) may be tested using the formula:

$$Z = \frac{\left| R - \left(\dfrac{2n_1 n_2}{n_1 + n_2} + 1 \right) \right| - 0.5}{\sqrt{\dfrac{2n_1 n_2 (2n_1 n_2 - n_1 - n_2)}{(n_1 + n_2)^2 (n_1 + n_2 - 1)}}}$$

The data from Table 100 above are cast in runs

$$\underbrace{A\ A\ A}_{1}\ \underbrace{B\ B}_{2}\ \underbrace{A\ A\ A\ A}_{3}\ \underbrace{B}_{4}\ \underbrace{A}_{5}\ \underbrace{B\ B\ B}_{6}\ \underbrace{A}_{7}\ \underbrace{B\ B\ B\ B}_{8}$$

$R = 8$

Entering the table in Appendix 22 at $n_1 = 10\ n_2 = 10$, we determine the critical value of R at the 5% level. In the two-sample runs test, we are interested only in the lower value of R shown for different values of n_1 and n_2. We find that $R \leq 6$ is required to reject H_0. Our obtained value exceeds this. We therefore accept the null hypothesis and conclude that the aggression scores of the two groups of nursery school children are from the same population.

Suppose now that the data in Table 101 refer to the responses of two groups of subjects to a Dogmatism Scale, the groups having been constituted as Radicals and Conservatives on another attitudinal measure. We wish to know whether there is a significant difference between the two sets of scores, the null hypothesis (H_0) in this case being that radical and conservative subjects' scores do not differ.

The scores are first arranged in a single ordered series from highest to lowest, the group from which they are derived is identified and the number of runs is computed.

$$\begin{array}{ccccccccccccc}
192 & 174 & 171 & 168 & 154 & 152 & 149 & 147 & 141 & 140 & 137 & 136 & 133 \\
\underbrace{Y \quad Y \quad Y \quad Y \quad Y \quad Y \quad Y \quad Y}_{1} & & & & & & & & \underbrace{X \quad X}_{2} & & \underbrace{Y}_{3} & \underbrace{X}_{4} & \underbrace{Y}_{5}
\end{array}$$

$$\begin{array}{cccccccccccccc}
121 & 115 & 104 & 100 & 99 & 98 & 97 & 93 & 91 & 85 & 82 & 81 & 80 & 65 \\
\underbrace{X \quad X \quad X}_{6} & & & \underbrace{Y}_{7} & \underbrace{X \quad X \quad X}_{8} & & & \underbrace{Y \quad Y}_{9} & & \underbrace{X \quad X}_{10} & & \underbrace{Y}_{11} & \underbrace{X \quad X}_{12}
\end{array}$$

Table 101   Radical and conservative subjects' scores on a scale of dogmatism

Group X (Radicals) (n₁)	Group Y (Conservatives) (n₂)
140	147
85	174
98	100
141	93
121	91
65	152
82	133
80	137
97	149
99	154
104	168
115	171
136	192
	81

Substituting our values:

$$n_1 = 13, \quad n_2 = 14, \quad R = 12$$

into the formula:

$$Z = \frac{\left|12 - \left(\frac{2(13)(14)}{13+14} + 1\right)\right| - 0.5}{\sqrt{\frac{2(13)(14)(2(13)(14) - (13) - (14))}{(13+14)^2(13+14-1)}}} = 0.78$$

Entering the table in Appendix 3(b) we see that a critical value of ± 1.96 is significant at the 5% level. Our obtained value fails to reach this level. We therefore accept the null hypothesis (H_0) and conclude that radical and conservative subjects' dogmatism scores are not significantly different.

The problem of ties is important in the Wald–Wolfowitz runs test. Where the ties are all in one sample or the other then no problem arises. If the ties are in both samples however, the researcher must 'mix' them in all possible ways. If the results agree by all three ways then, again, there is no problem. Supposing however that the results differ. The probability of each X must then be determined and the average of these taken by way of testing the null hypothesis. By way of example, suppose that in two sets of scores, one score of 22 appears three times, twice in the X distribution and once in the Y distribution. It could be arranged in the set of scores in three ways:

22 22 22   or   22 22 22   or   22 22 22

X X Y   X Y X   Y X X

In this series of ties, then, the problem must be solved three times in order to determine the different number of runs that the various combinations could cause.

Let's look at the question of breaking ties in a concrete example.

Suppose that having assigned subjects randomly to experimental and control groups a researcher uses a games/simulation technique to teach a specific skill with the E group $(n = 15)$ and a lecture method with the C group $(n = 16)$. He then post-tests both groups on skill-acquisition with the following results set out so as to determine the number of runs.

First Run

Score
$$\underbrace{\overset{\text{C C C C C C C C C C C C C}}{\text{5 6 6 8 9 9 9 9 10 12 12 12 12}}}_{1} \, \underbrace{\overset{\text{E}}{\text{13}}}_{2}$$

Score
$$\underbrace{\overset{\text{C C}}{\text{13 13}}}_{3} \, \underbrace{\overset{\text{E E E E}}{\text{14 15 15 15}}}_{4} \, \underbrace{\overset{\text{C}}{\text{15}}}_{5} \, \overset{\text{E E E E E E E}}{\text{18 19 19 19 21 22 22}}$$

Score
$$\underbrace{\overset{\text{E E E}}{\text{22 22 25}}}_{6}$$

Substituting our values,

$$n_1 = 15, \quad n_2 = 16 \quad \text{and} \quad R = 6$$

Look at the scores at 13 however. In the first run they have been ordered as

E C C

13 13 13˙

What is the effect of ordering them as

$$\begin{array}{ccc} C & C & E \\ 13 & 13 & 13 \end{array} ?$$

Second Run

$$\underset{\displaystyle 1}{\underline{\begin{array}{cccccccccccccccc} C & C & C & C & C & C & C & C & C & C & C & C & C & C & C \\ 5 & 6 & 6 & 8 & 9 & 9 & 9 & 9 & 10 & 12 & 12 & 12 & 12 & 13 & 13 \end{array}}}$$

Score

$$\underset{\displaystyle 2 \qquad\qquad 3 \qquad\qquad 4}{\underline{\begin{array}{ccccccccccccccc} E & E & E & E & C & E & E & E & E & E & E & E & E & E & E \\ 13 & 14 & 15 & 15 & 15 & 15 & 18 & 19 & 19 & 19 & 21 & 22 & 22 & 22 & 22 & 25 \end{array}}}$$

Score

We see that R is reduced to 4.

The safest procedure we can use in such cases is to compute the number of runs using all possible ways. When all the orderings lead to the same decision, that is rejection or non-rejection of the null hypothesis, then the researcher may safely adhere to this decision. What of the case, however, when the various ways of breaking ties leads to some R values which are significant and some which are not? Siegel (1956) suggests that one should take the *average p value* as the obtained probability in these circumstances. Blalock (1979), on the other hand, talks of resolving the problem by flipping a coin, although the safest procedure, he suggests, is probably to withhold judgement. When a large number of tied ranks occur the Wald–Wolfowitz test is not appropriate.

In our example above when

$$R = 6 \qquad Z = 3.65$$
$$R = 4 \qquad Z = 4.37$$

The decision in both cases is to reject the null hypothesis and conclude that the experimental condition results in skill acquisition scores that are significantly different from those gained by control group subjects.

16.11 Using the FISHER EXACT PROBABILITY TEST

The Fisher Exact Probability Test is a useful non-parametric technique for analysing data at the nominal or ordinal level of measurement when the two independent samples are small in size. The test provides exact probability values of events as extreme as, or more extreme than, those observed. It is used when the scores from two independent random samples all fall into one or other of two mutually exclusive classes. That is to say, every subject in both groups receives one of two possible scores. In our presentation, the Fisher exact probability test is restricted to Ns as large as 30 where neither of the row marginal totals exceeds $N = 15$.

Table 102 Strike/non-strike decisions in 23 trade union delegates

	Strike	Non-strike	Total
Male	11	1	(12)
Female	3	8	(11)
Total	(14)	(9)	23

Suppose, by way of example, that a political scientist researching decision-making in a manufacturing industry observes the following voting result in a small union committee consisting of 23 male and female delegates. He wishes to determine the significance of the difference in the proportions voting for strike/non-strike action, the null hypothesis (H_o) in this case being that the population proportions of males and females favouring strike action are the same. The data are set out in Table 102.

PROCEDURE FOR COMPUTING THE FISHER EXACT PROBABILITY TEST

1   Cast the data in a 2×2 contingency table, as set out below.

A	B	$(A+B)$
C	D	$(C+D)$
$(A+C)$	$(B+D)$	N

2   Compute the marginal totals $(A + B)$, and $(C + D)$.

3   Enter the tables in Appendix 23 at the appropriate value for B and where $(A + B)$ and $(C + D)$ also correspond with the observed data. In the columns next to B are the critical values of D required for significance. If the obtained D is equal to or less than the critical value, then reject H_0. In the event of none of the B values corresponding to the obtained B, use the value of A instead and substitute the C value for the D.

Using the data from our example set out in Table 102 above $B = 1$, $(A + B) = 12$ and $(C + D) = 11$ we find that there is no critical value of B corresponding to the obtained value of $B = 1$.

We therefore enter the appropriate table at $A = 11$ and find that a critical value of $C < 3$ is significant at the 0.5% level. The obtained value for C is 3. The political scientist should therefore reject the null hypothesis and conclude that the population proportions of males and females supporting strike action are not the same.

Sometimes a researcher applying the Fisher exact probability test to data where $N \leq 30$ discovers that one of the row marginal totals exceeds 15. Look, for example, at the data below. Suppose our political scientist finds $A + B = 16$. What is he to do?

	Strike	Non-Strike	
Male	A 12	B 4	$A + B = 16$
Female	C 3	D 8	$C + D = 11$
	(15)	(12)	

The data may be re-cast so as to permit the use of the tables in Appendix 23.

	Male	Female	
Strike	A 12	B 3	$A + B \doteq 15$
No-strike	C 4	D 8	$C + D = 12$
	(16)	(11)	

16.12 The analysis of 2 × 2 contingency tables

Elsewhere in the text (pages 127, 134 and 164) we have made passing reference to 2 × 2 contingency tables. Because of the frequency with which social and life scientists deal with two dichotomous variables cast in 2 × 2 tables, we now outline ways of analysing such data in greater detail.

Recall that, in our example of the chi-square one sample test in Section 12.1, observed frequencies of workers' choice of transportation set out in Table 43 are distributed between categories in *one row*. When this is the case we refer to a *one-way classification*. Often, however, two npminal level observations are obtained by researchers, as in the case of an ornithologist recording bird sightings by species *and* habitat or an epidemiologist classifying smokers by their age *and* daily consumption. When researchers arrange observed frequencies in two or more rows they refer to their data as *a two-way classification*. Tables of such data are called *contingency tables*. The simplest of contingency tables is one which has only two nominal categories of each variable; it is called a *2 × 2 table*.

Chi square test for a 2 × 2 table

A form of the chi square test for a 2 × 2 table is shown in Table 102.1.

Table 102.1   2 × 2 contingency table

		Variable A		
		Category 1	Category 2	
Variable B	Category 1	a	b	$a+b$
	Category 2	c	d	$c+d$
		$a+c$	$b+d$	$N = a+b+c+d$

$$\chi^2 = \frac{N(ad-bc)^2}{(a+b)(c+d)(a+c)(b+d)}$$

On page 127 we discuss an adjustment called Yates' correction that is applied in the case of a chi-square test of data cast in a 2 × 2 contingency table. Briefly, it involves subtracting 0.5 from the positive discrepancies (observed − expected) and adding 0.5 to the negative discrepancies before these values are squared. In the adjusted formula below, the term $|ad-bc|$ means 'the *absolute value* of $(ad-bc)$', that is to say, the numerical value of the expression irrespective of its sign.

$$\chi^2 = \frac{N(|ad-bc| - 0.5N)^2}{(a+b)(c+d)(a+c)(b+d)}$$

By way of example:

An audience researcher for a television company finds that of the 101 men she interviews, 86 watch a particular programme and 15 do not. Among the women she interviews, 32 watch and 12 do not. Can the researcher conclude that there is a statistically significant difference between the proportion of male and female viewers? The frequencies are set out in Table 102.2.

Table 102.2   Male and female viewers of TV programme X

	Watch programme	Do not watch programme	Totals
Men	a 86	b 15	$a + b$ 101
Women	c 32	d 12	$c + d$ 44
	$a + c$ 118	$b + d$ 27	$a + b + c + d$ 145

$$\chi^2 = \frac{145[(86 \times 12) - (32 \times 15) - 72.5)]^2}{101 \times 44 \times 118 \times 27}$$

$$= 2.355$$

The critical value of chi square in Appendix 4 at $(r - 1)(c - 1) = 1$d.f. is 3.84. The audience researcher must therefore conclude that there is no significant difference between the proportion of male and female viewers.

Fisher exact test for a 2 × 2 table

In Section 16.11 we present an example of the use of this non-parametric test for analysing data at the nominal or ordinal levels of measurement when the two independent samples are small in size. In the format set out in Section 16.11, the Fisher exact test is restricted to Ns as large as 30 where neither of the row marginal totals exceeds $N = 15$. Everitt (1977) however, presents a formula for use when Ns are larger than 30 and recommends the use of the Fisher exact test in 2 × 2 tables as an alternative to chi square when expected frequencies are less than 5. The hypothetical example that follows is based on Everitt's (1977) exposition.

In Table 102.3 the 40 driving test results of candidates examined by Inspector X are classified on the basis of their gender and success/failure in the test. The inspector's supervisor wishes to test if there is an association between gender and success in the inspector's results.

Table 102.3   Driving test success and failure in a sample of male and female testees

	Male (a)	Female (b)	(a + b)
Fail	2	6	8
Pass	(c) 18	(d) 14	(c + d) 32
	(a + c) 20	(b + d) 20	(a + b + c + d) 40

In the formula set out below, P refers to the smallest of the frequencies a, b, c, d. In Table 102.3 the smallest frequency is 2. Assuming that the two variables are independent, the probability (P) of obtaining any specific arrangement of the frequencies a, b, c, d, *keeping in mind that the marginal totals are to be regarded as fixed,* is given by:

$$P = \frac{(a + b)! \ (c + d)! \ (a + c)! \ (b + d)!}{a! \ b! \ c! \ d! \ N!}$$

Recalling the footote on page 129, $a!$ is a *factorial*, that is to say, the product of a and all the whole numbers less than it down to unity. For example:

$$8! = 8 \times 7 \times 6 \times 5 \times 4 \times 3 \times 2 \times 1 = 40{,}320$$

From Table 102.3,

$$P_2 = \frac{8! \times 32! \times 20! \times 20!}{2! \times 6! \times 18! \times 14! \times 40!}$$

The calculation of P is undertaken using the factorial function of a scientific calculator.

$$P = 0.095760$$

Following Everitt's exposition, we must return to Table 102.3 and, *keeping in mind that the marginal frequencies are to be taken as fixed,* we ask in what ways can the frequencies in the body of the table be rearranged so as to represent *more extreme discrepancies* between men and women with respect to their success or failure in the driving test. Two more extreme possibilities present themselves:

A				B		
1	7	8		0	8	8
19	13	32		20	12	32
20	20	40		20	20	40

Substituting these values in the formula in turn,

$$P = \frac{(a + b)! \ (c + d)! \ (a + c)! \ (b + d)!}{a! \ b! \ c! \ d! \ N!}$$

we obtain:

$$P_1 = 0.020160 \text{ for } \mathbf{A}$$

and

$$P_0 = 0.001638 \text{ for } \mathbf{B}$$

We may now conclude that the probability of obtaining the result in Table 102.3, or one more suggestive of a departure from independence is given by:

$$P = P_2 + P_1 + P_0$$
$$= 0.09576 + 0.020160 + 0.001638$$
$$= 0.117558$$

$P = 0.117558$ represents the probability of obtaining among the 8 driving test failures 2 or fewer who are men when the hypothesis of the equality of the proportions of men and women who fail is true. Put differently, a discrepancy between male and female failures such as that obtained might be expected to occur by chance about one in ten times. Because that value is larger than the commonly used levels of significance (0.05 or 0.01) the supervisor might well conclude that the driving inspector's records of passes and failures provide no evidence of gender influence.

Everitt draws attention to an important distinction between the Fisher exact test and the chi square test. In the case of the Fisher test, a significant result indicates a departure from the null hypothesis in a *specific direction*. In our hypothetical example of driving test successes and failures, the Fisher test is employed to decide whether the proportions of male and female failures are equal or whether the proportion of men who fail is *less* than the proportion of women. In a word, the Fisher exact test is *one-tailed*. By contrast, using chi square on the same data tests departure from the equality of proportions hypothesis in *either direction*; the chi square test is *two-tailed*. (See also Section 10.14.)

Pooling information from several 2 × 2 tables

Where a researcher has generated a number of 2 × 2 tables, all relating to the same investigation, he or she may wish to combine these so as to undertake an overall assessment of the association between the row and column variables under scrutiny. It goes without saying that common sense must prevail in any strategy employed in pooling data. As Everitt observes, combining data that do not accurately reflect the information in the original tables is inadmissible.

What techniques, then, are available to the researcher?

The commonly used method of aggregating the data into a single 2 × 2 table and computing a chi square statistic is legitimate when corresponding proportions in the separate tables are alike. Where this is not the case the procedure is to be avoided. Equal caution is advised by Everitt in respect of the procedure of aggregating the chi-square values calculated separately for each table. Such a method, he observes, takes no account of the direction of the differences between the proportions in the separate tables and, in consequence, lacks the power to identify a difference that appears consistently in the same direction in all or in most of the individual tables.

What follows is an outline of a technique that Everitt identifies for combining information from several 2 × 2 tables. The method takes account of problems arising when samples and proportions do vary. The procedure involves weighting the results from the separate 2 × 2 tables and is based on a weighted mean of the differences between proportions.

Returning to our hypothetical example of driving test results, let us suppose that the supervisor now has available three sets of data from a particular driving test inspector, gathered over several months. The data are set out in Table 102.4. Clearly, separate chi-square analyses of the 2 × 2 tables fail to reveal any significant association between gender and test results. But what if the supervisor were to combine the information contained in the separate tables?

Table 102.4   Driving test success and failure in three samples of male and female testees

		Male	Female	
Sample A	Fail	2	6	$\chi^2 = 1.406$ ns.
	Pass	18	14	
Sample B	Fail	12	20	$\chi^2 = 2.101$ ns.
	Pass	60	51	
Sample C	Fail	17	30	$\chi^2 = 2.845$ ns.
	Pass	75	70	

Pooling the data in Table 102.4

(a)   By *summing the chi-square values for each 2 × 2 table* (i.e. Sample A, Sample B and Sample C a chi-square value of $\chi^2 = 6.352$ is obtained. At 3 degrees of freedom* the result is non-significant.

(b) By *aggregating the data from Samples A, B and C into a single 2 × 2 table* and computing the chi-square statistic.

	Male	Female
Fail	31	56
Pass	184	135

$\chi^2 = 7.496$

d.f. $= 1$

significant at $p = 0.01$

Here is evidence of a significant association between the number of successes and failures in the driving test and the gender of the testees. The procedure is justified to the extent that there is a broad similarity in the corresponding proportions in the three samples:

Proportions of Failures

Sample A	male	0.100
	female	0.300
Sample B	male	0.166
	female	0.282
Sample C	male	0.185
	female	0.300

(c) *Pooling data by Cochran's method.* Cochran's method is appropriate when differences in sample sizes are extreme and where corresponding proportions are dissimilar. The test statistic Y is given by:

$$Y = \frac{(w_1\,d_1) + (w_2\,d_2) + (w_3\,d_3) \ldots}{[(w_1 P_1 Q_1) + (w_2 P_2 Q_2) + (w_3 P_3 Q_3) \ldots]^{\frac{1}{2}}}$$

where:

n_{11} and n_{12} are the sample sizes in the two groups in the first 2 × 2 table.

n_{21} and n_{22} are the sample sizes in the two groups in the second 2 × 2 table.

n_{31} and n_{32} are the sample sizes in the two groups in the third 2 × 2 table etc . . .

p_{11} and p_{12} are the observed proportions in the two samples in the first 2 × 2 table.

p_{21} and p_{22} are the observed proportions in the two samples in the second 2 × 2 table.

p_{31} and p_{32} are the observed proportions in the two samples in the third 2 × 2 table . . .

$$P_1 = \frac{(n_{11}\,p_{11} + n_{12}\,p_{12})}{(n_{11} + n_{12})} \text{ and } Q_1 = (1 - P_1)$$

---

* The sum of g chi-square variables each with one degree of freedom is itself distributed as chi square with g degrees of freedom (Everitt, 1977, page 27).

$$P_2 = \frac{(n_{21} P_{21} + n_{22} P_{22})}{(n_{21} + n_{22})} \text{ and } Q_2 = (1 - P_2)$$

$$P_3 = \frac{(n_{31} P_{31} + n_{32} P_{32})}{(n_{31} + n_{32})} \text{ and } Q_3 = (1 - P_3)$$

etc . . .

$$d_1 = (p_{11} - p_{12})$$

$$d_2 = (p_{21} - p_{22})$$

$$d_3 = (p_{31} - p_{32})$$

etc . . .

$$w_1 = \frac{(n_{11} n_{12})}{(n_{11} + n_{12})}$$

$$w_2 = \frac{(n_{21} n_{22})}{(n_{21} + n_{22})}$$

$$w_3 = \frac{(n_{31} n_{32})}{(n_{31} + n_{32})}$$

etc . . .

From the data set out in Table 102.4, we derive the various quantities necessary to compute the Cochran statistic Y:

Sample A	n_{11}	= 20	n_{12}	= 20
	p_{11}	= 0.100	p_{12}	= 0.300
	P_1	= 0.200	Q_1	= 0.800
	d_1	= 0.200	w_1	= 10.00
Sample B	n_{21}	= 72	n_{22}	= 71
	p_{21}	= 0.166	p_{22}	= 0.282
	P_2	= 0.224	Q_2	= 0.776
	d_2	= 0.116	w_2	= 35.748
Sample C	n_{31}	= 92	n_{32}	= 100
	p_{31}	= 0.185	p_{32}	= 0.300
	P_3	= 0.245	Q_3	= 0.755
	d_3	= 0.115	w_3	= 47.916

$$Y = \frac{[(10 \times 0.200) + (35.748 \times 0.116) + (47.916 \times 0.115)]}{(10 \times 0.200 \times 0.800) + (35.748 \times 0.224 \times 0.776) + (47.916 \times 0.245 \times 0.755)^{\frac{1}{2}}}$$

$$= 2.85$$

In Appendix 3(b) we interpolate the table at 2.85 where this value corresponds to a probability of $p = 2 \times 0.0022 = 0.0044$ (two-tailed test).

The Cochran test reveals a significant association between driving test outcomes and the gender of the testees examined by our hypothetical driving test inspector. The greater sensitivity of the Cochran test is well illustrated when the result is compared with that obtained in the chi-square analysis based on the aggregated data from the three 2 × 2 tables.

McNemar test for a 2 × 2 table

In Section 14.3 (page 176) we illustrate a before-and-after design with hypothetical data to do with changes in the sociability ratings awarded to children attending a Child Guidance Clinic following an intensive course of group therapy. Essentially, the *same* children are rated on two occasions. When such one-to-one matching is employed, the usual chi square test for a 2 × 2 table is not applicable. The McNemar test for correlated proportions is appropriate for comparing frequencies in samples matched on variables such as age, gender, IQ, social class, etc., and where, as in the Child Guidance Clinic example above, each subject acts as his or her own control.

Take, for example, the hypothetical data set out in Table 102.5. Of the 90 students sitting an examination, 30 fail both questions and 25 pass both questions. Of the remainder, 12 pass Q1 but fail Q2 while 23 pass Q2 but fail Q1. The tutor setting the examination wishes to ascertain whether Q1 and Q2 are of equal difficulty. If such is the case, a candidate is as likely to have passed only on Q1 as to have passed only on Q2. If the questions' degree of difficulty varies, a candidate is more likely to have passed on one rather than the other. It follows from our respective posing of the null and alternative hypotheses above that we are not interested in those candidates who have either passed or failed both Q1 and Q2. Our interest in the data in Table 102.5 focuses on cells a and d.

Table 102.5 Examination success and failures on Q1 and Q2

	Fail Q2	Pass Q2
Pass Q1	(a) 12	(b) 25
Fail Q1	(c) 30	(d) 23

The McNemar test is given by:

$$\chi^2 = \frac{(|a - d| - 1)^2}{a + d}$$

where -1 is a correction factor for continuity (see Siegel, 1956, page 64 for a fuller discussion).

$$\chi^2 = \frac{(|12 - 23| - 1)^2}{12 + 23}$$

$$= \frac{100}{35}$$

$$= 2.86$$

From the table in Appendix 4 we see that at d.f. = 1 a critical value of 3.84 is significant at the 5% level. Our obtained value fails to reach that level. The tutor must conclude that for these candidates Q1 and Q2 appear to be of equal difficulty.

The Gart test for order effects

There are occasions in a two-sample situation when a researcher wishes to test both for a difference between two treatments and for the possibility of an order effect. An example suggested by Leach (1979, page 137) illustrates how the Fisher exact test can be applied twice to the same set of data, first in a way that tests for an *order effect* and second in a way that tests for a *treatment effect*.

Suppose that a social psychologist wishes to find out whether order effects are important in forming impressions of others. Seventeen subjects are presented with two conflicting descriptions of an imaginary person X. One of those descriptions (the E block in Table 102.6) shows X behaving in an *extrovert* way while the other (the I block) depicts X as *introverted*. The experimental design provides for half of the subjects to read character descriptions in the order I–E and the other half in the order E–I. The investigator wishes to test the hypothesis that subjects' impressions as revealed by a post-treatment questionnaire tend to be significantly influenced by the last block read, that is to say, there is an order effect.

Seventeen subjects read the two descriptions of person X, 9 in I–E order, 8 in the E–I order. Their post-treatment characterization of X by questionnaire is scored by the researcher as revealing (a) a predominant influence of the I block, (b) a predominant influence of the E block, or (c) an equal influence of both I and E.

In our hypothetical example, let us suppose that 3 subjects' characterizations fall into the (c) category, that is, deriving equally from I and E. In the Gart test for order effects these are discarded. Table 102.6 sets out the results of the 14 subjects who are scored as predominantly influenced by I or by E.

Table 102.6   Order and treatment effects in subjects' characterizations

		Predominant influence I block	Predominant influence E block	
Order of presentation	I–E	5	3	(8)
	E–I	4	2	(6)
		(9)	(5)	(14)

		Predominant influence I block	Predominant influence E block	
Order of presentation	I–E	10	0	(10)
	E–I	1	3	(4)
		(11)	(3)	(14)

Source: adapted from Leach, 1979, page 138.

Order effect (Table 102.6)

Following procedures for computing the Fisher exact test set out on page 218, from the data in Table 102.6, $b = 3$. $a + b = 8$, $c + d = 6$. Entering the tables in Appendix 23 we see that there is no critical value of b coresponding to the obtained value of $b = 3$. We therefore enter the appropriate table at $a = 5$ level and find that a critical value of $c \leq 0$ is significant at the 0.05 level. The obtained value for c is 4. We cannot reject the null hypothesis and must conclude that there is no effect of order in the presentation of the block.

Treatment effect (Table 102.6)

Again, from the procedures for computing the Fisher exact test set out on page 218 and from the data in Table 102.6, $b = 0$, $a + b = 10$ and $c + d = 4$. Entering the tables in Appendix 23 we find that there is no critical value of B corresponding to the obtained value of $b = 0$. We therefore enter the appropriate table at $a = 10$ and find that a critical value of $c \le 1$ is significant at the .05 level. The obtained value for c is 1. We therefore reject the null hypothesis and conclude that there is a treatment effect, the data in Table 102.6 suggesting that the I block is the more salient of the two in determining the composite impression, with 13 of the 14 characterizations showing its potency.

Applying the G-test in 2 × 2 contingency tables

We refer readers again to the hypothetical example set out on page 220. An audience researcher for a television company finds that of the 101 men she interviews, 86 watch a particular programme and 15 do not. Among the women she interviews, 32 watch the programme and 12 do not. Can the researcher conclude that there is a statistically significant difference between the proportions of men and women viewers?

In using the G-test for analysing 2×2 and $k \times n$ contingency tables we do not have to distinguish between *observed* and *expected* frequencies. We symbolize the observed frequencies therefore simply as f. Observed frequencies are multiplied by the natural logarithm of themselves and the products summed ($\sum f . \ln f$). The operation is easily accomplished with a scientific calculator.

Example

Obtain $\sum f . \ln f$ for the frequencies 5 and 8.

Operation	Readout
$\boxed{5}$ × $\boxed{\ln}$ $\boxed{M+}$	8.04718956
$\boxed{8}$ × $\boxed{\ln}$ $\boxed{M+}$	16.63553233
$\boxed{MR}$	24.68272189

Thus, $(5.\ln5) + (8.\ln8) = 24.683$ to three decimal places. Note that the operation $\boxed{M+}$ executes both the multiplication of f by $\ln f$ and adds the product to the sum accumulated in the memory.

The procedure for using the G-test with a 2×2 contingency table is shown in connection with the audience research data outlined above.

Step 1 Display the observed frequencies in a contingency table showing row, column and grand totals.

	Watch programme	Do not watch programme	Totals
Men	a 86	b 15	$a + b$ 101
Women	c 32	d 12	$c + d$ 44
Totals	$a + c$ 118	$b + d$ 27	$a + b + c + d$ 145

Step 2   Calculate $\Sigma f.\ln f$ for all observed frequencies:

(86.ln86) + (15.ln15) + (32.ln32) + (12.ln12) = 564.417

Step 3   Calculate $\Sigma f.\ln f$ for the grand total $(a + b + c + d)$:

145.ln145 = 721.626

Step 4   Calculate $\Sigma f.\ln f$ for all row and column totals:

(101.ln101) + (44.ln44) + (118.ln118) + (27.ln27) = 1284.560

Step 5   Add the numbers obtained in Steps 2 and 3 and subtract the number obtained in Step 4:

Step 2 + Step 3 – Step 4 = 564.417 + 721.626 – 1284.560 = 1.483

Step 6   Double this number to give G:

$G = 1.483 \times 2 = 2.966$

Step 7   Calculate Williams' correction factor from:

$$\text{Correction factor} = 1 + \frac{\left[\left(\frac{n}{a+b}\right) + \left(\frac{n}{c+d}\right) - 1\right]\left[\left(\frac{n}{a+c}\right) + \left(\frac{n}{b+d}\right) - 1\right]}{6n}$$

$$= 1 + \frac{\left[\left(\frac{145}{101}\right) + \left(\frac{145}{44}\right) - 1\right]\left[\left(\frac{145}{118}\right) + \left(\frac{145}{27}\right) - 1\right]}{6 \times 145}$$

$$= 1 + \frac{20.891}{870} = 1.024$$

Step 8   Divide G by the correction factor to obtain $G_{adj.}$

$G_{adj.} = 2.966/1.024 = 2.896$

Step 9   Compare $G_{adj.}$ with chi square in Appendix 4 at $(r - 1)(c - 1) = 1$d.f. The value is below the critical value of 3.84 at p = 0.05. The audience researcher should conclude that there is no sigificant difference between the proportions of male and female viewers.

Comparing the value of $G_{adj.}$ with that of the calculated test statistic in the chi square test (with Yates' correction applied), the latter is 2.355, lower than the value of $G_{adj.}$.

16.13   A median test for 2 × 2 tables (Conover, 1971)

A special application of the chi square test can be used to explore whether several samples come from populations having the same median. The technique assumes that each sample is a random sample, that the samples are independent of one another and that the measurement scale is at least ordinal.

To test whether several (k) populations have the same median, a random sample is drawn from each population. A 2 × k contingency table is constructed and the number of values in each sample above and below the median are entered as shown in the diagram below. The χ^2 test is then applied to the 2 × k table.

	Sample 1	Sample 2	$\cdots$	Sample k	Totals
Above the median (a)	a_1	a_2	$\cdots$	a_k	A
Below the median (b)	b_1	b_2	$\cdots$	b_k	B
Totals	n_1	n_2	$\cdots$	n_k	N

χ^2 is given by:

$$\chi^2 = \frac{N^2}{AB}\left[\frac{\left(a_1 - \frac{n_1 A}{N}\right)^2}{n_1} + \frac{\left(a_2 - \frac{n_2 A}{N}\right)^2}{n_2} + \cdots + \frac{\left(a_k - \frac{n_k A}{N}\right)^2}{n_k}\right]$$

By way of illustration we cite an example suggested by Conover (1971, page 169).

An agriculturalist experiments with four different methods of growing barley, the four methods being randomly assigned to 34 plots of land and the yield per acre is then computed for each of the plots. The hypothetical data are set out in Table 102.7.

Table 102.7 Yields of barley per acre under four methods of treatment

	Method		
1	2	3	4
183	190	201	178
191	191	200	182
194	181	191	181
189	183	193	177
189	184	196	179
196	183	195	181
191	188	194	180
192	191		181
190	189		
	184		

To determine whether there is a difference in yields as a result of the method employed, the agriculturalist decides to use the median test, assuming that a difference in population medians can be interpreted as a difference in the efficacy of the methods applied. The null hypothesis is that all four methods have the same median yield per acre.

A combined sample median (a *grand median*) is computed from the 34 sets of values (see pages 27–28). The grand median is 189. The values given in Table 102.7 are allocated *above* or *below* the grand median in Table 102.8.

Table 102.8 Classification of barley yields above or below grand median

	Method				Totals
	1	2	3	4	
Above median (a) > 189	6	3	7	0	(16)
Below median (b) ≤ 189	3	7	0	8	(18)
	(9)	(10)	(7)	(8)	(34)

$$\chi^2 = \frac{N^2}{AB} \left[\frac{\left(a_1 - \frac{n_1 A}{N}\right)^2}{n_1} + \frac{\left(a_2 - \frac{n_2 A}{N}\right)^2}{n_2} + \cdots + \frac{\left(a_k - \frac{n_k A}{N}\right)^2}{n_k} \right]$$

$$= \frac{(34)^2}{(16)(18)} \left[\frac{\left(6 - \frac{(9)(16)}{34}\right)^2}{9} + \frac{\left(3 - \frac{(10)(16)}{34}\right)^2}{10} + \frac{\left(7 - \frac{(7)(16)}{34}\right)^2}{7} + \frac{\left(0 - \frac{(8)(16)}{34}\right)^2}{8} \right]$$

$$= 4.01 \times (0.34 + 0.29 + 1.96 + 1.77) = 17.5$$

Unless there are many values equal to the grand median, a should generally be roughly equal to b. In this event, Conover (1971, page 170) provides a short computation of the median test where χ^2 is given by:

$$\chi^2 = \frac{(a_1 - b_1)^2}{n_1} + \frac{(a_2 - b_2)^2}{n_2} + \cdots + \frac{(a_k - b_k)^2}{n_k}$$

Substituting the values from Table 102.8:

$$\chi^2 = \frac{(6 - 3)^2}{9} + \frac{(3 - 7)^2}{10} + \frac{(7 - 0)^2}{7} + \frac{(0 - 8)^2}{8}$$

$$= 17.6$$

Entering the table in Appendix 4 at d.f. = 3 the obtained value is seen to exceed the critical value of $\chi^2 = 11.34$ $p = 0.01$. The agriculturalist should reject the null hypothesis and conclude that there is a significant difference in the yields as a result of the four different methods employed. Should that investigator have wanted to, a one-way analysis of variance could have been applied to the data (see pages 181 to 185).

16.14   The analysis of $k \times n$ contingency tables

Combining $k \times n$ tables

In Section 16.12 we set out several ways of pooling data cast in 2×2 tables. The need also occurs to combine data in larger contingency tables, where, for example, researchers collect data from different gender groups, different age groups or different social classes and wish to combine the separate tables in order to obtain more sensitive tests of association between variables than are available from analyses of each table individually (see Section 17.5). We should remind readers at this point of Everitt's (1977, page 48) caveat that 'the combination of data from different investigations or samples should be considered only when differential effects from one investigation to another can be ruled out'.

Suppose that the data set out in Table 95 concerning the option choices of male and female social science students represent the results of an inquiry at University X. The researcher also has data available from Universities Y and Z which offer an identical range of option choices. She now wishes to combine each separate table in order to obtain a more sensitive test of association between gender and option choice. The hypothetical data are set out in Table 102.9 below.

Table 102.9   Gender and option choices at Universities X, Y and Z

UNIVERSITY X

Students	Constitutional Law	Demography	Statistics	Criminology	Economics	
Men	10	11	12	14	13	(60)
Women	20	10	8	30	12	(80)
	(30)	(21)	(20)	(44)	(25)	(140)

UNIVERSITY Y

Students	Constitutional Law	Demography	Statistics	Criminology	Economics	
Men	9	17	18	5	4	(53)
Women	21	12	9	13	6	(61)
	(30)	(29)	(27)	(18)	(10)	(114)

UNIVERSITY Z

Students	Constitutional Law	Demography	Statistics	Criminology	Economics	
Men	11	11	22	10	11	(65)
Women	30	13	10	29	10	(92)
	(41)	(24)	(32)	(39)	(21)	(157)

University X

Computing chi square from Table 102.9, the *expected frequencies* are:

Men	12.8	9.0	8.6	18.9	10.7
Women	17.2	12.0	11.4	25.1	14.3

and $\dfrac{(O - E)^2}{E}$ =

Men	0.62	0.44	1.34	1.27	0.49
Women	0.45	0.33	1.01	0.96	0.37

$$\chi^2 = \sum \frac{(O - E)^2}{E} = 7.28$$

Degrees of freedom are given by d.f. = $(r - 1)(c - 1) = (2 - 1)(5 - 1) = 4$. From the table in Appendix 4 we find that at d.f. = 4 a value of 9.49 is required for significance at the 0.05 level. Since our value fails to reach this we accept the null hypothesis and conclude that there is no significant difference between men and women at University X in respect of their selection of social science options.

University Y

Computing chi square from Table 102.9, the *expected frequencies* are:

Men	13.9	13.5	12.6	8.4	4.6
Women	16.1	15.5	14.4	9.6	5.4

and $\dfrac{(O - E)^2}{E}$ =

Men	1.73	0.91	2.31	1.38	0.08
Women	1.49	0.79	2.02	1.20	0.07

$$\chi^2 = \sum \frac{(O - E)^2}{E} = 11.98$$

From the table in Appendix 4 we find that at d.f. = 4 a value of 9.49 is required for significance at the 0.05 level and 13.28 at the .01 level. Our obtained value exceeds that at the .05 level. We therefore reject the null hypothesis and conclude that there is a significant difference between men and women at University Y in respect of their selection of social science options. Looking at the expected frequencies and at the results of the $(O - E)^2 / E$ calculations, it seems that there is a tendency for men to be under-represented in Constitutional Law and Criminology and over-represented in Statistics. Conversely, women seem to be under-represented in Statistics and over-represented in Criminology.

University Z

Computing chi square from Table 102.9, the *expected frequencies* are:

Men	17.0	9.9	13.2	16.1	8.7
Women	24.0	14.1	18.8	22.9	12.3

and $(O - E)^2 =$
$$\frac{}{E}$$

Men	2.12	0.12	5.87	2.31	0.61
Women	1.50	0.09	4.12	1.62	0.43

$$x^2 = \sum \frac{(O - E)^2}{E} = 18.79$$

From the table in Appendix 4 we find that at d.f. = 4 a value of 13.28 is required for significance at the 0.01 level. Our obtained value exceeds that. We therefore reject the null hypothesis and conclude that there is a sigificant difference between men and women at University Z in respect of their social science options. Looking at the expected frequencies and at the results of the $(O - E)^2 / E$ calculations, it seems that there is a tendency for men to be under-represented in Constitutional Law and Criminology and over-represented in Statistics. Women, on the other hand, seem to be under-represented in Statistics and over-represented in Constitutional Law and Criminology.

Universities X, Y and Z taken together

Students	Constitutional Law	Demography	Statistics	Criminology	Economics	Options
Men	30	39	52	29	28	(178)
Women	71	35	27	72	28	(233)
	(101)	(74)	(79)	(101)	(56)	(411)

Computing chi square from Table 102.9, the *expected frequencies* are:

Men	43.7	32.0	34.2	43.7	24.3
Women	57.3	42.0	44.8	57.3	31.7

and $\dfrac{(O - E)^2}{E} =$

Men	4.29	1.53	9.26	4.94	0.56
Women	3.28	1.17	7.07	3.77	0.43

$$\chi^2 = \sum \frac{(O - E)^2}{E} = 36.30$$

From the table in Appendix 4 we find that at d.f. = 4 a value of 13.28 is required for significance at the 0.01 level. Our obtained value greatly exceeds that. We therefore reject the null hypothesis and conclude that there is a highly significant difference between men and women at Universities X, Y and Z in respect of their social science options choices.

We summarize the findings below:

$$\begin{array}{lll}
\text{University X} & \chi^2 = 7.28 & \text{ns.} \\
\text{University Y} & \chi^2 = 11.98 & .05 \\
\text{University Z} & \chi^2 = 18.79 & .01 \\
\hline
\text{Combined data (X, Y, Z)} & \chi^2 = 36.30 & .001
\end{array}$$

The analysis based on the pooled data leads to a more significant result than the analysis of each table separately. A scrutiny of the $(O - E)^2 / E$ values for Constitutional Law, Statistics and Criminology in each of the separate and then the combined data shows the weight of these three subject options in distinguishing between the preferences of male and female students.

Identifying sources of association in $k \times n$ tables

The hypothetical data in Section 17.1 (page 242) show that although a significant difference is found among the means of professional, skilled manual and unskilled workers on a particular attribute, it is not known if all three means of the social class groups differ significantly from one another. To find that out, a further statistical test is called for.

A similar problem occurs in isolating sources of significant association in $k \times n$ contingency tables. For example, in a significant overall chi-square test on a $k \times n$ table, a researcher may wish to make additional comparisons of cells in order to identify the specific part(s) of the table that contribute to the overall finding of non-independence of the variables.

The overall chi-square statistic for a contingency table can be partitioned into as many components as the table has degrees of freedom (Everitt, 1977, page 41), each component chi-square value corresponding to a particular 2×2 table deriving from the original table and each component independent of the rest. Thus, it is possible to explore departures from independence and identify those categories that are responsible for the significant overall chi-square value.

By way of example, the hypothetical data set out in Table 102.10 concern the employment status of juveniles classified on the basis of their secondary school truancy records.

Table 102.10 Employment status and history of truancy at secondary school

Employment Status	Truancy Record			
	Non-Truant	Occasional Truant	Persistent Truant	
Employed	13	14	6	(33)
Unemployed	19	18	26	(63)
	(32)	(32)	(32)	(96)

The partitioning procedures for a $k \times n$ table that we now outline follow those set out in Everitt (1977) pages 41–44.

First, a chi-square test of the association of the data in Table 102.10 gives $\chi^2 = 5.26$, which narrowly fails to reach the 0.05 level of statistical significance at d.f. = 2. An examination of the table, however, suggests that whilst the relationship between employment status and truancy among the non-truants and the occasional truants is very similar, there appears to be a distinct relationship between unemployment and persistent truancy. And here, warns Everitt, the temptation to combine the non-truant and occasional truant data and then compute a chi-square test on the resultant 2 × 2 table is strictly contrary to good statistical practice. 'Dipping in the brantub' again after an overall chi-square test has failed to yield a significant result is totally unjustified. Had the researcher decided *in advance of examining the data* that a combination of non-truant and occasional truant data was called for, this would have constituted a perfectly legitimate procedure. We proceed on the basis that such was the good intention of the investigator.

Using the notation in Table 102.11, the overall chi square (χ^2) is partitioned into two components $(\chi_1^2$ and $\chi_2^2)$ each having one degree of freedom.

Table 102.11

a_1	a_2	•	•	•	•	a_c	A
b_1	b_2	•	•	•	•	b_c	B
•	•	•	•	•	•	•	•
n_1	n_2	•	•	•	•	n_c	N

Source: Everitt (1977) page 42.

From the 2 × 3 Table 102.10 two fourfold tables are derived.

(1)   (2)

a_1	a_2
b_1	b_2

$(a_1 + a_2)$	a_3
$(b_1 + b_2)$	b_3

Substituting the data from Table 102.10:

(1)   (2)

13	14
19	18

27	6
37	26

Chi-square formulae for the 2×2 tables above are given by:

$$\chi_1^2 = \frac{N^2 \, (a_1 \, b_2 - a_2 \, b_1)^2}{AB \, n_1 \, n_2 \, (n_1 + n_2)}$$

$$\chi_2^2 = \frac{N^2 \, [b_3 \, (a_1 + a_2) - a_3 \, (b_1 + b_2)]^2}{ABn_3 \, (n_1 + n_2) \, (n_1 + n_2 + n_3)}$$

Substituting the values from Table 102.10 we compute:

$$\chi_1^2 = \frac{96^2 \times (13 \times 18 - 14 \times 19)^2}{33 \times 63 \times 32 \times 32 \times 64} = 0.069$$

$$\chi_2^2 = \frac{96^2 \times (26 \times 27 - 6 \times 37)^2}{33 \times 63 \times 32 \times 64 \times 96} = 5.194$$

By way of check:

$$\chi^2 = \chi_1^2 + \chi_2^2 = 0.069 + 5.194 = 5.263$$

Both components of χ^2 have one degree of freedom.

It can be seen that whereas χ_1^2 is not significant, χ_2^2 is significant at the 0.05 level. Thus, by partitioning the overall *non-significant* chi square from the 2×3 contingency table, a more sensitive test of association between the variables has been achieved. The researcher may now conclude that whereas non-truants and occasional truants do not differ in respect of their employment status, the two groups combined do, in fact, differ significantly from persistent truants.

Standardizing the margins

Looking at the data in Table 102.10 we can see at a glance that it is in the *persistent truant column* that the least frequency of *employed* and the greatest frequency of *unemployed* occur in the respective rows. In arriving at its instant decision, our eye, as it were, takes in the frequencies in the particular cells and 'weighs them' in relation to the column and row marginal totals. In very large contingency tables, however, the eye is much more likely to be fooled by the absolute magnitude of the frequencies displayed so that 'taking the margins into account' becomes very difficult indeed unless we are able to use some systematic aid to eye and mind. Just such an aid is proposed by Mosteller (1968) in a technique called *standardizing the margins*. Though it looks somewhat tedious, the procedure is readily accomplished with the help of a scientific calculator.

    Standardizing the margins involves making all the *row margins* equal and at the same time making all the *column margins* equal. We illustrate the procedure using the data from Table 102.10.

Table 102.12   Truancy record and employment status

	Non-Truant	Occasional Truant	Persistent Truant	Total
Employed	13	14	6	(33)
Unemployed	19	18	26	(63)
Total	(32)	(32)	(32)	(96)

PROCEDURE

1. Begin by dividing each frequency in Table 102.12 by its column total. Table 102.13 records the results of this procedure.

Table 102.13

	Non-Truant	Occasional Truant	Persistent Truant	Total
Employed	.406	.437	.188	1.031
Unemployed	.594	.563	.812	1.969
Total	1.000	1.000	1.000	3.000

It can be seen that the column margins have been equalized but the row margins remain to be equalized.

2. To equalize the row margins we now divide each of the new values in Table 102.13 by its row total, which gives us the data set out in table 102.14.

Table 102.14

	Non-Truant	Occasional Truant	Persistent Truant	Total
Employed	.394	.424	.182	1.000
Unemployed	.302	.286	.412	1.000
Total	.696	.710	.594	2.000

We can see from Table 102.14 that we have now succeeded in equalizing the row margins but we now find that the column totals are no longer equal. We therefore proceed by dividing each of the new values in Table 102.14 by its new column margin total. The data are set out in table 102.15.

Table 102.15

	Non-Truant	Occasional Truant	Persistent Truant	Total
Employed	.566	.597	.306	1.469
Unemployed	.434	.403	.694	1.531
Total	1.000	1.000	1.000	3.000

Once again we have managed to equalize the column margin totals whilst the row margin totals remain unequal. We continue as before by dividing each of the new values in Table 102.15 by its new row margin total. The results are set out in Table 102.16.

Table 102.16

	Non-Truant	Occasional Truant	Persistent Truant	Total
Employed	.385	.406	.208	0.999
Unemployed	.283	.263	.453	0.999
Total	.668	.669	.661	1.998

The row margin totals have now been equalized but the column margin totals are not quite equal as yet. We again divide each of the values in Table 102.16 by its column margin total and set out the results in Table 102.17.

Table 102.17

	Non-Truant	Occasional Truant	Persistent Truant	Total
Employed	.576	.607	.315	1.498
Unemployed	.424	.393	.685	1.502
Total	1.000	1.000	1.000	3.000

Table 102.17 shows column margin totals equalized and row margin totals equalized within rounding error.

3.   One final step remains in order to highlight our result. Suppose that there were no association between degree of truancy and employment status; in that event we should expect each value in Table 102.17 to be 0.500. In constructing our final Table (102.18), we have subtracted 0.500 from each obtained value in Table 102.17 to demonstrate the degree to which each category of truant is over- or under-represented in terms of its employment status.

239

Table 102.18

	Non-Truant	Occasional Truant	Persistent Truant	Total
Employed	+ 0.076	+ 0.107	– 0.185	– 0.002
Unemployed	– 0.076	– 0.107	+ 0.185	+ 0.002
Total	0.000	0.000	0.000	0.000

Table 102.18 shows that persistent truancy is more negatively associated with employment and more positively associated with unemployment than is the case in the non-truant and occasional truant groups. The procedure of *standardizing the margins* serves to highlight the differences among the three truancy groupings. The differential values set out in Table 102.18 are consistent with the differential values for $(O - E)^2 / E$ in a χ^2 analysis of the data from Table 102.12:

	Non-Truant	Occasional Truant	Persistent Truant
Employed	.364	.818	2.273
Unemployed	.190	.428	1.190

$\chi^2 = 5.263$

CHAPTER 17
DESIGN 6

Multi-group research design: more than two groups, one single variable

Group 1 subjects A, B, C, D	Group 2 subjects E, F, G, H	Group 3 subjects J, K, L, M	Group 4 subjects N, O, P, Q
X_1 X_a	X_2 X_e	X_3 X_j	X_4 X_n
X_b	X_f	X_k	X_o
X_c	X_g	X_l	X_p
X_d	X_h	X_m	X_q

17.1   Using ONE-WAY ANALYSIS OF VARIANCE, INDEPENDENT SAMPLES (Fixed effects, completely randomized model)

The one-way analysis of variance for independent samples is the most commonly used technique for examining the differences between two or more group means. It involves testing the difference between means of random samples taken from the population(s) of specific interest. The reader is advised to refer to page 181 for a brief explanation of the rationale behind analysis of variance techniques before proceeding to the example that follows.

Suppose that as part of a households and families survey, conducted in a small community, we wished to determine if the length of time between marriage and the birth of the first child was influenced by the social class of the father.

From a population of families with two or more children, thirty fathers, ten from each of three broad social class groupings, were selected at random and the intervals between marriage and the birth of their first child were recorded (see Table 103).

The null hypothesis (H_0) in this case is that the interval between marriage and the first child birth is not affected by the social class of the father.

The assumptions of analysis of variance are that our data are at the interval or ratio level of measurement, that our samples are drawn at random, and that the variances of the sample populations

241

Table 103 Birth intervals (months): by social class

Independent variable		
SOCIAL CLASS OF FATHER		
Professional	**Skilled manual**	**Unskilled**
31	18	7
32	19	9
34	21	12
36	22	12
37	24	13
38	25	13
41	26	16
43	27	12
43	26	15
44	27	16
Mean 37.9	Mean 23.5	Mean 12.5

are equal. It is common practice in using analysis of variance to check this last assumption by means of a simple test of homogeneity of variance. This involves computing individual variances in respect of the subgroups in each of our cells and then dividing the largest variance by the smallest. The result of this division is treated as an F value and interpreted in the usual way by means of an F table. If this obtained F value is not statistically significant we may proceed on the assumption that the variances of the sampled populations are homogeneous and complete the analysis of variance. Our test of the homogeneity of variance in respect of the data in our example is set out below.

HOMOGENEITY OF VARIANCE CHECK

Variance is given by the formula:

$$\text{Var} = \frac{\sum X^2 - \frac{(\sum X)^2}{n}}{n - 1}$$

where n = number of scores in sample.

1 Square each score (X^2) and total the scores ($\sum X$) and squares of scores ($\sum X^2$) for each group. Enter in a separate data table. (Table 104).

2 Calculate variance for each group.

Group 1.

$$\text{Var} = \frac{\sum X_1^2 - \frac{(\sum X_1)^2}{n_1}}{n_1 - 1}$$

$$= \frac{14{,}565 - \frac{(379)^2}{10}}{10 - 1} = \frac{14{,}565 - 14{,}364.1}{9}$$

$$= 22.32$$

242

Table 104 One-way analysis of variance for independent samples

Professional (Group 1)		Skilled manual (Group 2)		Unskilled (Group 3)	
X_1	X_1^2	X_2	X_2^2	X_3	X_3^2
31	961	18	324	7	49
32	1024	19	361	9	81
34	1156	21	441	12	144
36	1296	22	484	12	144
37	1369	24	576	13	169
38	1444	25	625	13	169
41	1681	26	676	16	256
43	1849	27	729	12	144
43	1849	26	676	15	225
44	1936	27	729	16	256
$\sum X_1 = 379$	$\sum X_1^2 = 14{,}565$	$\sum X_2 = 235$	$\sum X_2^2 = 5621$	$\sum X_3 = 125$	$\sum X_3^2 = 1637$

Group 2.

$$\text{Var} = \frac{\sum X_2^2 - \frac{(\sum X_2)^2}{n_2}}{n_2 - 1}$$

$$= \frac{5621 - \frac{(235)^2}{10}}{10 - 1} = \frac{5621 - 5522.5}{9}$$

$$= 10.94$$

Group 3.

$$\text{Var} = \frac{\sum X_3^2 - \frac{(\sum X_3)^2}{n_3}}{n_3 - 1}$$

$$= \frac{1637 - \frac{(125)^2}{10}}{10 - 1} = \frac{1637 - 1562.5}{9}$$

$$= 8.28$$

3 Compute F_{max}.

$$F_{max} = \frac{\text{Largest variance}}{\text{Smallest variance}}$$

$$= \frac{22.32}{8.28} = 2.70$$

4 Consult the table in Appendix 6 for d.f. = $n_1 - 1 = 9$ and d.f. $n_3 - 1 = 9$ to determine the significance of F_{max}.

We see that our obtained value of 2.70 is smaller than the value in the table at the .05 level $(F = 3.176)$. We therefore accept the null hypothesis that there is no significant difference among our subgroup variances. We may therefore proceed with our one-way analysis of variance.

The reader should note that although a check on the homogeneity of variance is recommended, when samples are sufficiently large and contain equal numbers, the results of an analysis of variance will not be greatly affected by unequal variances in subgroups.

VARIANCE ANALYSIS

The total variance in this independent sample, randomized design can be partitioned into:

Systematic effects
BETWEEN GROUPS (Treatments or Conditions) variance is the variance between group means caused by the independent variable, which in this case is social class.

Error effects
WITHIN GROUPS variance is the variance due to subject differences and uncontrolled factors.

$$F = \frac{\text{Between groups variance}}{\text{Within groups variance}}$$

1 Calculate the Grand Total (Correction Factor).

$$GT = \frac{(\sum X)^2}{N} = \frac{(\sum X_1 + \sum X_2 + \sum X_3)^2}{N}$$

$$= \frac{(379 + 235 + 125)^2}{30}$$

$$= 18204.03$$

where

$$\sum X = \text{sum of all raw scores}$$

and

$$N = \text{total number of scores.}$$

2 Compute the TOTAL sum of squares (SS_{TOTAL}).

$$SS_{\text{TOTAL}} = \sum X^2 - GT = [\sum X_1^2 + \sum X_2^2 + \sum X_3^2] - GT$$

$$= 14{,}565 + 5621 + 1637 - 18{,}204.03$$

$$= 3618.97$$

where

$$\sum X^2 = \text{sum of squares of all raw scores.}$$

3 Compute the BETWEEN GROUPS sum of squares (SS_{BETWEEN}).

$$SS_{\text{BETWEEN}} = \frac{(\sum X_1)^2}{n_1} + \frac{(\sum X_2)^2}{n_2} + \frac{(\sum X_3)^2}{n_3} - GT$$

$$= \frac{(379)^2}{10} + \frac{(235)^2}{10} + \frac{(125)^2}{10} - 18{,}204.03$$

$$= 21{,}449.1 - 18{,}204.03 = 3245.07$$

4 Compute the WITHIN GROUPS sum of squares (SS_{WITHIN}).

For Group 1:

$$SS_{W_1} = \sum X_1^2 - \frac{(\sum X_1)^2}{n_1}$$

$$= 14{,}565 - \frac{(379)^2}{10}$$

$$= 14{,}565 - 14{,}364.1 = 200.90$$

For Group 2:

$$SS_{W_2} = \sum X_2^2 - \frac{(\sum X_2)^2}{n_2}$$

$$= 5621 - \frac{(235)^2}{10}$$

$$= 5621 - 5522.50 = 98.50$$

For Group 3:

$$SS_{W_3} = \sum X_3^2 - \frac{(\sum X_3)^2}{n_3}$$

$$= 1637 - \frac{(125)^2}{10}$$

$$= 1637 - 1562.50 = 74.50$$

The sum of these gives the total WITHIN GROUPS sum of squares

$$SS_{\text{WITHIN}} = 200.90 + 98.50 + 74.50$$

$$= 373.90$$

Alternatively, the WITHIN GROUPS sum of squares for this example can be computed using

$$SS_{\text{WITHIN}} = SS_{\text{TOTAL}} - SS_{\text{BETWEEN}}$$

$$= 3618.97 - 3245.07 = 373.90$$

5 Determine the degrees of freedom for each sum of squares.

$$\text{d.f. for } SS_{\text{TOTAL}} = (N - 1) = (30 - 1) = 29$$

where N = number of scores overall.

$$\text{d.f. for } SS_{\text{BETWEEN}} = (k - 1) = (3 - 1) = 2$$

where k = number of groups.

$$\text{d.f. for } SS_{\text{WITHIN}} = (N - k) = (30 - 3) = 27$$

6 Estimate the variances.

$$\text{Var} = \frac{\text{Sum of squares}}{\text{d.f.}}$$

$$\text{Var}_{\text{BETWEEN}} = \frac{SS_{\text{BETWEEN}}}{\text{d.f.}_{\text{BETWEEN}}} = \frac{3245.07}{2} = 1622.54$$

$$\text{Var}_{\text{WITHIN}} = \frac{SS_{\text{WITHIN}}}{\text{d.f.}_{\text{WITHIN}}} = \frac{373.9}{27} = 13.85$$

7 Compute F.

$$F = \frac{\text{Between groups variance}}{\text{Within groups variance}} = \frac{1622.54}{13.85}$$

$$= 117.15$$

8 Enter results in an analysis of variance summary table.

Table 105 Analysis of variance table

Source of variation	Sum of squares	d.f.	Variance	F
Between groups	3245.07	2	1622.54	117.15
Within groups	373.90	27	13.85	
Total	3618.97	29		

9 Determine the significance of F using the table in Appendix 6.

Our obtained value of $F = 117.15$ is larger than the values in Appendix 6 at both the .05 ($F = 3.3541$) and the .01 ($F = 5.568$) levels for d.f.'s = 2 and 27. We therefore reject the null hypothesis and conclude that there is a significant difference among groups means.

Although we have found a significant difference among the means of the groups, we do not know if *all three* means differ significantly from one another. The F test tells us that *at least* two differ, but does not identify which two.

In order to find out which of the means differ, we must apply a further statistical test to our data. This is called a Tukey Test.

PROCEDURE FOR COMPUTING THE TUKEY TEST FOR COMPARING MEANS

1 Construct a table of sample mean differences.

2 Compute T.

$$T = (q) \times \sqrt{\frac{\text{Var}_{\text{WITHIN}} \text{ (Error Variance)}}{N}}$$

where

N = number in each group or the number of scores from which each mean is calculated.

The q value in the above formula is found by consulting Appendix 7 and determining the value corresponding to the number of means (n in the Tukey table) and the degrees of freedom for the denominator of our prior F test. That denominator is, of course, the within groups variance and its appropriate degrees of freedom are read as v in the Tukey table.

Table 106 Comparison of sample mean differences

	Comparison means	M_1	M_2	M_3
Group 1	$M_1 = 37.9$		$M_1 - M_2 = 14.4$	$M_1 - M_3 = 25.4$
Group 2	$M_2 = 23.5$			$M_2 - M_3 = 11.0$
Group 3	$M_3 = 12.5$			

We see that in our example, q at the 0.05 level = 3.51 (interpolated) when $n = 3$ and v 27. Thus:

$$T_{0.05} = 3.51 \sqrt{\frac{13.85}{10}} = 4.13$$

If the T value of 4.13 is smaller than the difference between two means, then the means are significantly different. Referring to our table of mean differences, we see that there is a significant difference among all the groups.

We therefore can conclude that the interval between marriage and first child birth is dependent upon the social class of the father.

The Scheffé S test

An alternative to the Tukey Test is given by Scheffé and is particularly applicable to groups of unequal sizes. Basically what the Scheffé S test does is to compute the limits of a confidence interval (I) for each difference between means.

Applying the Scheffé test* to our data in Table 105, the formula is given by:

$$I = S\sqrt{(\text{Variance}_{\text{WITHIN}})(Wg)}$$

PROCEDURE FOR COMPUTING THE SCHEFFÉ TEST FOR COMPARING MEANS

1   Let's deal first with $\sqrt{(\text{Variance}_{\text{WITHIN}})(Wg)}$
From Table 105, VarianceWITHIN = 13.85

2   $Wg = \dfrac{1}{n} + \dfrac{1}{n}$

---

*   We follow the exposition of the Scheffe S Test outlined in H. Harrison Clarke (1972) *Advanced Statistics*. N.J. Englewood Cliffs: Prentice-Hall Inc.

Practical Statistics for Students

From Table 103,

$$\frac{1}{n} + \frac{1}{n} = \frac{1}{10} + \frac{1}{10} = \frac{1}{5}$$

Thus:

$$\sqrt{(\text{Variance}_{\text{WITHIN}})(Wg)} = \sqrt{\frac{13.85}{5}} = 1.66$$

3 $S = \sqrt{(k-1)(F_{.05})}$ or $\sqrt{(k-1)(F_{.01})}$

where k is the number of columns and $F_{.05}$ and $F_{.01}$ are the F ratios for significance at the 5% and 1% levels that are obtained from the table in Appendix 6.

At the 5% level $S = \sqrt{(k-1)(F_{.05})} = \sqrt{(3-1)(3.35)} = 2.59$

At the 1% level $S = \sqrt{(k-1)(F_{.01})} = \sqrt{(3-1)(5.49)} = 3.31$

4 Computing the Scheffé S test formula:

At the 5% level, $I = S\sqrt{(\text{Variance}_{\text{WITHIN}})(Wg)} = (2.59)(1.66) = 4.30$

At the 1% level, $I = S\sqrt{(\text{Variance}_{\text{WITHIN}})(Wg)} = (3.31)(1.66) = 5.49$

5 Where the confidence interval (I) for a given level is smaller than the difference between two means, then the means are significantly different. Referring to our table of mean differences (Table 106) we see that there is a significant difference among all groups. We therefore conclude (as we did in the case of the Tukey Test applied to these data), that the interval between marriage and first birth is dependent upon the social class of the father.

17.2 Using the KRUSKAL-WALLIS ONE-WAY ANALYSIS OF VARIANCE BY RANKS

A psychologist wishes to contrast the need for achievement (n'ach) test scores of boys drawn from different school backgrounds. He randomly selects twenty-four sixteen-year-old students from Public, Independent Day, Grammar, Comprehensive, and Secondary Modern Schools and rank orders their scores as shown in Table 107. He wishes to find out whether n'ach is associated with experience of certain school background, the null hypothesis (H_o) in this case being that there is no association between level of n'ach and school experience.

Table 107 Rank order on secondary school students' need for achievement scores

Type of School				
A	**B**	**C**	**D**	**E**
Public	**Independent day**	**Grammar**	**Comprehensive**	**Secondary modern**
2	1	6	7	12
3	4	9	11	17
5	8	10	15	21
13	14	16	19	23
	18	20	22	24

KRUSKAL–WALLIS ONE–WAY ANALYSIS OF VARIANCE

The Kruskal-Wallis one-way analysis of variance by ranks is appropriate to the problem. The assumptions made in using this test are that the populations from which the samples are drawn have similar distributions, that the samples are drawn at random and that they are independent of each other.

H, the statistic used in the Kruskal-Wallis test is given by the formula:

$$H = \frac{12K}{N(N+1)} - 3(N+1)$$

where

N = the number of cases in all the samples combined.

K = the total of the squared sum of the ranks in each of the samples divided by the respective number of cases in each of the samples. That is to say, where R_A, R_B, R_C, R_D and R_E are the sum of the ranks for each of the samples, and n_A, n_B, n_C, n_D, and n_E are the number of cases in each of the samples, then:

$$K = \frac{R_A^2}{n_A} + \frac{R_B^2}{n_B} + \frac{R_C^2}{n_C} + \frac{R_D^2}{n_D} + \frac{R_E^2}{n_E}$$

PROCEDURE FOR COMPUTING THE KRUSKAL-WALLIS ONE-WAY ANALYSIS OF VARIANCE BY RANKS

1 Rank all the cases irrespective of school origin from 1 to 24.

2 Sum the ranks for each sample R_A, R_B, R_C, R_D, R_E

3 Compute K.

In Table 108 the data are set out and R and n computed for each of the five samples.

Table 108 Sum of the ranks on secondary school students' n'ach scores

A	B	C	D	E
2	1	6	7	12
3	4	9	11	17
5	8	10	15	21
13	14	16	19	23
	18	20	22	24
$R_A = 23$	$R_B = 45$	$R_C = 61$	$R_D = 74$	$R_E = 97$
$n_A = 4$	$n_B = 5$	$n_C = 5$	$n_D = 5$	$n_E = 5$

$$K = \frac{R_A^2}{n_A} + \frac{R_B^2}{n_B} + \frac{R_C^2}{n_C} + \frac{R_D^2}{n_D} + \frac{R_E^2}{n_E}$$

$$= \frac{(23)^2}{4} + \frac{(45)^2}{5} + \frac{(61)^2}{5} + \frac{(74)^2}{5} + \frac{(97)^2}{5} = 4258.45$$

Substituting into our formula:

$$H = \frac{12(4258.45)}{24(24+1)} - 3(24+1) = \frac{51,101.4}{600} - 75$$

$$H = 10.17$$

Testing the significance of H for large samples

Where there are more than 5 cases in each of the samples, or, as in our example, there are more than 3 samples, the significance of H is obtained from the chi square table in Appendix 4.

Degrees of freedom are given by:

$$\text{d.f.} = k - 1 \text{ where } k \text{ is the number of samples.}$$

In the present example d.f. = 5 – 1 = 4.

From the table in Appendix 4 we see that a value of 9.49 (d.f. = 4) is significant at the 5% level. Our obtained value for H exceeds this. We therefore reject the null hypothesis and conclude that there is a relationship between 'need for achievement' level and school background.

Testing the significance of H for small samples

Where $k = 3$ and the number of cases in each of the three samples is 5 or fewer, then the significance of H can be tested by reference to the table in Appendix 24.

The power efficiency of the Kruskal-Wallis test relative to the parametric F test is reported as 95.5% (Gibbons 1976).

CORRECTING FOR TIES

Because the value of H is influenced by ties a correction for tied ranks is desirable.

To correct for the effect of ties H is computed and then divided by:

$$1 - \frac{\Sigma T}{N^3 - N}$$

where

$$T = t^3 - t \text{ (when } t \text{ is the number of tied observations in a tied group of scores)}$$

and

$$N = \text{the number of cases in all the samples combined.}$$

Correcting for ties has the effect of increasing the value of H. In consequence, it makes the result more significant than it would have been if left uncorrected.

The procedure for correcting for ties is set out below in an example. The hypothetical data in Table 109 refer to burglary rates in 30 medium-sized villages randomly selected in the north, south, east and west of Great Britain. The burglary rates for the four regional locations have been ranked from high to low (low rankings indicating low burglary rates).

Table 109 Burglary rates in 30 randomly-selected villages in Great Britain: computation

North		South		East		West	
Rate	Rank	Rate	Rank	Rate	Rank	Rate	Rank
4.5	12	6.2	21.5	4.5	12	3.9	4
4.5	12	5.2	18.5	3.7	1	4.1	5
6.2	21.5	7.1	29.5	4.3	9	5.2	18.5
4.8	15.5	6.8	24.5	5.2	18.5	4.5	12
6.8	24.5	4.8	15.5	3.8	2.5	7.0	27
4.2	7	7.0	27	4.5	12	4.2	7
		6.3	23	4.2	18.5	3.8	2.5
		7.0	27	7.1	29.5	4.2	7
$R_1 = 92.5$		$R_2 = 186.5$		$R_3 = 103$		$R_4 = 83$	

Our first task in correcting for ties is to determine how many groups of ties occur and how many scores are tied in each group. Look at Table 109. The first tie occurs between two burglary rates in east and west locations (both rates equalling 3.8). Both rates are assigned the rank of 2.5. t in this case = the number of tied observations which equals 2. For this occurrence:

$$T = t^3 - t = 8 - 2 = 6$$

The next tie occurs between three burglary rates in north and west locations (the three rates equalling 4.2). Here the rank of 7 is assigned. For this occurrence:

$$T = t^3 - t = 27 - 3 = 24.$$

Looking at the data in Table 109 we find that 9 groups of ties occur. We set out t for each set of ties and the computed value of T below.

t	2	3	5	2	4	2	2	3	2
T	6	24	120	6	60	6	6	24	6

We may now calculate the correction for ties:

$$1 - \frac{\sum T}{N^3 - n} = 1 - \frac{(6 + 24 + 120 + 6 + 60 + 6 + 6 + 24 + 6)}{(30)^3 - 30}$$

$$= 1 - \frac{258}{26,970} = .990$$

We can now calculate H by first computing K

$$K = \frac{R_A^2}{N_A} + \frac{R_B^2}{N_B} + \frac{R_C^2}{N_C} + \frac{R_D^2}{N_D}$$

$$= \frac{(92.5)^2}{6} + \frac{(186.5)^2}{8} + \frac{(103)^2}{8} + \frac{(83)^2}{8} = 7961.07$$

251

Practical Statistics for Students

Substituting into our formula:

$$H = \frac{12(7961.07)}{30(30+1)} - 3(30+1) = 102.723 - 93 = 9.723$$

H is now corrected for ties: $\dfrac{9.723}{0.99} = 9.82$

TESTING THE SIGNIFICANCE OF H

The significance of our combined value for H is tested by means of the χ^2 table in Appendix 4. Degrees of freedom are given by:

d.f. $= k - 1$ where k is the number of samples.

In the present example d.f. $= 4 - 1 = 3$.

From the table in Appendix 4 we see that a value of 7.82 (d.f. $= 3$) is significant at the 5% level. Our obtained value for H exceeds this. We therefore reject the null hypothesis and conclude that there is a relationship between burglary rates and geographical location.

17.3 Using a k-sample slippage test (Conover, 1971)

An alternative to the Kruskal-Wallis one-way analysis of variance by ranks is suggested by Conover (1971). The test is easy and quick to compute and has the advantage that ties caused by several observations equalling one another present little difficulty when using the procedure. The term 'slippage' in the heading refers to the central purpose of the test, which is to ascertain whether one or more population distribution functions have 'slipped to the right' of others, that is to say, whether they contain larger numbers than others.

The slippage test relies heavily on the largest values in each of the samples. It is powerful, Conover points out, as long as it remains the case that the populations having the larger means also have larger variances. In this event, the extreme values of a sample are sensitive indicators of differences in means. Where this does not hold true, as for example when the variances of populations remain equal despite some means being larger than others, or where variances decrease as means become larger, then the slippage test is rather insensitive to differences in means and the Kruskal-Wallis test is recommended instead. One further limitation is that the test may be employed only when all samples contain the same number of observations.

The slippage test makes the following assumptions:

that each sample is a random sample from some population,
that the k samples are independent of one another,
that the random variables are continuous,
that the data are, at least, at the ordinal level of measurement,
and that either the k population distribution functions are identical or that some of the populations contain larger observations than other populations.

By way of example, suppose that a Department for Education and Employment directive requires standardized tests of mathematical achievement to be administered in six East Midlands secondary schools in order to publicize the performances of their respective students. Twelve pupils are randomly selected from the fifth-form classes of each of the six secondary schools. Their scores are set out in Table 109.1.

The null hypothesis is that the mathematical achievement scores of the fifth-form classes in the six schools are identically distributed.

Table 109.1   Mathematical achievement scores in six secondary schools

| | | Schools | | | | | |
		A	B	C	D	E	F
	1	132	199	148	102	87	131
	2	141	172	58	117	95	135
	3	205	247*	53	229*	55	204
	4	226	193	74	123	166*	219*
	5	109	241	115	194	82	110
Pupils	6	180	121	137	223	57	181
	7	233*	160	87	177	70	209
	8	93	220	79	138	132	97
	9	197	215	191*	212	149	201
	10	168	93	171	206	117	143
	11	157	112	142	159	127	156
	12	143	101	69	162	90	142

PROCEDURES

1   Place an asterisk next to the greatest score in each of the six samples.

2   Underline the smallest asterisked score <u>once</u> and the greatest asterisked score <u>twice</u>.

3   Add up the number of scores in the sample with the greatest asterisked score (i.e. Sample B) that exceed the smallest asterisked score (i.e. 166 in Sample E). The obtained value is 7.

4   Enter the tables in Appendix 29 where k = *the number of samples,* that is, 6 and n = *the number of observations,* that is, 12.

5   We see that a critical value of 6 is significant at p = 0.05. Our obtained value exceeds the critical value and is therefore significant at less than p = 0.05.

We must reject the null hypothesis and conclude at least one of the six secondary schools is associated with larger achievement scores than at least one of the remaining five.

Conover points to the convenient way in which further multiple comparisons can be undertaken when the *k*-sample slippage test rejects the null hypothesis as in our example above. The 'largest' sample (i.e. Sample B in Table 109.1) is simply discarded and the analysis is repeated on the remaining $k - 1$ samples, interpolating the table in Appendix 29 at k = 5 and n = 12. Where the null hypothesis is again rejected, the procedure is repeated on the remaining $k - 2$ samples, and so on, until the null hypothesis is accepted.

Readers may wish to verify the following results from further multiple comparisons of the data in Table 109.1.

Comparing Sample A (233*) with Sample E (166*) results in an obtained value of 6, significant at less than p = 0.05 where k = 5 and n = 12.

Comparing Sample D (229*) with Sample E (166*) results in an obtained value of 6, significant at less than p = 0.05 where k = 4 and n = 12.

Comparing Sample F (219*) with Sample E (166*) results in an obtained value of 5, significant at less than p = 0.10 but greater than p = 0.05 where k = 3 and n = 12.

Comparing Sample C (191*) with Sample E (166*) results in an obtained value of 2, which is not statistically significant where k = 2 and n = 12.

17.4   Using the JONCKHEERE TREND TEST

The Jonckheere trend test may be used as an extension of the Kruskal-Wallis one-way analysis of variance by ranks (see page 248). It is appropriate when the researcher is dealing with three or more conditions where, in light of previous experience or some theory, he can predict the ordering of results in the various experimental treatments. The Jonckheere Trend Test allows the researcher to evaluate a predicted trend across the scores of different groups without ranking them. Like Page's L trend test (see page 195), the scores are arranged in the predicted order from the lowest on the left to the highest on the right.

Suppose for example that a communication research worker wishes to find out whether there is a predicted order of difference in the rates of correct information recall from four versions of a television commercial about a new car. The researcher assigns different subjects to each of the four conditions and runs the experiment.

The scores in Table 110 represent the correct responses of 24 subjects to the four TV clips in which technicality of language has been increased from Condition 1 (low technicality) to Condition 2 (medium technicality), Condition 3 (high technicality) and Condition 4 (extremely high technicality). As required in the Jonckheere trend test, the scores in the table of results have been arranged in the order predicted by the researcher, that is, from Condition 4 (the lowest) on the left to Condition 1 (the highest) on the right. The null hypothesis (H_o) in this case is that differences in correct recall between the 4 groups are random.

Table 110   Correct recall responses by four groups of different subjects to four levels of technicality in a TV advertisement

Group 4 extremely high technicality	Group 3 high technicality	Group 2 medium technicality	Group 1 low technicality
15	21	30	19
19	18	20	30
17	16	22	32
20	21	22	25
14	25	18	24
16	19	21	27
$M = 16.8$	20	22.2	26.2

PROCEDURE FOR COMPUTING JONCKHEERE'S TREND TEST

1   Next to each score, beginning in the left-hand column, write in brackets *the number of scores in all the columns to the right of it that are larger than the one under consideration.* By way of example, in Table 111, look at the score of 15 in the column, 'extremely high technicality'. In the brackets next to it there is the figure 18 which represents the number of scores in the columns to the right of the score 15 which are larger than it. Now look at the score 21 at the top of the column 'high technicality'. The figure 8 in the brackets next to that score represents the number of scores in the columns to the right of it that are larger than it.

Notice that the final column in Table 111, 'low technicality', has no scores to the right of it. Notice too that in computing the number of scores that are larger than a score being considered we do *not* include ties.

2   Sum the figures in brackets. Call this sum *P*.

Table 111   Jonckheere's Trend Test: computation

Group 4 extremely high technicality			Group 3 high technicality			Group 2 medium technicality			Group 1 low technicality
	(*P*)	(*Q*)		(*P*)	(*Q*)		(*P*)	(*Q*)	
15	(18)	(0)	21	(8)	(3)	30	(1)	(4)	19
19	(13)	(3)	18	(11)	(0)	20	(5)	(1)	30
17	(17)	(1)	16	(12)	(0)	22	(5)	(1)	32
20	(12)	(5)	21	(8)	(3)	22	(5)	(1)	25
14	(18)	(0)	25	(4)	(7)	18	(6)	(0)	24
16	(17)	(0)	19	(10)	(1)	21	(5)	(1)	27

From Table 111:

$$P = 18 + 13 + 17 + 12 + 18 + 17 + 8 + 11 + 12 + 8 + 4 + 10 + 1 + 5 + 5 + 5 + 6 + 5$$

$$= 175$$

3   Next to each score, beginning in the left-hand column write in brackets *the number of scores in all the columns to the right of it that are smaller than the one under consideration*. By way of example, in Table 111, look at the score of 15 in the column, 'extremely high technicality'. In the brackets next to it there is the figure 0 which represents the number of scores in the columns to the right of the score 15 which are smaller than it. Now, look at the score 21 at the top of the column 'high technicality'. The figure 3 in the brackets next to that score represents the number of scores in the columns to the right of it that are smaller than it.

Notice that the final column in Table 111, 'low technicality' has no scores to the right of it. Notice too that in computing the number of scores that are smaller than a score being considered we do *not* include ties.

Sum the figures in brackets. Call this sum *Q*. From Table 111:

$$Q = 0 + 3 + 1 + 5 + 0 + 0 + 3 + 0 + 0 + 3 + 7 + 1 + 4 + 1 + 1 + 1 + 0 + 1$$
$$= 31.$$

4   Compute *S* from the formula $S = P - Q$.

$$S = 175 - 31 = 144$$

Testing the significance of S

The significance of our obtained value may be tested by entering the table in Appendix 26 at the appropriate intersection of *C* and *n* where *C* = the number of conditions and *n* = the number of subjects in each condition. If the obtained value of *S* is *equal to or larger than* the critical value found there, then the null hypothesis may be rejected. Entering the table in Appendix 26 at *C* = 4, *n* = 6 we find that the critical value is 92 at the 1% level. Our obtained value exceeds this. We therefore reject the null hypothesis and conclude that the differences in rate of correct recall are not random but that there is a trend towards higher recall at lower levels of language technicality.

Correcting for ties

When more than a quarter of the scores in the Jonckheere trend test are tied we need to use a different approach to testing the significance of S. It involves computing the variance of S by means of a particular formula and then deriving a Z score. This may look somewhat complicated but it is quite straightforward as the following example illustrates.

Suppose that a psychology student wishes to test the hypothesis that children's spelling ability improves with age. She obtains the following data to do with thirty pupils divided equally into three sets, 10 seven-year-olds, 10 eight-year-olds and 10 nine-year-olds. The scores in Table 112 refer to the number of correctly-spelt words out of 10 that each child obtains in a test. The figures in the brackets (P) and (Q) are derived by means of the computation procedures set out on page 255. From Table 112:

$$P = 19 + 19 + 19 + 15 + 15 + 15 + 15 + 9 + 4 + 10 + 9 + 9 + 9 + 7 + 7 + 7 + 7 + 3$$
$$= 198$$

$$Q = 0 + 0 + 0 + 1 + 1 + 1 + 1 + 5 + 11 + 16 + 0 + 0 + 0 + 0 + 1 + 1 + 1 + 1 + 3 + 7$$
$$= 50$$

$$S = P - Q = 198 - 50 = 148$$

Before computing the variance of S by means of the formula set out below, let's first cast the data in Table 112 in a form to allow us ease of computation.

Table 112 Correctly-spelt words by pupils of different ages: computation of S

7-year-olds			8-year-olds			9-year-olds
P		**Q**	**P**		**Q**	
4	(19)	(0)	4	(10)	(0)	5
4	(19)	(0)	5	(9)	(0)	6
4	(19)	(0)	5	(9)	(0)	6
5	(15)	(1)	5	(9)	(0)	7
5	(15)	(1)	6	(7)	(1)	7
5	(15)	(1)	6	(7)	(1)	7
5	(15)	(1)	6	(7)	(1)	7
6	(9)	(5)	6	(7)	(1)	8
7	(4)	(11)	7	(3)	(3)	8
8	(0)	(16)	8	(0)	(7)	8

Table 113 Ordered contingency table for the data contained in Table 112

	Correctly-spelt words					
	4	5	6	7	8	t
Age 7 yrs	3	4	1	1	1	10
8 yrs	1	3	4	1	1	10
9 yrs	0	1	2	4	3	10
u	4	8	7	6	5	$30 = n$

where t = the row marginal totals in the contingency table
and u = the column marginal totals in the contingency table.

$$\text{Variance of } S = \frac{2(n^3 - \sum t^3 - \sum u^3) + 3(n^2 - \sum t^2 - \sum u^2) + 5n}{18}$$

$$+ \frac{(\sum t^3 - 3\sum t^2 + 2n)(\sum u^3 - 3\sum u^2 + 2n)}{9n(n-1)(n-2)}$$

$$+ \frac{(\sum t^2 - n)(\sum u^2 - n)}{2n(n-1)}$$

We may compute the various values required in the formula:

$$n = 30, \quad n^2 = 900 \quad \text{and} \quad n^3 = 27{,}000$$
$$\sum t^2 = 10^2 + 10^2 + 10^2 = 300$$
$$\sum t^3 = 10^3 + 10^3 + 10^3 = 3000$$
$$\sum u^2 = 4^2 + 8^2 + 7^2 + 6^2 + 5^2 = 190$$
$$\sum u^3 = 4^3 + 8^3 + 7^3 + 6^3 + 5^3 = 1260$$

Substituting in our variance of S formula we have:

$$\text{Variance} = \frac{2(27{,}000 - 3000 - 1260) + 3(900 - 300 - 190) + 150}{18}$$

$$+ \frac{(3000 - 900 + 60)(1260 - 570 + 60)}{270(29)(28)}$$

$$+ \frac{(300 - 30)(190 - 30)}{60(29)}$$

$$= 2635.55$$

$$Z = \frac{S}{\text{S.D.}} = \frac{148}{\sqrt{2635.55}} = 2.88$$

Because our hypothetical psychology student is predicting that children's spelling ability improves with age a one-tailed test of significance is appropriate. From the table in Appendix 3(b) we see that a Z value of 2.88 has a one-tailed probability of 0.002. The student should reject the null hypothesis and conclude that improvement in spelling ability is associated with age.

17.5 Using CHI SQUARE IN $k \times n$ TABLES

Suppose we are interested in the response of first, second, and third-year university students to the statement, 'No part of the Senate's deliberations should be barred to elected representatives of the Students' Union'. We obtain the following responses, classified as Strongly Agree, Agree, Uncertain, Disagree, and Strongly Disagree, and now wish to determine whether or not student opinion differs significantly by year in university. Our null hypothesis (H_o) in this case is that student opinion and year in university are unrelated.

Table 114 1st, 2nd, and 3rd year students' responses to statement: observed frequencies

Student group	Strongly agree	Agree	Uncertain	Disagree	Strongly disagree	
1st Year	6	12	40	15	10	(83)
2nd Year	14	22	50	11	9	(106)
3rd Year	30	50	10	6	2	(98)
Totals	(50)	(84)	(100)	(32)	(21)	(287)

When frequencies are in discrete categories either at the nominal or the ordinal level of measurement, chi square is appropriate to our problem. Its computation in the present 3×5 table is essentially the same as for the Chi Square One Sample Test that we outline on page 126.

PROCEDURE FOR COMPUTING CHI SQUARE

1   Compute the expected frequencies (E) for each cell by multiplying the column and row totals for that cell and dividing by N.

2   Calculate chi square using the formula:

$$\chi^2 = \sum \frac{(O-E)^2}{E}$$

where O is the observed frequency in each cell and E is the expected frequency in each cell.

Table 115 1st, 2nd, and 3rd year students' responses to statement: expected frequencies

Student group	Strongly agree	Agree	Uncertain	Disagree	Strongly disagree
1st Year	14.5	24.3	28.9	9.3	6.1
2nd Year	18.5	31.0	36.9	11.8	7.8
3rd Year	17.0	28.7	34.2	10.9	7.1

$$\chi^2 = \sum \frac{(O-E)^2}{E} = 4.9 + 6.2 + 4.3 + 3.5 + 2.5 + 1.1 + 2.6$$
$$+ 4.7 + 0.0 + 0.2 + 9.9 + 15.8 + 17.1$$
$$+ 2.2 + 3.7$$
$$= 78.7$$

Degrees of freedom are given by $(c - 1)(r - 1)$ where $c =$ the number of columns and $r =$ the number of rows.

$$\text{d.f.} = (5 - 1)(3 - 1) = 8$$

Entering the table in Appendix 4 at d.f. $= 8$ we see that a critical value of 20.09 is significant at the 1 % level. Our obtained value exceeds this. We therefore reject the null hypothesis and conclude that there is a significant relationship between year in university and opinion about Student Union access to Senate's deliberations.

Collapsing tables in contingency analyses

On page 127 we refer to the stability of the chi square test and the problem of low expected frequencies. One way round the difficulty, we suggest, is to combine categories or cells, that is to say, to collapse the contingency table. But how does one set about such a procedure?

What follows is an account of the actual decisions that Canadian researchers had to make in respect of communications data gathered in the wake of an urban disaster. The investigators produced a table of tallies based on 'who tells whom' about the disaster. Each frequency in Table 115.1 represents a unit of observation called a *communication tie*, that is a *housewife* tells another *housewife*, a *student* tells a *low-status occupation member*, etc.

Table 115.1 Communications between occupational statuses

From	To					
	Housewives	Students	Low Status	Medium Status	High Status	
Housewives	8	2	7	7	2	(26)
Students	3	4	3	1	0	(11)
Low Status	5	2	10	6	3	(26)
Medium Status	7	2	8	17	7	(41)
High Status	4	0	5	6	6	(20)
	27	10	33	36	18	(124)

Source: Erickson and Nosanchuk, 1979, page 257.

With chi-square assumptions in mind, the researchers estimated the *mean expected value* for the data in Table 115.1 by calculating the mean number of cases per cell:

$$\frac{N}{r \times c} = \frac{124}{5 \times 5} = 5$$

They decided that this value was a 'bit small' and therefore undertook to do some collapsing. From the frequencies in Table 115.1 they identified what they described as 'the outstanding candidates for collapsing', namely the communication ties involving students since they were so few in number. But what should they collapse them with? And here, judgement came in. It made sense to the researchers to combine housewives with students as 'non-workers' on the ground that communications among those in the formal work world (low-, medium- and high-status occupations) were perhaps different from other communications and ought not to be mixed up with them. In the collapsed Table 115.2, the mean expected value is now 124/16, or about 8, which the researchers found sufficient justification to proceed with the chi-square analysis.

Table 115.2 Collapsed communication data between occupational statuses

From	To				
	Non-workers	Low Status	Medium Status	High Status	
Non-workers	17	10	8	2	(37)
Low status	7	10	6	3	(26)
Medium status	9	8	17	7	(41)
High status	4	5	5	6	(20)
	(37)	(33)	(36)	(18)	(124)

Collapsing categories should be undertaken with caution. Blalock (1979, page 327) warns that collapsing categories produces non-random errors of a peculiar nature that are not very well understood. The decision to collapse categories clearly requires some judgements as to the 'similarity among categories, which in turn,' Blalock observes, 'presupposes that the categories were ordered in some fashion. That being the case, the appropriate procedure is to treat the variables as ordinal and to retain as much measurement accuracy as possible.'

Controlling for variables

Statistical texts often cite agricultural experiments when discussing the control of variables. Typically, seed crops are systematically subjected to variations in soil condition, moisture, light, heat and fertilizer, and the individual or combined effects of these independent variables are explored in relation to the dependent variable, crop yields per hectare.

Social and life scientists, too, have recourse to the control of one or several variables when examining relationships within their data. The fictitious and improbable example that follows draws on Blalock's (1979, pages 315–325) account of controlling for variables.

A researcher suspects that teachers in a large secondary school tend to award high marks to white pupils irrespective of their ability or effort, and, conversely, to give low marks to black children except in the case of those black pupils who possess high ability and are perceived by their teachers to work hard at their studies. The data concerning the four variables (*ethnicity, ability, perceived effort* and *marks awarded*) are set out in the master table 115.3.

Table 115.3 Ethnicity, ability, effort and reward (1)

Ability	Marks Awarded	White		Black		Totals
		Perceived Effort				
		Tries hard	Makes little effort	Tries hard	Makes little effort	
Above Average	High	59	41	39	17	156
	Low	20	25	15	39	99
Below Average	High	40	23	6	2	71
	Low	23	13	33	55	124
Totals		142	102	93	113	450

A useful tip on the construction of a master table is given by Blalock (1979, page 319):

> Ordinarily, it will be most convenient to construct . . . a master table so that the dependent variable appears as the innermost column on the left-hand side and so that the independent variable of greatest interest appears as the lowest row in the top heading. This will result in sub-tables with those frequencies that are being directly compared.

Table 115.3 has been constructed with Blalock's advice in mind. It consists of four sub-tables each relating to *perceived effort* and *marks awarded*. All pupils in the top right sub-table are *black* and *above average ability*; all pupils in the bottom left sub-table are *white* and *below average ability*, and so forth.

Controlling for *ability*, that is, examining *high-ability pupils* and *low-ability pupils* separately, we can explore the relationship between teachers' awards and perceived pupil effort in the two ethnic groups (see Table 115.4).

Table 115.4 Chi-square tests on the four sub-tables (ability controlled)

59	41
20	25

$\chi^2 = 1.82$ ns.

39	17
15	39

$\chi^2 = 17.64$: $p = < .001$

40	23
23	13

$\chi^2 = 0.07$ ns.

6	2
33	55

$\chi^2 = 2.86$ ns.

Only in the top right sub-table (*black pupils* of *high ability* and *varying degrees of effort*) is *high effort* differentially rewarded.

In Table 115.5, we have recast the data in the master table so as to control for *effort*, that is, by examining those pupils who *try hard* and those who *make little effort* separately, we can explore the relationship between teachers' awards and pupils' ability in the two ethnic groups.

Table 115.5 Ethnicity, ability, effort and reward (2)

Teachers' Marks	Perceived Effort			
	Tries Hard		Makes Little Effort	
	Above Average Ability	Below Average Ability	Above Average Ability	Below Average Ability
	White Pupils			
High	59	40	41	23
Low	20	23	25	13
	Black Pupils			
High	39	6	17	2
Low	15	33	39	55

Table 115.6 Chi-square tests on the four sub-tables (effort controlled)

59	40
20	23

$\chi^2 = 1.58$ ns.

41	23
25	13

$\chi^2 = 0.0014$ ns.

39	6
15	33

$\chi^2 = 27.04$: $p = <.001$.

17	2
39	55

$\chi^2 = 12.70$: $p = <.001$

Only among black pupils are teachers' awards related to pupils' ability. This relationship, moreover, is at its strongest among *black students* who *try hard*.

So far, the analyses provide support for the hunches of our hypothetical researcher into racial prejudice in classroom settings. What of that investigator's final suspicion that teachers, in general, tend to award higher marks to white pupils than to black students?

The hypothesis is tested in a 2×2 contingency table that relates *ethnicity* to *high* and *low* marks.

		Ethnicity	
		White	Black
Marks	High	163	64
	Low	81	142

$\chi^2 = 55.64$: $p = <.001$

The data provide strong evidence in support of the researcher's hunch.

Interaction effects

The analyses of the separate sub-tables in Tables 115.4 and 115.6 reveal that, when *ability* and *effort* are controlled, the relationships between teachers' marks and pupils' effort and teachers' marks and pupils' ability are shown to be different for white and black pupils. In a word, *interaction* is present. Where interaction is found to be statistically significant we need to qualify the generalization(s) that we make about our findings. Thus:

> For white pupils, no relationships are identified between either teachers' marks and pupil ability or between teachers' marks and pupil effort. For black pupils however, significant relationships exist between teachers' marks and pupil ability and between teachers' marks and pupil effort.

Another approach to identifying interaction effects is through *differences in proportions*. On pages 200–201 we have set out procedures for testing the significance of the difference between two proportions in two independent samples. We now apply those procedures to the data contained in Table 115.5. This time our focus is upon the relationships between *high marks* and pupils' ethnicity, ability and effort.

Table 115.7 Teachers' high marks and pupils' ethnicity, ability and effort (proportions shown in brackets)

Ethnicity	Tries Hard		Makes Little Effort		Totals
	Above Average Ability	Below Average Ability	Above Average Ability	Below Average Ability	
White	59 (36%)	40 (25%)	41 (25%)	23 (14%)	163 (100%)
Black	39 (61%)	6 (9%)	17 (27%)	2 (3%)	64 (100%)

To illustrate the procedure, let us compare proportions in the first column of Table 115.7 where hardworking, bright pupils of black and white origin are awarded high marks by teachers.

The difference between the two proportions (61%) and (36%) is given by:

$$Z = \frac{(p_{s_1} - p_{s_2}) - 0}{\hat{\delta}\, p_{s_1} - p_{s_2}}$$

where

p_{s_1} = the proportion of able, hardworking white pupils given high marks by teachers

p_{s_2} = the proportion of able, hardworking black pupils given high marks by teachers

and

$$\hat{\delta}\, p_{s_1} - p_{s_2} = \sqrt{\hat{p}_u \hat{q}_u}\ \sqrt{\frac{N_1 + N_2}{N_1 N_2}}$$

where $\hat{\delta}\, p_{s_1} - p_{s_2}$ = the estimated standard deviation of the difference between the sample proportions

and $\hat{p}_u$ = a pooled estimate of the sample error, obtained as a weighted average of the sample proportions

where

$$\hat{p}_u = \frac{N_1 p_{s_1} + N_2 p_{s_2}}{N_1 + N_2}$$

and $\hat{q}_u = 1 - \hat{p}_u$

In computing the test statistic Z when comparing the proportions in our example, we must first obtain $\hat{p}_u$ and $\hat{\delta}\, p_{s_1} - p_{s_2}$

$$\hat{p}_u = \frac{N_1 p_{s_1} + N_2 p_{s_2}}{N_1 + N_2} = \frac{163(.36) + 64(.61)}{163 + 64} = .430$$

therefore $\hat{q}_u = 1 - \hat{p}_u = .570$

$$\hat{\delta}\, p_{s_1} - p_{s_2} = \sqrt{\hat{p}_u \hat{q}_u}\ \sqrt{\frac{N_1 + N_2}{N_1 N_2}} = \sqrt{(.43)(.57)}\ \sqrt{\frac{163 + 64}{(163)(64)}} = .073$$

$$Z = \frac{(p_{s_1} - p_{s_2})}{\hat{\delta} p_{s_1} - p_{s_2}} - 0 = \frac{.36 - .61}{.073} = -3.42: p = < .001$$

We conclude that there is a significant interaction effect with respect to the high marks that teachers award to the two groups of hardworking, able pupils, resulting in a greater proportion of high marks being given to black students rather than white.

Now let us look at the proportions in the second column of Table 115.7. Here we are comparing the proportions of high grades awarded to two groups of hardworking pupils of lower academic ability: one black group and one white. Computing the difference between the respective proportions, (white 25%) and (black 9%) gives a Z of 2.71, which is significant at $p = <.05$ and $>.01$. We must conclude that in the case of lower ability, hardworking black pupils, the significant interaction effect is in the opposite direction to that we have identified in respect of their high-ability, hardworking fellow students.

By way of summary, in the hypothetical example of teachers' grades and pupils' ethnicity, we have determined that there is a significant association between teachers' awards and pupils' ethnicity favouring white students. We have also shown that when, in addition to *ethnicity* and *awards*, we also consider the independent variables, *effort* and *ability*, complex interaction effects are found to occur.

The analyses underline the need to test for interaction effects when dealing with more than two variables. (See also, Multivariate Analysis, page 271ff).

17.6 Using different measures of association between ordered variables: GOODMAN AND KRUSKAL'S gamma, SOMER'S *d* and KENDALL'S tau *b*

On pages 66 to 74 we review a number of ways of thinking about association and we identify various measures that can be employed in quantifying a relationship between two variables. It's time now to show how three closely related measures of association describe three different forms of relationship between ordered variables. These measures are gamma (γ), Somer's *d* and tau $b(\tau_b)$ and they belong to that approach to association which requires us to think of forming all possible comparisons of one member of the population with another.

Suppose that a sociologist collects information on the annual rates of library book borrowing in a sample of residents of a community. Using *occupation* as a criterion he classifies his respondents by social class and casts his data in the form of the table shown below.

Table 116   Library book borrowing by social class

Number of library books borrowed per year (Y variable)	Social class (X variable)			
	I	II	IIIa + IIIb	IV and V
3 or more	20	12	3	1
2	7	10	6	4
1	2	10	12	6
None	1	2	5	20

Suppose now that the sociologist wishes to look at his data in three different ways.* First, he merely wants to determine whether there is a monotonic relationship of any kind between social class and library book borrowing. Second, he proposes to test the hypothesis that social class 'causes' library book borrowing. Third, he wishes to test a hunch that a one-to-one correspondence exists between social class category rank and library book borrowing category rank.

Which measure of association is most appropriate in each of these situations? Let's look at how each measure—Goodman and Kruskal's gamma (γ), Somer's d, and Kendall's tau b (τ_b)—describes a different form of relationship. We can then determine which measure is most appropriate to the sociologist's needs.

Goodman and Kruskal's gamma (γ)

Recall that on page 71 we introduce the reader to Goodman and Kruskal's gamma and to the formula:

$$\gamma = \frac{P-Q}{P+Q} \quad \text{or} \quad \frac{S}{P+Q}$$

where

P = the number of positive pairs

Q = the number of negative pairs

and

$S = P - Q$, or the 'preponderance of pairs' (whether positive or negative).

In our discussion (pages 69 to 70) leading up to the presentation of gamma we note that in only taking account of positive and negative pairs, gamma *ignores ties*.

Let's now compute gamma for the data in Table 116 using the diagrammatic presentation of the gamma calculation (pages 72 to 74) as our guide.

$P = 20(10 + 6 + 4 + 10 + 12 + 6 + 2 + 5 + 20) + 12(6 + 4 + 12 + 6 + 5 + 20)$

$\quad + 3(4 + 6 + 20) + 7(10 + 12 + 6 + 2 + 5 + 20) + 10(12 + 6 + 5 + 20)$

$\quad + 6(6 + 20) + 2(2 + 5 + 20) + 10(5 + 20) + 12(20) = 3741$

$Q = 1(7 + 10 + 6 + 2 + 10 + 12 + 1 + 2 + 5) + 3(7 + 10 + 2 + 10 + 1 + 2) + 12(7 + 2 + 1)$

$\quad + 4(2 + 10 + 12 + 1 + 2 + 5) + 6(2 + 10 + 1 + 2) + 10(2 + 1) + 6(1 + 2 + 5) + 12(1 + 2) + 10(1) = 613$

$$\gamma = \frac{P-Q}{P+Q} = \frac{3128}{4354} = .72$$

Gamma is a measure of association ranging from -1.0 to $+1.0$. In respect of the hypothetical data in Table 116 we may conclude that a value of $\gamma = .72$ suggests a strong positive association between 'higher' social class and library book borrowing. The absolute value of gamma indicates the proportion of pairs that are 'accounted for' by the social class (i.e. the X variable) and library borrowing (i.e. the Y variable) relationship. Gamma is roughly analagous to r^2, the coefficient of determination which we discuss on pages 86 to 87.

In ignoring ties altogether, gamma is, in a sense, the 'weakest' of the three measures of association and, as such, is most appropriate to the 'weakest' of our sociologist's research questions, that is, the

* We are indebted to F. J. Kohout (1974) *Statistics For Social Scientists*, N.Y. John Wiley & Sons, pages 235-236 and pages 435-436, for the example and the distinction drawn in the use of γ, Somer's d, and τ_b.

one which asks *is there a monotonic relationship of any kind between social class and library book borrowing?*

Somer's d

A more stringent measure of association would be one that takes account of ties and shows a decrease when ties occur. For example, we might ask for a measure of association that in addition to the *preponderance of pairs* in the gamma calculation takes account also of those pairs that are tied on Y but not on X.

Somer's d_{yx} is just such a measure. What it does is to add T_y to the denominator of the gamma measure, T_y being the number of pairs tied on Y but not on X.

Let's see how T_y is computed. Look at Table 116 again. Because T_y represents all pairs tied on Y but not on X we know that such pairs will be located in different cells that fall along the same *rows* of the table. Taking each cell in turn, beginning in the upper left-hand cell, multiply the frequency in that cell by all the cases *falling to the right of the cell in the same row*. Having done this for each cell, simply add all of the products to give T_y. For the data in Table 116:

$$T_y = 20(12+3+1)+12(3+1)+3(1)+7(10+6+4)+10(6+4)+6(4)+2(10+12+6)$$

$$+10(12+6)+12(6)+1(2+5+20)+2(5+20)+5(20) = 1120$$

Somer's d_{yx} is given by the formula:

$$d_{YX} = \frac{P-Q}{P+Q+T_Y}$$

Substituting our data from Table 116:

$$d_{YX} = \frac{3128}{4354+1120} = .57$$

By taking account of pairs tied on Y but not X, Somer's d_{yx} shows a substantial decrease in the degree of association when compared with gamma for the same data. Somer's d_{yx} is an *asymmetrical measure of association,* that is to say, it is appropriate when we wish to predict order on Y from order on X. It follows therefore that Somer's d_{yx} is the most appropriate measure for the sociologist seeking to test the hypothesis that social class (variable X) 'causes' library book borrowing (variable Y).

Notice that as an asymmetrical measure of association, Somer's d_{yx} takes no account of pairs that are tied on X but not on Y. If we wish to predict order on X from order on Y, we would need to use another Somer's d, (that is Somer's d_{xy}) to take account of those pairs that are tied on X but not on Y.

Somer's d_{xy} is given by the formula:

$$d_{XY} = \frac{P-Q}{P+Q+T_X}$$

T_x is computed as follows.

Because T_x represents all the pairs that are tied on X but not on Y we know that such pairs will be located in different cells along the same columns of the table. Taking each cell in turn, beginning in the upper left-hand cell, multiply the frequency in that cell by all the cases *falling below it in the same column*. Having done this for each cell, simply add all of the products to give T_x. For

266

the data in Table 116:

$$T_X = 20(7+2+1)+7(2+1)+2(1)+12(10+10+2)+10(10+2)+10(2)+3(6+12+5)$$

$$+6(12+5)+12(5)+1(4+6+20)+4(6+20)+6(20) = 1112$$

$$d_{XY} = \frac{3128}{4354+1112} = .57$$

Kendall's tau b (τ_b)

Somer's d_{yx} and d_{xy}, as we have seen, go only half way in taking ties into account. A still more stringent measure of association would include both pairs tied on X but not on Y and pairs tied on Y but not on X. There is such a measure, Kendall's tau b and it is given by the formula:

$$\tau_b = \frac{P-Q}{\sqrt{(P+Q+T_X)(P+Q+T_Y)}}$$

where P, Q, T_X and T_Y have the same meaning as above.
Substituting our data from Table 116:

$$\tau_b = \frac{3128}{\sqrt{(5466)(5474)}} = .57^*$$

Because the interpretation of Kendall's tau b can be phrased in terms of a one-to-one correspondence between category ranks, tau is the most appropriate measure of association to be used in testing our sociologist's last hunch, namely that a one-to-one correspondence exists between the category rankings on social class and on library book borrowing.

By way of summary:

> When the researcher is concerned with ordinal level data cast in a contingency table to which gamma, Somer's d, and tau b are appropriate, his choice of one of these three measures must be made on the basis of *the type of relationship specified in his research hypothesis*. Thus, if his concern is merely with a weak monotonic relationship, gamma is the best choice. If his concern is with a strictly monotonic asymmetric relationship, Somer's d is the best choice. If his concern is with a category rank linear relationship tau b is his best choice. (Kohout, 1974 p. 231)

17.7 Using GUTTMAN'S LAMBDA (λ_{yx}) as an asymmetrical measure of association for nominal level data

Recall that in our discussion of measures of association (pages 66 to 74) we use the hypothetical example of the school backgrounds of commissioned and non-commissioned officers in a Regiment of Guards to introduce readers to proportional reduction in error statistics. We continue now in military vein, outlining Guttman's lambda (λ_{yx}), a proportional reduction in error statistic for use with nominal level data.

The rationale[†] for lambda involves us in 'guessing' the category of the dependent variable in which cases will fall. Look at Table 117.

---

*   The reader should note that although the results of Somer's d_{yx}, Somer's d_{xy}, and Kendall's tau b for our hypothetical data are the same, this will not generally be true.

†   Our outline draws on that in F. J. Kohout (1974) *Statistics For Social Scientists*, N.Y. John Wiley, pages 237-243.

Table 117 1000 soldiers categorized by rank

(*Y* variable)

Y_1 Private	Y_2 NCO	Y_3 Officer	
700	200	100	$N = 1000$

Suppose we select individuals randomly from the distribution of cases shown in Table 117. It's clear that the best single guess we can make about the category membership of these randomly-selected individuals is category Y_1 that is, the *modal category*, the one with the largest number of cases in it.

If we continue to draw individuals on a random basis it's equally clear that in the long run, our 'best guess' of Category Y_1 will be wrong 300 times out of 1000. Similarly, we will be wrong 800 out of 1000 times if we guess Category Y_2 and 900 times out of 1000 if we choose Category Y_3.

With no other information available to us, our best single guess is modal Category Y_1.

Suppose now that we are given additional information about the soldiers to do with their academic achievement prior to army service.

We can now categorize each of our 1000 cases in terms of this second variable (X) as shown in Table 118 below.

Let's see how the additional information on second variable X enables us to improve our guessing. What we do is this. We find the *modal category* Y (ie. the Y cell containing the largest number of cases within each X-category). From Table 118 we see that:

for X_1 the best guess is Y_1

for X_2 the best guess is Y_2

for X_3 the best guess is Y_3

Our choice of these three cells (Y_1, Y_2, and Y_3) results in 600 + 100 + 90 correct guesses, that is 790 correct guesses out of 1000. Recall that prior to information on the X-categories, our best guess was 700 out of 1000. With information on the X-categories we have reduced our errors by 90. P.R.E. (Proportional reduction in error) statistics use this quantity as a criterion for measuring the strength

Table 118 1000 soldiers categorized by rank and examination successes

		X_1 No examination qualifications	X_2 'O' levels and C.S.E.	X_3 'A' levels and above	
Y-variable categories	Y_1 Private	600	80	20	700
	Y_2 NCO	50	100	50	200
	Y_3 Officer	0	10	90	100
					(1000)

of the relationship between X and Y. The P.R.E. statistic, Lambda (λ_{yx}) may be defined as:

$$\lambda_{yx}^* = \frac{\text{Amount of reduction in error}}{\text{Amount of original error when } X \text{ is not known}}$$

Substituting our data from Table 118

$$\lambda_{yx} = \frac{90}{300} = .30$$

Kohout (1974) suggests the following formula for computing Lambda:

$$\lambda_{yx} = \frac{\Sigma L_{yx} - L_y}{N - L_y}$$

where

N = the total number of cases in the table

L_y = the number of cases in the modal Y-category ignoring X

L_{yx} = the number of cases in the largest cell within a given X-category.

Substituting from Table 118.

$$\lambda_{YX} = \frac{\Sigma L_{YX} - L_Y}{N - L_Y} = \frac{(600 + 100 + 90) - 700}{1000 - 700} = \frac{90}{300} = .30$$

Lambda λ_{yx} is an *asymmetric measure of association.* It shows how much error is reduced when we guess Y categories from knowledge of X, but *not the other way round.* Suppose, however, that we wish to guess X categories from knowledge of Y, that is to say, Y being the independent variable. In this event, lambda λ_{yx} is given by the formula:

$$\lambda_{yx} = \frac{\Sigma L_{xy} - L_x}{N - L_x}$$

Lambda is easy to compute and readily understood. However it has the undesirable property that it may take on the value of zero in cases where other measures of association will not be zero and where one would not refer to the X and Y variables as being uncorrelated or statistically independent. Take the following example.

Table 119 Example

		X-variable		
		X_1	X_2	
Y-variable	Y_1	300	400	700
	Y_2	200	100	300
		500	500	1000

If we compute λ_{yx} for the data in Table 119 we have the following result:

$$\lambda_{YX} = \frac{\Sigma L_{YX} - L_Y}{N - L_Y} = \frac{(300 + 400) - 700}{1000 - 700} = \frac{0}{300} = 0$$

---

* The subscript, YX means that Y, the dependent variable is to be guessed from our knowledge of X, the independent variable.

Suppose, instead, we decide to use Yule's Q (see example on page 165) as our measure of association. It transpires that for the data in Table 119 $Q = -.45$.

Suppose, instead, we were to use phi coefficient ϕ (see example on page 164) as our measure of association. It transpires that for the data in Table 119 $\phi = -.22$.

The problem with lambda (λ_{rx}) is this. Where the modal Y-category is *extremely large* in comparison with any other Y-category, then the 'individual' modal Y-categories within each X-category tend to correspond to the 'overall' modal Y-category. As Kohout observes, unless at least one of the modal Y-categories within an X-category is different from the 'overall' modal Y-category, then lambda must necessarily be zero. The problem can be avoided by dividing the Y-variable into categories in such a way that the marginal frequencies are as nearly equal as possible. Should it not be possible, theoretically, to subdivide the Y-variable, then in 2×2 contingency tables, Q (pages 165 to 166) or ϕ (pages 164 to 165) may be used as an alternative measure of association. In tables larger than 2×2, the percentage difference (page 163) is recommended. Alternatively, Kohout suggests, where it makes sense, then the roles of X and Y may be reversed and λ_{xy} may be computed instead of λ_{rx}.

CHAPTER 18
DESIGN 7

Multivariate analysis

Introduction

It's often the case that researchers are interested in the effect of two variables on a third, especially when they believe that the two variables may influence each other. Take, for example, the suggestion of Bryman and Cramer (1990) that there may be an interaction between the gender of a patient and the type of treatment given for a depressive illness. Women, they hypothesize, might respond more positively to psychotherapy where they are helped to discuss their feelings while men might react more favourable to treatment by an antidepressant drug. In this hypothetical example, the anticipation is that the type of treatment will interact with gender in alleviating the depressive condition. Variables in comparison investigations such as the one above are termed *factors*; studies which seek to explore the effects of two or more factors are known as *factorial studies*.

18.1   Multivariate analysis in contingency tables

Increasingly in the social and life sciences, researchers wish to explore variables that are classified in *multiple* rather than *two-dimensional* formats, as, for example, in the hypothetical data set out in Table 119.1. Everitt (1977) provides a useful account of methods for analysing multidimensional tables and has shown, incidentally, the erroneous conclusions that can result from the practice of analysing multidimensional data by summing over variables to reduce them to two-dimensional formats. In this section we too illustrate the misleading conclusions that can arise when researchers employ bivariate rather than multivariate analysis. The outline that follows draws closely on an exposition by Whiteley (1983).

Multidimensional data: some words on notation

The hypothetical data in Table 119.1 refer to a survey of voting behaviour in a sample of men and women in Britain.

Table 119.1   Sex, voting preference and social class: a three-way classification table

	Middle class		Working class	
	Conservative	Labour	Conservative	Labour
Men	81	29	40	131
Women	99	21	40	109

Source: adapted from Whiteley (1983)

In Table 119.1

the row variable (sex) is represented by i;

the column variable (voting preference) is represented by j;

the layer variable (social class) is represented by k.

The number in any one cell in Table 119.1 can be represented by the symbol n_{ijk} that is to say, the score in row category i, column category j, and layer category k, where $i = 1$ (men), 2 (women); $j = 1$ (Conservative), 2 (Labour); $k = 1$ (middle class), 2 (working class). It follows therefore that the numbers in Table 119.1 can also be represented as in Table 119.2.

Table 119.2   Sex, voting preference and social class: a three-way notational classification

	Middle class		Working class	
	Conservative	Labour	Conservative	Labour
Men	n_{111}	n_{121}	n_{112}	n_{122}
Women	n_{211}	n_{221}	n_{212}	n_{222}

Source: adapted from Whiteley (1983)

Thus,

$n_{121} = 29$ (men, Labour, middle class)

and

$n_{212} = 40$ (women, Conservative, working class)

Three types of marginals can be obtained from Table 119.2 by:

(1) *Summing over two variables to give the marginal totals for the third.* Thus n_{++k} = summing over sex and voting preference to give *social class,* for example:

$n_{111} + n_{121} + n_{211} + n_{221} = 230$ middle class

$n_{112} + n_{122} + n_{212} + n_{222} = 320$ working class

and n_{+j+} = summing over sex and social class to give *voting preference* and n_{i++} = summing over voting preference and social class to give *sex.*

(2) *Summing over one variable to give the marginal totals for the second and third variables.* Thus:

n_{+11} = 180 (middle-class Conservative)

n_{+21} = 50 (middle-class Labour)

n_{+12} = 80 (working-class Conservative)

n_{+22} = 240 (working-class Labour)

(3) *Summing over all three variables to give the grand total.* Thus:

n_{+++} = 550 = N

Using the chi square test in a three-way classification table

Whiteley shows how easy it is to extend the 2 × chi square test to the three-way case. The probability that an individual taken from the sample at random in Table 119.1 will be a woman is:

$$P_{2++} = \frac{n_{2++}}{n_{+++}} = \frac{269}{550} = 0.49$$

and the probabiity that a respondent's voting preference will be *Labour* is:

$$P_{+2+} = \frac{n_{+2+}}{n_{+++}} = \frac{290}{550} = 0.53$$

and the probability that a respondent will be *working class* is:

$$P_{++2} = \frac{n_{++2}}{n_{+++}} = \frac{320}{550} = 0.58$$

To determine the *expected probability* of an individual being a *woman, Labour supporter* and *working class* we assume that these variables are statistically independent (that is to say, there is no relationship between them) and simply apply the multiplication rule of probability theory:

$$P_{222} = (P_{2++})\ (P_{+2+})\ (P_{++2}) = (0.49)\ (0.53)\ (0.58) = 0.15$$

This can be expressed in terms of the expected frequency in cell n_{222} as:

$$N\ (P_{2++})\ (P_{+2+})\ (P_{++2}) = 550(0.49)\ (0.53)\ (0.58) = 82.8$$

Similarly, the expected frequency in cell n_{112} is:

$$N(P_{1++})\ (P_{+1+})\ (P_{++2})\ \text{where}$$

$$P_{1++} = \frac{n_{1++}}{n_{+++}} = \frac{281}{550} = 0.51$$

$$\text{and } P_{+1+} = \frac{n_{+1+}}{n_{+++}} = \frac{260}{550} = 0.47$$

$$\text{and } P_{++2} = \frac{n_{++2}}{n_{+++}} = \frac{281}{550} = 0.58$$

Thus $N(P_{1++})\ (P_{+1+})\ (P_{++2}) = 550\ (.51)\ (0.47)\ (0.58) = 76.5$

Table 119.3 gives the expected frequencies for the data shown in Table 119.1.

Table 119.3 Expected frequencies in sex, voting preference and social class

| | Middle class | | Working class | |
	Conservative	Labour	Conservative	Labour
Men	55.4	61.7	76.5	85.9
Women	53.4	59.5	74.3	82.8

Source: adapted from Whiteley (1983)

With the *observed* frequencies and the *expected* frequencies to hand, chi square is calculated in the usual way:

$$\chi^2 = \Sigma \frac{(O - E)^2}{E} = 158.59$$

Degrees of freedom

As Whiteley observes, degrees of freedom in a three-way contingency table are more complex than in a 2 × 2 classification. Essentially, however, degrees of freedom refer to the freedom with which the researcher is able to assign values to the cells, given fixed marginal totals. This can be computed by first determining the degrees of freedom for the marginals.

Each of the variables in our example (sex, voting preference and social class) contains two categories. It follows therefore that we have $(2 - 1)$ degrees of freedom for each of them, given that the marginal for each variable is fixed. Since the grand total of all the marginals (i.e. the sample size) is also fixed, it follows that one more degree of freedom is also lost. We subtract these fixed numbers from the total number of cells in our contingency table. In general therefore:

degrees of freedom (d.f.) = the number of cells in the table – 1 (for *N*) – the number of cells fixed by the hypothesis being tested.

Thus, where r = rows, c = columns and l = layers:

d.f. $= rcl - (r - 1) - (c - 1) - (l - 1) - 1$

$= rcl - r - c - l + 2$

that is to say, d.f. $= rcl - r - c - l + 2$ when we are testing the hypothesis of the *mutual independence of the three variables.*

In our example:

d.f. $= (2)\,(2)\,(2) - 2 - 2 - 2 + 2 = 4$

From chi-square tables we see that the critical value of χ^2 with 4 degrees of freedom is 9.49 at $p = 0.05$. Our obtained value greatly exceeds that number. We reject the null hypothesis and conclude that sex, voting preference and social class are significantly interrelated.

Having rejected the null hypothesis with respect to the *mutual independence* of the three variables, the researcher's task now is to identify which variables cause the null hypothesis to be rejected. We cannot simply assume that because our chi square test has given a significant result, it therefore follows that there are significant associations between all three variables. It may be the case, for example, that an association exists between two of the variables whilst the third is completely independent. What we need now is a test of *partial independence*. Whiteley shows the following three such possible tests in respect of the data in Table 119.1.

First, that sex is independent of social class and voting preference

$$p_{ijk} = (p_i) \, (p_{jk})$$

Second, that voting preference is independent of sex and social class

$$p_{ijk} = (p_j) \, (p_{ik})$$

And third, that social class is independent of sex and voting preference

$$p_{ijk} = (p_k) \, (p_{ij})$$

The following example shows how to construct the expected frequencies for the first hypothesis. We can determine the probability of an individual being, say, *woman, Labour* and *working class,* assuming the first hypothesis as follows:

$$p_{222} = (p_{2++}) \, (p_{+22}) = \frac{(n_{2++}) \, (n_{+22})}{(N) \quad (N)}$$

$$p_{222} = \frac{(269) \, (240)}{(550) \, (550)} = 0.213$$

$$E_{222} = N(p_{2++}) \, (p_{+22}) = 550 \, \frac{(269) \, (240)}{(550) \, (550)} = 117.4$$

That is to say, assuming that sex is independent of social class and voting preference, the expected number of female, working-class Labour supporters is 117.4.

When we calculate the expected frequencies for each of the cells in our contingency table in respect of our first hypothesis $(p_{ijk}) = (p_i) \, (p_{jk})$, we obtain the results shown in Table 119.4.

$$\chi^2 = \frac{\Sigma \, (O - E)^2}{E} = 4.78$$

Degrees of freedom is given by:

$$\text{d.f.} = rcl - (cl - 1) - (r - 1) - 1$$
$$= rcl - cl - r + 1 = 8 - 4 - 2 + 1 = 3$$

Whiteley (1983) observes:

Note that we are assuming c and l are interrelated so that once, say, p_{+11} is calculated, then p_{+12}, p_{+21} and p_{+22} are determined so we have only 1 degree of freedom; that is to say, we lose $(cl - 1)$ degrees of freedom in calculating that relationship.

From chi-square tables we see that the critical value of χ^2 with 3 degrees of freedom is 7.81 at $p = 0.05$. Our obtained value is less than this. We therefore accept the null hypothesis and conclude *that there is no relationship between sex on the one hand and voting preference and social class on the other.*

Suppose now that instead of casting our data into a three-way classification as shown in Table 119.1, we had simply used a 2 × 2 contingency table and that we had sought to test the null hypothesis that there is no relationship between *sex and voting preference.* The data are shown in Table 119.5.

Table 119.4 Expected frequencies assuming that sex is independent of social class and voting preference

| | Middle class | | Working class | |
	Conservative	Labour	Conservative	Labour
Men	91.9	25.6	40.9	122.6
Women	88.1	24.4	39.1	117.4

Source: adapted from Whiteley (1983)

When we compare chi square from these data our obtained value is $\chi^2 = 4.48$. Degrees of freedom are given by $(r - 1)(c - 1) = (2 - 1)(2 - 1) = 1$. From chi square tables we see that the critical value of χ^2 with 1 degree of freedom is 3.84 at $p = 0.05$. Our obtained value exceeds this. We reject the null hypothesis and conclude that sex is significantly associated with voting preference.

Table 119.5 Sex and voting preference: a two-way classification table

	Conservative	Labour
Men	121	160
Women	139	130

Source: adapted from Whiteley (1983)

But how can we explain the differing conclusions that we have arrived at in respect of the data in Tables 119.1 and 119.5?

These examples illustrate an important and general point, Whiteley observes. In the *bivariate* analysis (Table 119.5) we concluded that there was a significant relationship between sex and voting preference. In the *multivariate* analysis (Table 119.1) that relationship was found to be non-significant when we controlled for social class.

The lesson is plain: use a multivariate approach to the analysis of contingency tables wherever the data allow.

Factorial designs—the effect of two independent variables on the dependent variable
(a) No repeated measures on factors

| | | Independent variable B (Factor B) | |
		Level B_1	Level B_2
Independent variable A (Factor A)	Level A_1	Group 1 X_A subjects X_B A, B, C X_C	Group 3 X_G subjects X_H G, H, J X_J
	Level A_2	Group 2 X_D subjects X_E D, E, F X_F	Group 4 X_K subjects X_L K, L, M X_M

18.2 Using TWO-WAY ANALYSIS OF VARIANCE—single observation on separate groups

Two-way analysis of variance techniques are used to estimate the effect of two independent variables (factors) on a dependent variable. The hypothetical data presented in Table 120 allow us to illustrate the use of a two-way technique for single observations on separate groups.

The reader is advised to refer to page 181 for a brief explanation of the rationale behind analysis of variance techniques before proceeding to the present example.

We have purposely included only a small number of scores in each of the cells in our example in order to avoid lengthy and cumbersome calculations. It should be remembered, however, that in order to make a valid analysis the same assumptions that govern a one-way analysis of variance have to be met.

Suppose that we wish to compare psychological status (anxiety) between high- and low-stress groups as measured by a social adjustment scale in immigrant and non-immigrant cultures. After administering the social adjustment scale to a large inner-city population, equal numbers of people with stressful and non-stressful life events are selected at random from each culture and given a test designed to measure anxiety.

Two-way analysis of variance permits us to answer a number of questions concerning our samples and their anxiety levels.

Do immigrants differ significantly from non-immigrants in their anxiety levels? Do stress levels affect anxiety? Is there an interaction effect of stress level and culture in respect of the results? In other words, do high-stress immigrants show more anxiety than low-stress immigrants and so on?

The null hypotheses in respect of each of the questions we have raised are that neither stress level, nor culture, nor any interaction of these two factors affects anxiety. Our data are set out in Table 120.

Table 120 Stress, culture and anxiety levels

		Independent variable B (Factor B)	
		Immigrants	Non-immigrants
Independent Variable A (Factor A)	High stress	(Group 1) 9.1 9.4 9.6 9.8	(Group 2) 8.2 8.8 8.7 9.0
	Low stress	(Group 3) 7.5 7.1 7.2 7.1	(Group 4) 7.3 7.2 7.5 7.0

The total variance in this single observation design can be partitioned into:

SYSTEMATIC EFFECTS

BETWEEN GROUPS (treatment or conditions) variance is the variance between means caused by the independent variables. These effects can be separated as follows:

(a) FACTOR A—the effect of stress level on the variance.

(b) FACTOR B—the effect of culture on the variance.

(c) INTERACTION A × B—the combined effects of stress level and culture on the variance.

ERROR EFFECTS

WITHIN GROUPS (CELLS) variance is the variance due to subject differences and uncontrolled factors.

PROCEDURE FOR COMPUTING TWO-WAY ANALYSIS OF VARIANCE (SINGLE OBSERVATIONS ON SEPARATE GROUPS)

1 Square each score (X^2) and total both the scores (ΣX) and the squares of scores (ΣX^2) for each group. Enter in separate data Table 121.

2 Compute the grand total GT (Correction Factor).

$$GT = \frac{(\Sigma X)^2}{N_T}$$

where N_T = Total number of scores.

$$GT = \frac{(130.5)^2}{16} = 1064.39$$

Table 121 Two-way analysis of variance: computational procedures

		Factor B				Total X for rows
		Immigrants (Column 1)		Non-immigrants (Column 2)		
		X_1	X_1^2	X_2	X_2^2	
	High stress (Row 1)	9.1	82.81	8.2	67.24	17.30
		9.4	88.36	8.8	77.44	18.20
		9.6	92.16	8.7	75.69	18.30
		9.8	96.04	9.0	81.00	18.80
		37.9	359.37	34.7	301.37	72.60
Factor A		X_3	X_3^2	X_4	X_4^2	
	Low stress (Row 2)	7.5	56.25	7.3	53.29	14.80
		7.1	50.41	7.2	51.84	14.30
		7.2	51.84	7.5	56.25	14.70
		7.1	50.41	7.0	49.00	14.10
		28.9	208.91	29.00	210.38	57.90
	Total X for columns	66.8		63.7		

ΣX TOTAL = 37.9 + 34.7 + 28.9 + 29 = 130.50

ΣX^2 TOTAL = 359.37 + 301.37 + 208.91 + 210.38 = 1080.03

3 Compute the TOTAL sum of squares (SS_{TOTAL})

$$SS_{TOTAL} = \Sigma X^2 - GT$$
$$= 1080.03 - 1064.39 = 15.64$$

4 Compute the BETWEEN GROUPS sum of squares $(SS_{BETWEEN})$.

$$SS_{BETWEEN} = \left[\frac{(\Sigma X_1)^2}{N_1} + \frac{(\Sigma X_2)^2}{N_2} + \frac{(\Sigma X_3)^2}{N_3} + \frac{(\Sigma X_4)^2}{N_4} \right] - GT$$

$$= \left[\frac{(37.9)^2}{4} + \frac{(34.7)^2}{4} + \frac{(28.9)^2}{4} + \frac{(29)^2}{4} \right] - 1064.39$$

$$= [359.103 + 301.023 + 208.803 + 210.25] - 1064.39$$

$$= 14.789$$

5 Compute the sum of squares for FACTOR A (SS_{FA}).

$$SS_{FA} = \frac{(\text{Sum of row 1})^2}{\text{Number in row 1}} + \frac{(\text{sum of row 2})^2}{\text{Number in row 2}} - GT$$

$$= \frac{(72.6)^2}{8} + \frac{(57.9)^2}{8} - 1064.39$$

$$= 658.845 + 419.051 - 1064.39$$

$$= 13.506$$

6 Compute the sum of squares for FACTOR B (SS_{FB}).

$$SS_{FB} = \frac{(\text{Sum of column 1})^2}{\text{Number in column 1}} + \frac{(\text{Sum of column 2})^2}{\text{Number in column 2}} - GT$$

$$= \frac{(66.8)^2}{8} + \frac{(63.7)^2}{8} - 1064.39$$

$$= 557.78 + 507.211 - 1064.39$$

$$= 0.601$$

7 Compute the sum of squares for INTERACTION $(SS_{A \times B})$.

$$SS_{A \times B} = SS_{BETWEEN} - SS_{FA} - SS_{FB}$$
$$= 14.789 - 13.506 - 0.601$$
$$= 0.682$$

8 Compute the WITHIN GROUPS (CELLS) sum of squares (SS_{WITHIN})-

$$SS_{WITHIN} = SS_{TOTAL} - SS_{BETWEEN}$$
$$= 15.64 - 14.789$$
$$= 0.851$$

9 Determine the degrees of freedom for each sum of squares.

$$\text{d.f. for } SS_{TOTAL} = (N - 1) = (16 - 1) = 15$$

where N = total number of scores

$$\text{d.f. for } SS_{BETWEEN} = (G - 1) = (4 - 1) = 3$$

279

where G = number of groups

$$\text{d.f. for } SS_{FB} = (C-1) = (2-1) = 1$$

where C = number of columns

$$\text{d.f. for } SS_{FA} = (r-1) = (2-1) = 1$$

where r = number of rows

$$\text{d.f. for } SS_{A \times B} = (r-1)(C-1) = (2-1)(2-1) = 1$$

$$\text{d.f. for } SS_{WITHIN} = (N - rC) = 16 - (2)(2) = 12$$

10 Estimate the variances.

$$\text{Var} = \frac{\text{Sum of squares}}{\text{d.f.}}$$

$$\text{Var}_{\text{FACTOR A}} = \frac{SS_{FA}}{\text{d.f.}_{FA}} = \frac{13.506}{1} = 13.506$$

$$\text{Var}_{\text{FACTOR B}} = \frac{SS_{FB}}{\text{d.f.}_{FB}} = \frac{0.601}{1} = 0.601$$

$$\text{Var}_{A \times B} = \frac{SS_{A \times B}}{\text{d.f.}_{A \times B}} = \frac{0.682}{1} = 0.682$$

$$\text{Var}_{\text{WITHIN}} = \frac{SS_{\text{WITHIN}}}{\text{d.f.}_{\text{WITHIN}}} = \frac{0.851}{12} = 0.071$$

11 Compute the F values for the main effects.

$$F_{\text{FACTOR A}} = \frac{\text{Var}_{\text{FACTOR A}}}{\text{Var}_{\text{WITHIN}}} = \frac{13.506}{0.071} = 190.225$$

$$F_{\text{FACTOR B}} = \frac{\text{Var}_{\text{FACTOR B}}}{\text{Var}_{\text{WITHIN}}} = \frac{0.601}{0.071} = 8.465$$

$$F_{A \times B} = \frac{\text{Var}_{A \times B}}{\text{Var}_{\text{WITHIN}}} = \frac{0.682}{0.071} = 9.606$$

12 Enter the results in an Analysis of Variance Table.

Table 122 Analysis of variance table

Source of variation	Sum of squares	d.f.	Variance	F
(Between groups)	(14.789)	(3)		
Factor A	13.506	1	13.506	190.225
Factor B	0.601	1	0.601	8.465
Interaction	0.682	1	0.682	9.606
Within groups	0.851	12	0.071	
Total	15.640	15		

13   Refer to the table in Appendix 6 to determine the significance of the F values.

Our first null hypothesis is that culture has no effect upon anxiety. The obtained F value for factor B, 8.465 exceeds the F of 4.7472 in the table for 1 and 12 degrees of freedom at the .05 level. We therefore reject the null hypothesis and conclude that immigrants have different anxiety levels from non-immigrants.

Our second null hypothesis is that stress level has no effect upon anxiety scores. Our obtained value of 190.225 exceeds the F of 9.3302 found in the table at the 0.01 level for 1 and 12 degrees of freedom. We therefore reject the null hypothesis and conclude that high-stress subjects have higher anxiety results than low-stress subjects.

Finally, our third null hypothesis is that there is no interaction effect on the anxiety results. That is to say, the combined effects of stress level and culture do not affect the results. The obtained value of $F = 9.606$ for interaction (A $\times$ B) exceeds the value of 4.7472 found in the table at the 0.05 level for 1 and 12 degrees of freedom. We therefore reject the null hypothesis and conclude that the combined effects of stress level and culture significantly affect the anxiety scores.

Since the F test has shown that there are significant differences between means, we must now use a Tukey test to compare means.

We already know that the means for immigrants and non-immigrants are significantly different. We also know that the means for high-stress and low-stress groups are different. There is no need to apply further tests on them.

On the other hand, although we have obtained a significant F value for interaction, we do not know how many of the interaction means (the cell means) are significantly different; all we know is that at least two of them are. The Tukey test can assist us here.

PROCEDURE FOR COMPUTING THE TUKEY TEST

1   Calculate the means for each cell or group and construct a table of sample mean differences.

Table 123   Sample mean differences

	Comparison means	M_2	M_3	M_4
Group 1 High-stress immigrants	$M_1 = 9.48$	$M_1 - M_2 = 0.8$**	$M_1 - M_3 = 2.25$**	$M_1 - M_4 = 2.23$**
Group 2 High-stress non-immigrants	$M_2 = 8.68$		$M_2 - M_3 = 1.45$**	$M_2 - M_4 = 1.43$**
Group 3 Low-stress immigrants	$M_3 = 7.23$			$M_3 - M_4 = -0.02$
Group 4 Low-stress non-immigrants	$M_4 = 7.25$			

** Significant at 0.01 level

2 Compute T.

$$T = (q)\sqrt{\frac{\text{Var}_{\text{WITHIN}} \text{ (Error variance)}}{N}}$$

where

N = number in each group or the number of scores from which mean is calculated.

The (q) value in the formula is found by consulting Appendix 7 and determining the value corresponding to the number of means (n in the Tukey table) and the degrees of freedom for the denominator of our prior F test. That denominator is, of course, the within groups variance and its appropriate degrees of freedom which are read as v in the Tukey table.

We see that in our example, q at the 0.05 level = 4.2 when n = 4 and r = 12. Thus:

$$T_{0.05} = 4.2\sqrt{\frac{0.07}{4}} = 4.2\,(0.13) = 0.56$$

also, q at the 0.01 level = 5.5 when n = 4 and r = 12. Thus:

$$T_{0.01} = 5.5\sqrt{\frac{0.07}{4}} = 5.5\,(0.13) = 0.72$$

If the T value is smaller than the absolute difference between two means, then the means are significantly different at that level.

Referring to our table of mean differences (Table 123) we can draw the following conclusions:

(*a*) High-stress immigrants score higher on anxiety than high-stress non-immigrants.

(*b*) High-stress immigrants score higher on anxiety than low-stress immigrants.

(*c*) High-stress non-immigrants score higher on anxiety than low-stress immigrants.

(*d*) Low-stress immigrants do not score significantly higher on anxiety than low-stress non-immigrants.

CHAPTER 19
DESIGN 8

Factorial designs—the effect of two independent variables on the dependent variable
(b)   Repeated measures on ONE factor

		Independent Variable B (Factor B)	
		Level B_1	**Level B_2**
Independent variable A (*Factor A*)	Level A_1	Group$_1$ X_{A_1} X_{B_1} X_{C_1}	Group$_1$ X_{A_2} X_{B_2} X_{C_2}
	Level A_2	Group$_2$ X_{E_1} X_{F_1} X_{G_1}	Group$_2$ X_{E_2} X_{F_2} X_{G_2}

19.1   Using TWO-WAY ANALYSIS OF VARIANCE—repeated observations on one factor

When we wish to determine the effects of two independent variables on a dependent variable and the experimental design involves repeated measures of the same subjects on one of the independent variables, then the two-way analysis of variance we outlined in page 277 is inappropriate. The following hypothetical example shows the changes that have to be made as a result of different partitioning of the total variance.

Suppose that two groups of subjects, one containing males and the other containing females, are selected at random and tested on a particular skills test on two separate occasions. On one occasion the subjects are administered a hormone-based drug and on the other, no drug is administered. In order to control for the practice effects, half of the subjects from each group take the drug on the first testing occasion and the other half take the drug on the second testing situation.

283

By applying a two-way analysis of variance we can answer the following questions. Do the males score higher than the females on this particular skills test? Does taking the drug have an effect on skill test scores? Is there a difference in skill test scores due to the combined effects (interaction) of gender and drug/no drug?

Our data are set out in Table 124 below.

Table 124 Skill test results

		Independent variable 2 (Factor 2)	
		Drug (Column 1)	No drug (Column 2)
Independent variable 1		23	23
		19	22
		24	25
		22	24
	Group 1	24	23
	Males (row 1)	18	17
		17	20
		20	19
		18	17
		23	21
(Factor 1)		10	12
		12	12
		10	13
		7	10
	Group 2	13	13
	Females (row 1)	11	11
		10	14
		9	12
		13	13
		11	14

The assumptions governing the two-way analysis of variance (repeated observations on one factor) are similar to those for the one-way analysis of variance detailed on page 241.

In this repeated measures design, the total variance can be partitioned as follows:

SYSTEMATIC EFFECTS

1   The variance caused by the main effects of the independent variables, namely:

(a)   FACTOR 1—the effects of gender on the variance.

(b)   FACTOR 2—the effects of drug or no drug on the variance.

(c)   INTERACTION (Gender × Drug)—the combined effects of factors 1 and 2 on the variance.

ERROR EFFECTS

2   BETWEEN SUBJECTS ERROR (WITHIN GROUPS) variance due to the subject differences and uncontrolled factors.

3   WITHIN SUBJECTS ERROR variance caused by individual variations on the two testing situations.

PROCEDURE FOR COMPUTING TWO-WAY ANALYSIS OF VARIANCE (REPEATED MEASURES ON ONE FACTOR)

1   Square each score (X^2) and total the scores (ΣX) and the squares of scores (ΣX^2) for each cell. Enter these in a separate data table.

Table 125   Skill test results: computational procedures for analysis of variance

		Factor 2				Totals for subjects	(Totals)^2 for subjects
		Drug		No drug			
	Group 1	X_1	X_1^2	X_2	X_2^2	(X_1+X_2)	$(X_1+X_2)^2$
		23	529	23	529	46	2116
		19	361	22	484	41	1681
		24	576	25	625	49	2401
		22	484	24	576	46	2116
	Males	24	576	23	529	47	2209
		18	324	17	289	35	1225
		17	289	20	400	37	1369
		20	400	19	361	39	1521
		18	324	17	289	35	1225
		23	529	21	441	44	1936
		208	4392	211	4523	419	17,799
Factor 1	Group 2	X_3	X_3^2	X_4	X_4^2	(X_3+X_4)	$(X_3+X_4)^2$
		10	100	12	144	22	484
		12	144	12	144	24	576
		10	100	13	169	23	529
		7	49	10	100	17	289
		13	169	13	169	26	676
	Females	11	121	11	121	22	484
		10	100	14	196	24	576
		9	81	12	144	21	441
		13	169	13	169	26	676
		11	121	14	196	25	625
		106	1154	124	1552	230	5356

Column Total   314   5546   335   6102

$$\Sigma X_{TOTAL} = 208 + 211 + 106 + 124 = 649$$
$$\Sigma X^2_{TOTAL} = 4392 + 4523 + 1154 + 1552 = 11,621$$

2   Compute GT (Correction factor).

$$GT = \frac{(\Sigma X)^2}{N_T}$$

where N_r = total number of scores.

$$= \frac{(649)^2}{40} = 10,530.025$$

Practical Statistics for Students

3 Compute the TOTAL sum of squares (SS_{TOTAL}).

$$SS_{TOTAL} = \sum X_{TOTAL}^2 - GT$$
$$= 11,621 - 10,530.025$$
$$= 1090.975$$

4 Compute the sum of squares for FACTOR 1 (SS_{F1}).

$$SS_{F1} = \sum \frac{(\text{Sum of row})^2}{\text{Number in rows}} - GT$$
$$= \frac{(419)^2}{20} + \frac{(230)^2}{20} - 10,530.025$$
$$= 8778.05 + 2645 - 10,530.025$$
$$= 893.025$$

5 Compute the sum of squares for FACTOR 2 (SS_{F2}).

$$SS_{F2} = \sum \frac{(\text{Sum of column})^2}{\text{Number in column}} - GT$$
$$= \frac{(314)^2}{20} + \frac{(335)^2}{20} - 10,530.025$$
$$= 4929.8 + 5611.25 - 10,530.025$$
$$= 11.025$$

6 Compute the INTERACTION sum of squares $(SS_{1 \times 2})$.

$$SS_{(1 \times 2)} = \frac{(\sum X_1)^2}{n_1} + \frac{(\sum X_2)^2}{n_2} + \frac{(\sum X_3)^2}{n_3} + \frac{(\sum X_4)^2}{n_4} - SS_{F1} - SS_{F2} - GT$$
$$= \frac{(208)^2}{10} + \frac{(211)^2}{10} + \frac{(106)^2}{10} + \frac{(124)^2}{10} - 893.025 - 11.025 - 10,530.025$$
$$= 4326.4 + 4452.1 + 1123.6 + 1537.6 - 893.025 - 11.025 - 10,530.025$$
$$= 5.625$$

7 Compute BETWEEN SUBJECTS ERROR sum of squares (SS_{BSE}).

$$SS_{BSE} = \frac{\sum (X_1 + X_2)^2 + \sum (X_3 + X_4)^2}{\text{Number of columns}} - SS_{F1} - GT$$
$$= \frac{17,799 + 5356}{2} - 893.025 - 10,530.025$$
$$= 11,577.5 - 893.025 - 10,530.025$$
$$= 154.450$$

8 Compute WITHIN SUBJECTS ERROR sum of squares (SS_{WSE}).

$$SS_{WSE} = SS_{TOTAL} - SS_{F1} - SS_{F2} - SS_{(INT)} - SS_{BSE}$$
$$= 1090.975 - 893.025 - 11.025 - 5.625 - 154.450$$
$$= 26.85$$

9   Determine the degrees of freedom for each of the sums of squares.

$$\text{d.f. for } SS_{\text{TOTAL}} = (N_T - 1) = (40 - 1) = 39$$

where N_T = total number of scores.

$$\text{d.f. for } SS_{F1} = (r - 1) = (2 - 1) = 1$$

where r = number of rows.

$$\text{d.f. for } SS_{F2} = (C - 1) = (2 - 1) = 1$$

where C = number of columns.

$$\text{d.f. for } SS_{\text{(INT)}} = (r - 1)(C - 1) = (2 - 1)(2 - 1) = 1$$

$$\text{d.f. for } SS_{\text{BSE}} = r(n - 1) = 2(10 - 1) = 18$$

where n = number of subjects in each group.

$$\text{d.f. for } SS_{\text{WSE}} = r(n - 1)(C - 1) = 2(10 - 1)(2 - 1) = 18$$

10   Compute the variances for the main and error effects.

$$\text{Var}_{\text{Factor 1}} = \frac{SS_{F1}}{\text{d.f.}_{F1}} = \frac{893.025}{1} = 893.025$$

$$\text{Var}_{\text{Factor 2}} = \frac{SS_{F2}}{\text{d.f.}_{F2}} = \frac{11.025}{1} = 11.025$$

$$\text{Var}_{\text{Interaction}} = \frac{SS_{\text{INT}}}{\text{d.f.}_{\text{INT}}} = \frac{5.625}{1} = 5.625$$

$$\text{Var}_{\text{BSE}} = \frac{SS_{\text{BSE}}}{\text{d.f.}_{\text{BSE}}} = \frac{154.45}{18} = 8.581$$

$$\text{Var}_{\text{WSE}} = \frac{SS_{\text{WSE}}}{\text{d.f.}_{\text{WSE}}} = \frac{26.850}{18} = 1.492$$

11   Compute F values for main effects.

$$F_{F1} = \frac{\text{Var}_{F1}}{\text{Var}_{\text{BSE}}} = \frac{893.025}{8.581} = 104.070$$

$$F_{F2} = \frac{\text{Var}_{F2}}{\text{Var}_{\text{WSE}}} = \frac{11.025}{1.492} = 7.389$$

$$F_{\text{INT}} = \frac{\text{Var}_{\text{INT}}}{\text{Var}_{\text{WSE}}} = \frac{5.625}{1.492} = 3.77$$

12   Enter the results in an analysis of variance table.
Consult the table in Appendix 6 to determine the significance of the F values.
Our obtained value for Factor 1, $F = 104.07$ exceeds the F of 8.29 in the table at the 0.01 level for d.f. = 1, 18. We therefore reject the null hypothesis and conclude that the males scored differently from the females.
Similarly, our obtained value for Factor 2, $F = 7.389$ exceeds the F of 4.41 at the 0.05 level. Again we reject the null hypothesis and conclude that the specific drug has an effect upon skill performance in the test situation.

Table 126 Analysis of variance table

Source of variation	Sum of squares	d.f.	Variance	F
Factor 1	893.025	1	893.025	104.07
Factor 2	11.025	1	11.025	7.389
Interaction	5.625	1	5.625	3.77
Within Subjects Error	26.85	18	1.492	
Between Subjects Error	154.45	18	8.581	
Total	1090.975	39		

Our obtained value for the Interaction, $F = 3.77$ is less than the 4.41 value in the table at the 0.05 level. We therefore accept the null hypothesis and conclude that the combined effects of gender and drugs do not affect skill test performance.

Factorial designs—the effect of two independent variables on the dependent variable

(c)   **Repeated measures on BOTH factors**

		Independent variable B (Factor B)	
		Level B_1	Level B_2
Independent variable A	Level A_1	Group 1	Group 1
(Factor A)	Level A_2	Group 1	Group 1

Or

Group 1 subjects	Independent variable A (Factor A)			
	Level A_1		Level A_2	
	Independent variable B (Factor B)			
	Level B_1	Level B_2	Level B_1	Level B_2
M	$X_M(A_1 B_1)$	$X_M(A_1 B_2)$	$X_M(A_2 B_1)$	$X_M(A_2 B_2)$
N	$X_N(A_1 B_1)$	$X_N(A_1 B_2)$	$X_N(A_2 B_1)$	$X_N(A_2 B_2)$
O	$X_O(A_1 B_1)$	$X_O(A_1 B_2)$	$X_O(A_2 B_1)$	$X_O(A_2 B_2)$
P	$X_P(A_1 B_1)$	$X_P(A_1 B_2)$	$X_P(A_2 B_1)$	$X_P(A_2 B_2)$
Q	$X_Q(A_1 B_1)$	$X_Q(A_1 B_2)$	$X_Q(A_2 B_1)$	$X_Q(A_2 B_2)$
R	$X_R(A_1 B_1)$	$X_R(A_1 B_2)$	$X_R(A_2 B_1)$	$X_R(A_2 B_2)$

20.1 Using TWO-WAY ANALYSIS OF VARIANCE—repeated observations on both factors

If we wish to determine the effects of two independent variables on a dependent variable and only one random sample is selected, the two-way analysis of variance is appropriate. It involves the analysis of repeated measures of the same subjects on both independent variables.

In order to ease the computational burden in the example we use hypothetical scores and a small sample. The reader should remember, however, that with authentic data the same assumptions apply to this analysis of variance technique as to those that have been described in earlier sections.

In an experiment to assess the psychological effect of sleep deprivation, subjects were selected at random and asked to perform standardized cycle ergometer tests. One involved light exercise and one involved heavy exercise over a three-day series which included 24 hours without sleep before day 2. The subjects were required to rate their perceived exertion on a scale 2–10 after each of the six bouts of exercise.

We wish to answer the following questions:

1 Is perceived exertion different after heavy and light exercise bouts?

2 Does perceived exertion for given work loads change with sleep deprivation?

3 Is there a combined effect of sleep deprivation and type of exercise on perceived exertion?

Our hypothetical data are set out in Table 127 below.

Table 127 Perceived exertion rates during heavy and light exercise after periods of sleep deprivation: Format A

			Independent Variable (D)		
		Subjects	Day 1 (D_1)	Day 2 (24 hours' loss) (D_2)	Day 3 (D_3)
Independent Variable (E)	Light Exercise (E_1)	S_1	3	5	3
		S_2	4	5	4
		S_3	3	5	3
		S_4	5	6	5
	Heavy Exercise (E_2)	S_1	5	7	5
		S_2	5	6	7
		S_3	7	7	6
		S_4	6	8	7

Although Table 127 includes all the relevant information needed for our analysis of variance, the form of presentation does not easily lend itself to the computational description that follows. The same information is therefore set out in a more suitable format in Table 128.

The total variance in this repeated measure design can be partitioned into:

1 BETWEEN SUBJECTS VARIANCE.

2 WITHIN SUBJECTS VARIANCE.

Table 128 Perceived exertion rates during heavy and light exercise after periods of sleep deprivation: Format B

Subjects	Light exercise (E_1)			Heavy exercise (E_2)		
	Day 1 (D_1)	Day 2 (D_2)	Day 3 (D_3)	Day 1 (D_1)	Day 2 (D_2)	Day 3 (D_3)
S_1	3	5	3	5	7	5
S_2	4	5	4	5	6	7
S_3	3	5	3	7	7	6
S_4	5	6	5	6	8	7

The within subjects variance has both systematic and error effects on the total variance. The main effects of the independent variables make up the systematic variance, whereas the error effects are caused by possible individual variations between the repeated measures. This within subjects variance can be subdivided as follows:

SYSTEMATIC EFFECTS

(a) Factor E—the effect of the differing levels of exercise on the variance.

(b) Factor D—the effect of sleep deprivation on the variance.

(c) Interaction (E × D)—the combined effects of exercise and sleep deprivation.

ERROR EFFECTS

(d) Exercise × Subjects WITHIN GROUPS ERROR—the effect due to the chance fluctuation of subjects' scores over the exercise conditions.

(e) Deprivation × Subjects WITHIN GROUPS ERROR—the effect due to chance fluctuation between scores over the separate days, i.e. different levels of deprivation.

(f) Interaction × Subjects WITHIN GROUPS ERROR.

PROCEDURE FOR COMPUTING THE TWO-WAY ANALYSIS OF VARIANCE—
REPEATED OBSERVATIONS ON BOTH FACTORS

Table 129 below will assist the reader to interpret the various formulae used in the computations.

1 Prepare a data table including the rows and columns totals.
Let

$$n = \text{number of subjects} = 4$$

$$q = \text{number of exercise treatments} = 2$$

$$r = \text{number of days (conditions)} = 3$$

2 Square each score and total the raw scores (ΣX) and squares of scores (ΣX^2).

$$\Sigma X = 15 + 21 + 15 + 23 + 28 + 25 = 127$$

$$\Sigma X^2 = 3^2 + 4^2 + \ldots + 6^2 + 7^2 = 721$$

Table 129 Interpretative layout of two-way analysis of variance

| Subjects | Light exercise E_1 | | | | Heavy exercise E_2 | | | | |
	D_1	D_2	D_3	Total	D_1	D_2	D_3	Total	Subject totals
S_1				S_1E_1				S_1E_2	$(S_1E_1 + S_1E_2)$
S_2				S_2E_1				S_2E_2	$(S_2E_1 + S_2E_2)$
S_3				S_3E_1				S_3E_2	$(S_3E_1 + S_3E_2)$
S_4				S_4E_1				S_4E_2	$(S_4E_1 + S_4E_2)$
Column totals	C_1	C_2	C_3		C_4	C_5	C_6		
Exercise totals	$E_1 = C_1 + C_2 + C_3$				$E_2 = C_4 + C_5 + C_6$				
Day totals	$D_1 = (C_1 + C_4)$			$D_2 = (C_2 + C_5)$		$D_3 = (C_3 + C_6)$			

Table 130 Perceived exertion rates, prepared for analysis of variance

| Subjects | Light exercise E_1 | | | | Heavy exercise E_2 | | | | |
	Day 1 (D_1)	Day 2 (D_2)	Day 3 (D_3)	Total	Day 1 (D_1)	Day 2 (D_2)	Day 3 (D_3)	Total	Subject totals
S_1	3	5	3	11	5	7	5	17	$11 + 17 = 28$
S_2	4	5	4	13	5	6	7	18	$13 + 18 = 31$
S_3	3	5	3	11	7	7	6	20	$11 + 20 = 31$
S_4	5	6	5	16	6	8	7	21	$16 + 21 = 37$
Column totals	15	21	15		23	28	25		$\Sigma X = 127$
Exercise totals	$E_1 = 15 + 21 + 15 = 51$				$E_2 = 23 + 28 + 25 = 76$				
Day totals	$D_1 = (15 + 23) = 38,$			$D_2 = (21 + 28) = 49,$				$D_3 = (15 + 25) = 40$	

3 Compute the Grand Total GT (Correction Factor).

$$GT = \frac{(\Sigma X)^2}{N_r}$$

where N_r = total number of scores.

$$= \frac{(127)^2}{24}$$

$$= 672.04$$

4 Compute the TOTAL sum of squares (SS_{TOTAL})

$$SS_{TOTAL} = \Sigma X^2 - GT$$
$$= 721 - 672.04 = 48.96$$

5   Compute the BETWEEN SUBJECTS sum of squares (SS_{BET})

$$SS_{BET} = [S] - GT$$

$$= \left[\frac{S_1^2}{qr} + \frac{S_2^2}{qr} + \frac{S_3^2}{qr} + \frac{S_4^2}{qr} \right] - GT$$

$$= \left[\frac{28^2}{(2)(3)} + \frac{31^2}{(2)(3)} + \frac{31^2}{(2)(3)} + \frac{37^2}{(2)(3)} \right] - 672.04$$

$$= [130.67 + 160.17 + 160.17 + 228.17] - 672.04$$

$$= [679.168] - 672.04$$

$$= 7.125$$

6   Compute the sum of squares for EXERCISE LEVELS (SS_{EX}).

$$SS_{EX} = [E] - GT$$

$$= \left[\frac{E_1^2}{rn} + \frac{E_2^2}{rn} \right] - GT$$

$$= \left[\frac{51^2}{(3)(4)} + \frac{76^2}{(3)(4)} \right] - GT$$

$$= [216.75 + 481.333] - 672.04$$

$$= [698.083] - 672.04$$

$$= 26.043$$

7   Compute the sum of squares for Deprivation (Days) (SS_D).

$$SS_D = [D] - GT$$

$$= \left[\frac{D_1^2}{qn} + \frac{D_2^2}{qn} + \frac{D_3^2}{qn} \right] - GT$$

$$= \left[\frac{(38)^2}{(2)(4)} + \frac{(49)^2}{(2)(4)} + \frac{(40)^2}{(2)(4)} \right] - 672.04$$

$$= [180.5 + 300.125 + 200] - 672.04$$

$$= [680.625] - 672.04$$

$$= 8.585$$

8   Compute the INTERACTION sum of squares (SS_{INT}).

$$SS_{INT} = [ED] - E - D + GT$$

$$= \left[\frac{C_1^2}{n} + \frac{C_2^2}{n} + \frac{C_3^2}{n} + \frac{C_4^2}{n} + \frac{C_5^2}{n} + \frac{C_6^2}{n} \right] - E - D + GT$$

$$= \left[\frac{(15)^2}{4} + \frac{(21)^2}{4} + \frac{(15)^2}{4} + \frac{(23)^2}{4} + \frac{(28)^2}{4} + \frac{(25)^2}{4} \right] - 698.08 - 680.63 + 672.04$$

$$= [56.25 + 110.25 + 56.25 + 132.25 + 196 + 156.25] - 689.08 - 680.63 + 672.04$$

$$= [707.25] - 698.08 - 680.63 + 672.04$$

$$= 0.582$$

9 Compute the EXERCISE×SUBJECTS ERROR sum of squares ($SS_{E×S}$)

$$SS_{E×S} = [ES] - E - S + GT$$

$$= \left[\frac{(S_1E_1)^2}{r} + \frac{(S_1E_2)^2}{r} + \cdots + \frac{(S_4E_2)^2}{r} \right] - E - S + GT$$

$$= \left[\frac{(11)^2}{3} + \frac{(13)^2}{3} + \frac{(11)^2}{3} + \frac{(16)^2}{3} + \frac{(17)^2}{3} + \frac{(18)^2}{3} + \frac{(20)^2}{3} + \frac{(21)^2}{3} \right] - 698.08 - 679.18 + 672.04$$

$$= [40.33 + 56.33 + 40.33 + 85.33 + 96.33 + 108 + 133.33 + 147] - 698.08 - 679.18 + 672.04$$

$$= [707.998] - 698.08 - 679.18 + 672.04$$

$$= 1.79$$

10 Compute the Deprivation (Days)×SUBJECTS ERROR sum of squares ($SS_{D×S}$)

$$SS_{D×S} = [DS] - D - S + GT$$

$$= \left[\frac{\sum (\text{Subjects' scores on each day})^2}{q} \right] - D - S + GT$$

$$= \left[\frac{(3+5)^2}{2} + \frac{(5+7)^2}{2} + \frac{(3+5)^2}{2} + \cdots + \frac{(5+6)^2}{2} + \frac{(6+8)^2}{2} + \frac{(5+7)^2}{2} \right] - 680.63 - 679.18 + 672.04$$

$$= [32 + 72 + 32 + 40.5 + 60.5 + 60.5 + 50 + 72 + 40.5 + 60.5 + 98 + 72]$$
$$- 680.63 - 679.18 + 672.04$$

$$= [690.5] - 680.625 - 679.168 + 672.04$$

$$= 2.747$$

11 Compute the INTERACTION×SUBJECTS ERROR sum of squares ($SS_{E×D×S}$)

$$SS_{E×D×S} = \sum X^2 - [ES] - [DS] - [ED] + [E] + [D] + [S] - GT$$

$$= 721 - 706.998 - 690.5 - 707.25 + 698.083 + 680.625 + 679.168 - 672.04$$

$$= 2.088$$

12 Determine the degrees of freedom for the main and error effects.

$$\text{d.f. for } SS_{TOTAL} = (N - 1) = (24 - 1) = 23$$

$$\text{d.f. for } SS_{BET} = (n - 1) = (4 - 1) = 3$$

$$\text{d.f. for } SS_{EX} = (q - 1) = (2 - 1) = 1$$

$$\text{d.f. for } SS_D = (r - 1) = (3 - 1) = 2$$

$$\text{d.f. for } SS_{INT} = (r - 1)(q - 1) = (2 - 1)(3 - 1) = 2$$

$$\text{d.f. for } SS_{E×S} = (q - 1)(n - 1) = (2 - 1)(4 - 1) = 3$$

$$\text{d.f. for } SS_{D×S} = (r - 1)(n - 1) = (3 - 1)(4 - 1) = 6$$

$$\text{d.f. for } SS_{D×E×S} = (n - 1)(q - 1)(r - 1) = (4 - 1)(2 - 1)(3 - 1) = 6$$

13   Compute variances for main and error effects.

$$\text{Var}_{\text{EX}} = \frac{SS_{\text{EX}}}{\text{d.f.}_{\text{EX}}} = \frac{26.043}{1} = 26.043$$

$$\text{Var}_{\text{D}} = \frac{SS_{\text{D}}}{\text{d.f.}_{\text{D}}} = \frac{8.585}{2} = 4.293$$

$$\text{Var}_{\text{INT}} = \frac{SS_{\text{INT}}}{\text{d.f.}_{\text{INT}}} = \frac{0.582}{2} = 0.291$$

$$\text{Var}_{\text{E}\times\text{S}} = \frac{SS_{\text{E}\times\text{S}}}{\text{d.f.}_{\text{E}\times\text{S}}} = \frac{1.79}{3} = 0.597$$

$$\text{Var}_{\text{D}\times\text{S}} = \frac{SS_{\text{D}\times\text{S}}}{\text{d.f.}_{\text{D}\times\text{S}}} = \frac{2.747}{6} = 0.458$$

$$\text{Var}_{\text{D}\times\text{E}\times\text{S}} = \frac{SS_{\text{E}\times\text{D}\times\text{S}}}{\text{d.f.}_{\text{E}\times\text{D}\times\text{S}}} = \frac{2.088}{6} = 0.348$$

14   Compute F for main effects.

$$F_{\text{EX}} = \frac{\text{Var}_{\text{EX}}}{\text{Var}_{\text{E}\times\text{S}}} = \frac{26.043}{0.597} = 43.623$$

$$F_{\text{D}} = \frac{\text{Var}_{\text{D}}}{\text{Var}_{\text{D}\times\text{S}}} = \frac{4.293}{0.458} = 9.373$$

$$F_{\text{INT}} = \frac{\text{Var}_{\text{INT}}}{\text{Var}_{\text{E}\times\text{D}\times\text{S}}} = \frac{0.291}{0.348} = 0.836$$

15   Enter the results in an analysis of variance table.

Table 131   Analysis of variance table

Source of variation	Sum of squares	d.f.	Variance	F
1. Between subjects	7.125	3		
2. Within subjects				
Exercise	26.043	1	26.043	43.623
Deprivation (Days)	8.585	2	4.293	9.373
Interaction	0.582	2	0.291	0.836
Ex × Subj. Error	1.79	3	0.597	
D × Subj. Error	2.747	6	0.458	
E × D × Subj. Error	2.088	6	0.348	
Total	48.96	23		

Consult the table in Appendix 6 to determine the significance of the F values.

The obtained value for the exercise levels effect, $F = 43.623$, exceeds the F value in the table for 1 and 3 d.f. We therefore reject the null hypothesis and conclude that type of exercise does affect perceived exertion rate.

The obtained value for the deprivation effect, $F = 9.373$, exceeds the value 5.1433 in the table at the 0.05 level for d.f. 2 and 6. We therefore reject the null hypothesis and conclude that sleep deprivation does have an effect on perceived exertion rate.

The obtained value for the interaction effect of exercise × deprivation, $F = 0.836$, does not exceed the F value of 5.1433 in the table at the 0.05 level for d.f. 2 and 6. We therefore accept the null hypothesis and conclude there is no interaction effect on perceived exertion rate.

Before closing our analysis, however, we must further investigate the sleep deprivation effect on perceived exertion rate. We found a significant difference between these levels but our F test does not tell us if all of the differences (i.e. 3) are significant, only that at least one significant difference exists. The differences between the means on day 1, 2 and 3 may now be examined using the Tukey test.

PROCEDURE FOR COMPUTING THE TUKEY TEST

1   Calculate the mean for each day, regardless of high or low intensity exercise.

$$\text{Mean for Day } 1 = 4.75$$
$$\text{Mean for Day } 2 = 6.13$$
$$\text{Mean for Day } 3 = 5.00$$

2   Construct a table of mean differences.

	Comparison means	M_1	M_2	M_3
Day 1	4.75		$M_1 - M_2 = -1.38^*$	$M_1 - M_3 = -0.25$
Day 2	6.13			$M_2 - M_3 = -1.13^*$
Day 3	5.00			

     * Significant at 0.05 level

3   Compute T.

$$T = (q)\sqrt{\frac{\text{Var}_{\text{WITHIN}} \ (\text{ERROR VARIANCE})}{N}}$$

where N = number of scores from which each mean is calculated, i.e. 8.

Consult Appendix 7 to determine the value corresponding to the number of means and the degrees of freedom for the denominator of our prior F test for deprivation (i.e. 6).

$$(q_{0.05}) = 4.34$$

$$T_{0.05} = 4.34\sqrt{\frac{0.458}{8}} = 4.34 \times 0.24 = 1.04$$

If the T value of 1.04 is smaller than the absolute differences between the two means, then the means are significantly different. Referring to our table of mean differences we can draw the following conclusions:

(a)   There is no difference between the perceived exertion rates on day 1 and day 3.

(b)   Perceived exertion rates on the same work loads were definitely higher on day two (after the sleep deprivation) than on the other days.

APPENDIX 1

Table of Random Numbers

23157	54859	01837	25993	76249	70886	95230	36744
05545	55043	10537	43508	90611	83744	10962	21343
14871	60350	32404	36223	50051	00322	11543	80834
38976	74951	94051	75853	78805	90194	32428	71695
97312	61718	99755	30870	94251	25841	54882	10513
11742	69381	44339	30872	32797	33118	22647	06850
43361	28859	11016	45623	93009	00499	43640	74036
93806	20478	38268	04491	55751	18932	58475	52571
49540	13181	08429	84187	69538	29661	77738	09527
36768	72633	37948	21569	41959	68670	45274	83880
07092	52392	24627	12067	06558	45344	67338	45320
43310	01081	44863	80307	52555	16148	89742	94647
61570	06360	06173	63775	63148	95123	35017	46993
31352	83799	10779	18941	31579	76448	62584	86919
57048	86526	27795	93692	90529	56546	35065	32254
09243	44200	68721	07137	30729	75756	09298	27650
97957	35018	40894	88329	52230	82521	22532	61587
93732	59570	43781	98885	56671	66826	95996	44569
72621	11225	00922	68264	35666	59434	71687	58167
61020	74418	45371	20794	95917	37866	99536	19378
97839	85474	33055	91718	45473	54144	22034	23000
89160	97192	22232	90637	35055	45489	88438	16361
25966	88220	62871	79265	02823	52862	84919	54883
81443	31719	05049	54806	74690	07567	65017	16543
11322	54931	42362	34386	08624	97687	46245	23245

APPENDIX 2

Table of factorials

N:	N!
1	1
2	2
3	6
4	24
5	120
6	720
7	5040
8	40320
9	362880
10	3628800
11	39916800
12	479001600
13	6227020800
14	87178291200
15	1307674368000
16	$2.0922789888 \times 10^{13}$
17	$3.55687428096 \times 10^{14}$
18	$6.402373705728 \times 10^{15}$
19	$1.216451004088 \times 10^{17}$
20	$2.432902008177 \times 10^{18}$
21	$5.109094217171 \times 10^{19}$
22	$1.124000727778 \times 10^{21}$
23	$2.585201673888 \times 10^{22}$
24	$6.204484017332 \times 10^{23}$
25	$1.551121004333 \times 10^{25}$
26	$4.032914611266 \times 10^{26}$
27	$1.088886945042 \times 10^{28}$
28	$3.048883446117 \times 10^{29}$
29	$8.841761993739 \times 10^{30}$
30	$2.652528598122 \times 10^{32}$

APPENDIX 3(a)

Percentage of scores under the Normal Curve from 0 to Z

z	0	1	2	3	4	5	6	7	8	9
0.0	0.0000	0.0040	0.0080	0.0120	0.0160	0.0199	0.0239	0.0279	0.0319	0.0359
0.1	0.0398	0.0438	0.0478	0.0517	0.0557	0.0596	0.0636	0.0675	0.0714	0.0754
0.2	0.0793	0.0832	0.0871	0.0910	0.0948	0.0987	0.1026	0.1064	0.1103	0.1141
0.3	0.1179	0.1217	0.1255	0.1293	0.1331	0.1368	0.1406	0.1443	0.1480	0.1517
0.4	0.1554	0.1591	0.1628	0.1664	0.1700	0.1736	0.1772	0.1808	0.1844	0.1879
0.5	0.1915	0.1950	0.1985	0.2019	0.2054	0.2088	0.2123	0.2157	0.2190	0.2224
0.6	0.2258	0.2291	0.2324	0.2357	0.2389	0.2422	0.2454	0.2486	0.2518	0.2549
0.7	0.2580	0.2612	0.2642	0.2673	0.2704	0.2734	0.2764	0.2794	0.2823	0.2852
0.8	0.2881	0.2910	0.2939	0.2967	0.2996	0.3023	0.3051	0.3078	0.3106	0.3133
0.9	0.3159	0.3180	0.3212	0.3238	0.3264	0.3289	0.3315	0.3340	0.3365	0.3389
1.0	0.3413	0.3438	0.3461	0.3485	0.3508	0.3531	0.3554	0.3577	0.3599	0.3621
1.1	0.3643	0.3665	0.3686	0.3708	0.3729	0.3749	0.3770	0.3790	0.3810	0.3830
1.2	0.3849	0.3869	0.3888	0.3907	0.3925	0.3944	0.3962	0.3980	0.3997	0.4015
1.3	0.4032	0.4049	0.4066	0.4082	0.4099	0.4115	0.4131	0.4147	0.4162	0.4177
1.4	0.4192	0.4207	0.4222	0.4236	0.4251	0.4265	0.4279	0.4292	0.4306	0.4319
1.5	0.4332	0.4345	0.4357	0.4370	0.4382	0.4394	0.4406	0.4418	0.4429	0.4441
1.6	0.4452	0.4463	0.4474	0.4484	0.4495	0.4505	0.4515	0.4525	0.4535	0.4545
1.7	0.4554	0.4564	0.4573	0.4582	0.4591	0.4599	0.4608	0.4616	0.4625	0.4633
1.8	0.4641	0.4649	0.4656	0.4664	0.4671	0.4678	0.4686	0.4693	0.4699	0.4706
1.9	0.4713	0.4719	0.4726	0.4732	0.4738	0.4744	0.4750	0.4756	0.4761	0.4767

Percentage of scores under the Normal Curve from 0 to Z (continued)

z	0	1	2	3	4	5	6	7	8	9
2.0	0.4772	0.4778	0.4783	0.4788	0.4793	0.4798	0.4803	0.4808	0.4812	0.4817
2.1	0.4821	0.4826	0.4830	0.4834	0.4838	0.4842	0.4846	0.4850	0.4854	0.4857
2.2	0.4861	0.4864	0.4868	0.4871	0.4875	0.4878	0.4881	0.4884	0.4887	0.4890
2.3	0.4893	0.4896	0.4898	0.4901	0.4904	0.4906	0.4909	0.4911	0.4913	0.4916
2.4	0.4918	0.4920	0.4922	0.4925	0.4927	0.4929	0.4931	0.4932	0.4934	0.4936
2.5	0.4938	0.4940	0.4941	0.4943	0.4945	0.4946	0.4948	0.4949	0.4951	0.4952
2.6	0.4953	0.4955	0.4956	0.4957	0.4959	0.4960	0.4961	0.4962	0.4963	0.4964
2.7	0.4965	0.4966	0.4967	0.4968	0.4969	0.4970	0.4971	0.4972	0.4973	0.4974
2.8	0.4974	0.4975	0.4976	0.4977	0.4977	0.4978	0.4979	0.4979	0.4980	0.4981
2.9	0.4981	0.4982	0.4982	0.4983	0.4984	0.4984	0.4985	0.4985	0.4986	0.4986
3.0	0.4987	0.4987	0.4987	0.4988	0.4988	0.4989	0.4989	0.4989	0.4990	0.4990
3.1	0.4990	0.4991	0.4991	0.4991	0.4992	0.4992	0.4992	0.4992	0.4993	0.4993
3.2	0.4993	0.4993	0.4994	0.4994	0.4994	0.4994	0.4994	0.4995	0.4995	0.4995
3.3	0.4995	0.4995	0.4995	0.4996	0.4996	0.4996	0.4996	0.4996	0.4996	0.4997
3.4	0.4997	0.4997	0.4997	0.4997	0.4997	0.4997	0.4997	0.4997	0.4997	0.4998
3.5	0.4998	0.4998	0.4998	0.4998	0.4998	0.4998	0.4998	0.4998	0.4998	0.4998
3.6	0.4998	0.4999	0.4999	0.4999	0.4999	0.4999	0.4999	0.4999	0.4999	0.4999
3.7	0.4999	0.4999	0.4999	0.4999	0.4999	0.4999	0.4999	0.4999	0.4999	0.4999
3.8	0.4999	0.4999	0.4999	0.4999	0.4999	0.4999	0.4999	0.4999	0.4999	0.4999
3.9	0.5000	0.5000	0.5000	0.5000	0.5000	0.5000	0.5000	0.5000	0.5000	0.5000

APPENDIX 3(b)

Probabilities associated with values as extreme as observed value of Z in the Normal Curve of Distribution

z	0.00	0.01	0.02	0.03	0.04	0.05	0.06	0.07	0.08	0.09
0.0	0.5000	0.4960	0.4920	0.4880	0.4840	0.4801	0.4761	0.4721	0.4681	0.4641
0.1	0.4602	0.4562	0.4522	0.4480	0.4443	0.4404	0.4364	0.0435	0.4286	0.4247
0.2	0.4207	0.4168	0.4129	0.4090	0.4052	0.4013	0.3974	0.3936	0.3897	0.3859
0.3	0.3821	0.3783	0.3745	0.3707	0.3669	0.3632	0.3594	0.3557	0.3520	0.3483
0.4	0.3446	0.3409	0.3372	0.3336	0.3300	0.3264	0.3228	0.3192	0.3156	0.3121
0.5	0.3085	0.3050	0.3015	0.2981	0.2946	0.2912	0.2877	0.2843	0.2810	0.2776
0.6	0.2743	0.2709	0.2676	0.2643	0.2611	0.2578	0.2546	0.2514	0.2483	0.2451
0.7	0.2420	0.2389	0.2358	0.2327	0.2296	0.2266	0.2236	0.2206	0.2177	0.2148
0.8	0.2119	0.2090	0.2061	0.2033	0.2005	0.1977	0.1949	0.1922	0.1894	0.1867
0.9	0.1841	0.1814	0.1788	0.1762	0.1736	0.1711	0.1685	0.1660	0.1635	0.1611
1.0	0.1587	0.1562	0.1539	0.1515	0.1492	0.1469	0.1446	0.1423	0.1401	0.1379
1.1	0.1357	0.1335	0.1314	0.1292	0.1271	0.1251	0.1230	0.1210	0.1190	0.1170
1.2	0.1151	0.1131	0.1112	0.1093	0.1075	0.1056	0.1038	0.1020	0.1003	0.0985
1.3	0.0968	0.0951	0.0934	0.0918	0.0901	0.0885	0.0869	0.0853	0.0838	0.0823
1.4	0.0808	0.0793	0.0778	0.0761	0.0749	0.0735	0.0721	0.0708	0.0694	0.0681
1.5	0.0668	0.0655	0.0643	0.0630	0.0618	0.0606	0.0594	0.0582	0.0571	0.0559
1.6	0.0548	0.0537	0.0526	0.0516	0.0505	0.0495	0.0485	0.0475	0.0465	0.0455
1.7	0.0446	0.0436	0.0427	0.0418	0.0409	0.0401	0.0392	0.0384	0.0375	0.0367
1.8	0.0359	0.0351	0.0344	0.0336	0.0329	0.0322	0.0314	0.0307	0.0301	0.0294
1.9	0.0287	0.0281	0.0274	0.0268	0.0262	0.0256	0.0250	0.0244	0.0239	0.0233

Read values of z to one decimal place down the left hand column *Column z*, Read across *Row z* for values to two decimal places. The probabilities contained in the table are *one-tailed*. For two-tailed tests, multiply by 2.

Probabilities associated with values as extreme as observed value of Z in the Normal Curve of Distribution (continued)

z	0.00	0.01	0.02	0.03	0.04	0.05	0.06	0.07	0.08	0.09
2.0	0.0228	0.0222	0.0217	0.0212	0.0207	0.0202	0.0197	0.0192	0.0188	0.0183
2.1	0.0179	0.0174	0.0170	0.0166	0.0162	0.0158	0.0154	0.0150	0.0146	0.0143
2.2	0.0139	0.0136	0.0132	0.0129	0.0125	0.0122	0.0119	0.0116	0.0113	0.0110
2.3	0.0107	0.0104	0.0102	0.0099	0.0096	0.0094	0.0091	0.0089	0.0087	0.0084
2.4	0.0082	0.0080	0.0078	0.0075	0.0073	0.0071	0.0069	0.0068	0.0066	0.0064
2.5	0.0062	0.0060	0.0059	0.0057	0.0055	0.0054	0.0052	0.0051	0.0049	0.0048
2.6	0.0047	0.0045	0.0044	0.0043	0.0041	0.0040	0.0039	0.0038	0.0037	0.0036
2.7	0.0035	0.0034	0.0033	0.0032	0.0031	0.0030	0.0029	0.0028	0.0027	0.0026
2.8	0.0026	0.0025	0.0024	0.0023	0.0023	0.0022	0.0021	0.0021	0.0020	0.0019
2.9	0.0019	0.0018	0.0018	0.0017	0.0016	0.0016	0.0015	0.0015	0.0014	0.0014
3.0	0.0013	0.0013	0.0013	0.0012	0.0012	0.0011	0.0011	0.0011	0.0010	0.0010
3.1	0.0010	0.0009	0.0009	0.0009	0.0008	0.0008	0.0008	0.0008	0.0007	0.0007
3.2	0.0007									
3.3	0.0005									
3.4	0.0003									
3.5	0.00023									
3.6	0.00016									
3.7	0.00011									
3.8	0.00007									
3.9	0.00005									
4.0	0.00003									

Examples
(i) The probability of a $z \geq 0.14$ on a one-tailed test is $p = 0.4443$
(ii) The probability of a $z \geq 1.98$ on a two-tailed test is $p = 2 \times (0.0239) = 0.0478$

APPENDIX 3(c)

Table of z values for r

r	z	r	z	r	z	r	z	r	z
.000	.000	.200	.203	.400	.424	.600	.693	.800	1.099
.005	.005	.205	.208	.405	.430	.605	.701	.805	1.113
.010	.010	.210	.213	.410	.436	.610	.709	.810	1.127
.015	.015	.215	.218	.415	.442	.615	.717	.815	1.142
.020	.020	.220	.224	.420	.448	.620	.725	.820	1.157
.025	.025	.225	.229	.425	.454	.625	.733	.825	1.172
.030	.030	.230	.234	.430	.460	.630	.741	.830	1.188
.035	.035	.235	.239	.435	.466	.635	.750	.835	1.204
.040	.040	.240	.245	.440	.472	.640	.758	.840	1.221
.045	.045	.245	.250	.445	.478	.645	.767	.845	1.238
.050	.050	.250	.255	.450	.485	.650	.775	.850	1.256
.055	.055	.255	.261	.455	.491	.655	.784	.855	1.274
.060	.060	.260	.266	.460	.497	.660	.793	.860	1.293
.065	.065	.265	.271	.465	.504	.665	.802	.865	1.313
.070	.070	.270	.277	.470	.510	.670	.811	.870	1.333
.075	.075	.275	.282	.475	.517	.675	.820	.875	1.354
.080	.080	.280	.288	.480	.523	.680	.829	.880	1.376
.085	.085	.285	.293	.485	.530	.685	.838	.885	1.398
.090	.090	.290	.299	.490	.536	.690	.848	.890	1.422
.095	.095	.295	.304	.495	.543	.695	.858	.895	1.447
.100	.100	.300	.310	.500	.549	.700	.867	.900	1.472
.105	.105	.305	.315	.505	.556	.705	.877	.905	1.499
.110	.110	.310	.321	.510	.563	.710	.887	.910	1.528
.115	.116	.315	.326	.515	.570	.715	.897	.915	1.557
.120	.121	.320	.332	.520	.576	.720	.908	.920	1.589
.125	.126	.325	.337	.525	.583	.725	.918	.925	1.623
.130	.131	.330	.343	.530	.590	.730	.929	.930	1.658
.135	.136	.335	.348	.535	.597	.735	.940	.935	1.697
.140	.141	.340	.354	.540	.604	.740	.950	.940	1.738
.145	.146	.345	.360	.545	.611	.745	.962	.945	1.783
.150	.151	.350	.365	.550	.618	.750	.973	.950	1.832
.155	.156	.355	.371	.555	.626	.755	.984	.955	1.886
.160	.161	.360	.377	.560	.633	.760	.996	.960	1.946
.165	.167	.365	.383	.565	.640	.765	1.008	.965	2.014
.170	.172	.370	.388	.570	.648	.770	1.020	.970	2.092
.175	.177	.375	.394	.575	.655	.775	1.033	.975	2.185
.180	.182	.380	.400	.580	.662	.780	1.045	.980	2.298
.185	.187	.385	.406	.585	.670	.785	1.058	.985	2.443
.190	.192	.390	.412	.590	.678	.790	1.071	.990	2.647
.195	.198	.395	.418	.595	.685	.795	1.085	.995	2.994

APPENDIX 4

χ² distribution			

d.f.	Probability (P)			
	0.050	0.025	0.010	0.001
1	3.841	5.024	6.635	10.828
2	5.991	7.378	9.210	13.816
3	7.815	9.348	11.345	16.266
4	9.488	11.143	13.277	18.467
5	11.071	12.833	15.086	20.515
6	12.592	14.449	16.812	22.458
7	14.067	16.013	18.475	24.322
8	15.507	17.535	20.090	26.125
9	16.919	19.023	21.666	27.877
10	18.307	20.483	23.209	29.588
11	19.675	21.920	24.725	31.264
12	21.026	23.337	26.217	32.909
13	22.362	24.736	27.688	34.528
14	23.685	26.119	29.141	36.123
15	24.996	27.488	30.578	37.697
16	26.296	28.845	32.000	39.252
17	27.587	30.191	33.409	40.790
18	28.869	31.526	34.805	42.312
19	30.144	32.852	36.191	43.820
20	31.410	34.170	37.566	45.315
21	32.671	35.479	38.932	46.797
22	33.924	36.781	40.289	48.268
23	35.173	38.076	41.638	49.728
24	36.415	39.364	42.980	51.179
25	37.653	40.647	44.314	52.620
26	38.885	41.923	45.642	54.052
27	40.113	43.194	46.963	55.476
28	41.337	44.461	48.278	56.892
29	42.557	45.722	49.588	58.302
30	43.773	46.979	50.892	59.703
40	55.759	59.342	63.691	73.402
50	67.505	71.420	76.154	86.661
60	79.082	83.298	88.379	99.607
80	101.879	106.629	112.329	124.839
100	124.342	129.561	135.807	149.449

APPENDIX 5

t Distribution

d.f.	\multicolumn{4}{c}{Level of significance for one-tailed test}

	0.05	0.025	0.01	0.005
	\multicolumn{4}{c}{Level of significance for two-tailed test}			
d.f.	0.10	0.05	0.02	0.01
1	6.314	12.706	31.821	63.657
2	2.920	4.303	6.965	9.925
3	2.353	3.182	4.541	5.841
4	2.132	2.776	3.747	4.604
5	2.015	2.571	3.365	4.032
6	1.943	2.447	3.143	3.707
7	1.895	2.365	2.998	3.499
8	1.860	2.306	2.896	3.355
9	1.833	2.262	2.821	3.250
10	1.812	2.228	2.764	3.169
11	1.796	2.201	2.718	3.106
12	1.782	2.179	2.681	3.055
13	1.771	2.160	2.650	3.012
14	1.761	2.145	2.624	2.977
15	1.753	2.131	2.602	2.947
16	1.746	2.120	2.583	2.921
17	1.740	2.110	2.567	2.898
18	1.734	2.101	2.552	2.878
19	1.729	2.093	2.539	2.861
20	1.725	2.086	2.528	2.845
21	1.721	2.080	2.518	2.831
22	1.717	2.074	2.508	2.819
23	1.714	2.069	2.500	2.807
24	1.711	2.064	2.492	2.797
25	1.708	2.060	2.485	2.787
26	1.706	2.056	2.479	2.779
27	1.703	2.052	2.473	2.771
28	1.701	2.048	2.467	2.763
29	1.699	2.045	2.462	2.756
30	1.697	2.042	2.457	2.750
40	1.684	2.021	2.423	2.704
60	1.671	2.000	2.390	2.660
120	1.658	1.980	2.358	2.617
∞	1.645	1.960	2.326	2.576

APPENDIX 6

APPENDIX 6

F Distribution
ν_1 = d.f. for the greater variance
ν_2 = d.f. for the lesser variance
(a)   0.05 level

ν_2 \ ν_1	1	2	3	4	5	6	7	8	9
1	161.45	199.50	215.71	224.58	230.16	233.99	236.77	238.88	240.54
2	18.513	19.000	19.164	19.247	19.296	19.330	19.353	19.371	19.385
3	10.128	9.5521	9.2766	9.1172	9.0135	8.9406	8.8867	8.8452	8.8323
4	7.7086	6.9443	6.5914	6.3882	6.2561	6.1631	6.0942	6.0410	5.9938
5	6.6079	5.7861	5.4095	5.1922	5.0503	4.9503	4.8759	4.8183	4.7725
6	5.9874	5.1433	4.7571	4.5337	4.3874	4.2839	4.2067	4.1468	4.0990
7	5.5914	4.7374	4.3468	4.1203	3.9715	3.8660	3.7870	3.7257	3.6767
8	5.3177	4.4590	4.0662	3.8379	3.6875	3.5806	3.5005	3.4381	3.3881
9	5.1174	4.2565	3.8625	3.6331	3.4817	3.3738	3.2927	3.2296	3.1789
10	4.9646	4.1028	3.7083	3.4780	3.3258	3.2172	3.1355	3.0717	3.0204
11	4.8443	3.9823	3.5874	3.3567	3.2039	3.0946	3.0123	2.9480	2.8962
12	4.7472	3.8853	3.4903	3.2592	3.1059	2.9961	2.9134	2.8486	2.7964
13	4.6672	3.8056	3.4105	3.1791	3.0254	2.9153	2.8321	2.7669	2.7444
14	4.6001	3.7389	3.3439	3.1122	2.9582	2.8477	2.7642	2.6987	2.6458
15	4.5431	3.6823	3.2874	3.0556	2.9013	2.7905	2.7066	2.6408	2.5876
16	4.4940	3.6337	3.2389	3.0069	2.8524	2.7413	2.6572	2.5911	2.5377
17	4.4513	3.5915	3.1968	2.9647	2.8100	2.6987	2.6143	2.5480	2.4443
18	4.4139	3.5546	3.1599	2.9277	2.7729	2.6613	2.5767	2.5102	2.4563
19	4.3807	3.5219	3.1274	2.8951	2.7401	2.6283	2.5435	2.4768	2.4227
20	4.3512	3.4928	3.0984	2.8661	2.7109	2.5990	2.5140	2.4471	2.3928
21	4.3248	3.4668	3.0725	2.8401	2.6848	2.5727	2.4876	2.4205	2.3660
22	4.3009	3.4434	3.0491	2.8167	2.6613	2.5491	2.4638	2.3965	2.3419
23	4.2793	3.4221	3.0280	2.7955	2.6400	2.5277	2.4422	2.3748	2.3201
24	4.2597	3.4028	3.0088	2.7763	2.6207	2.5082	2.4226	2.3551	2.3002
25	4.2417	3.3852	2.9912	2.7587	2.6030	2.4904	2.4047	2.3371	2.2821
26	4.2252	3.3690	2.9752	2.7426	2.5868	2.4741	2.3883	2.3205	2.2655
27	4.2100	3.3541	2.9604	2.7278	2.5719	2.4591	2.3732	2.3053	2.2501
28	4.1960	3.3404	2.9467	2.7141	2.5581	2.4453	2.3593	2.2913	2.2360
29	4.1830	3.3277	2.9340	2.7014	2.5454	2.4324	2.3463	2.2783	2.2229
30	4.1709	3.3158	2.9223	2.6896	2.5336	2.4205	2.3343	2.2662	2.2207
40	4.0847	3.2317	2.8387	2.6060	2.4495	2.3359	2.2490	2.1802	2.1240
60	4.0012	3.1504	2.7581	2.5252	2.3683	2.2541	2.1665	2.0970	2.0401
120	3.9201	3.0718	2.6802	2.4472	2.2899	2.1750	2.0868	2.0164	1.9688
∞	3.8415	2.9957	2.6049	2.3719	2.2141	2.0986	2.0096	1.9384	1.8799

F Distribution (continued)

10	12	15	20	24	30	40	60	120	∞
241.88	243.91	245.95	248.01	249.05	250.10	251.14	252.20	253.25	254.31
19.396	19.413	19.429	19.446	19.454	19.462	19.471	19.479	19.487	19.496
8.7855	8.7446	8.7029	8.6602	8.6385	8.6166	8.5944	8.5720	8.5594	8.5264
5.9644	5.9117	5.8578	5.8025	5.7744	5.7459	5.7170	5.6877	5.6381	5.6281
4.7351	4.6777	4.6188	4.5581	4.5272	4.4957	4.4638	4.4314	4.3085	4.3650
4.0600	3.9999	3.9381	3.8742	3.8415	3.8082	3.7743	3.7398	3.7047	3.6689
3.6365	3.5747	3.5107	3.4445	3.4105	3.3758	3.3404	3.3043	3.2674	3.2298
3.3472	3.2839	3.2184	3.1503	3.1152	3.0794	3.0428	3.0053	2.9669	2.9276
3.1373	3.0729	3.0061	2.9365	2.9005	2.8637	2.8259	2.7872	2.7475	2.7067
2.9782	2.9130	2.8450	2.7740	2.7372	2.6996	2.6609	2.6211	2.5801	2.5379
2.8536	2.7876	2.7186	2.6464	2.6090	2.5705	2.5309	2.4901	2.4480	2.4045
2.7534	2.6866	2.6169	2.5436	2.5055	2.4663	2.4259	2.3842	2.3410	2.2962
2.6710	2.6037	2.5331	2.4589	2.4202	2.3803	2.3392	2.2966	2.2524	2.2064
2.6022	2.5342	2.4630	2.3879	2.3487	2.3082	2.2664	2.2229	2.1778	2.1307
2.5437	2.4753	2.4034	2.3275	2.2878	2.2468	2.2043	2.1601	2.1141	2.0658
2.4935	2.4247	2.3522	2.2756	2.2354	2.1938	2.1507	2.1058	2.0589	2.0096
2.4499	2.3807	2.3077	2.2304	2.1898	2.1477	2.1040	2.0584	2.0107	1.9604
2.4117	2.3421	2.2686	2.1906	2.1497	2.1071	2.0629	2.0166	1.9681	1.9168
2.3779	2.3080	2.2341	2.1555	2.1141	2.0712	2.0264	1.9795	1.9302	1.8780
2.3479	2.2776	2.2033	2.1242	2.0825	2.0391	1.9938	1.9464	1.8963	1.8432
2.3210	2.2504	2.1757	2.0960	2.0540	2.0102	1.9645	1.9165	1.8657	1.8117
2.2967	2.2258	2.1508	2.0707	2.0283	1.9842	1.9380	1.8894	1.8380	1.7831
2.2747	2.2036	2.1282	2.0476	2.0050	1.9605	1.9139	1.8648	1.8128	1.7570
2.2547	2.1834	2.1077	2.0267	1.9838	1.9390	1.8920	1.8424	1.7896	1.7330
2.2365	2.1649	2.0889	2.0075	1.9643	1.9192	1.8718	1.8217	1.7684	1.7110
2.2197	2.1479	2.0716	1.9898	1.9464	1.9010	1.8533	1.8027	1.7488	1.6906
2.2043	2.1323	2.0558	1.9736	1.9299	1.8842	1.8361	1.7851	1.7306	1.6717
2.1900	2.1179	2.0411	1.9586	1.9147	1.8687	1.8203	1.7689	1.7138	1.6541
2.1768	2.1045	2.0275	1.9446	1.9005	1.8543	1.8055	1.7537	1.6981	1.6376
2.1646	2.0921	2.0148	1.9317	1.8874	1.8409	1.7918	1.7396	1.6835	1.6223
2.0772	2.0035	1.9245	1.8389	1.7929	1.7444	1.6928	1.6373	1.5766	1.5089
1.9926	1.9174	1.8364	1.7480	1.7001	1.6491	1.5943	1.5343	1.4673	1.3893
1.9105	1.8337	1.7505	1.6587	1.6084	1.5543	1.4952	1.4290	1.3519	1.2539
1.8307	1.7522	1.6664	1.5705	1.5173	1.4591	1.3940	1.3180	1.0214	1.0000

F Distribution
(b)   0.01 level

v_2 \ v_1	1	2	3	4	5	6	7	8	9
1	4052.2	4999.5	5403.4	5624.6	5763.6	5859.0	5928.4	5981.1	6022.5
2	98.503	99.000	99.166	99.249	99.299	99.333	99.356	99.374	99.388
3	34.116	30.817	29.457	28.710	28.237	27.911	27.672	27.489	27.345
4	21.198	18.000	16.694	15.977	15.522	15.207	14.976	14.799	14.659
5	16.258	13.274	12.060	11.392	10.967	10.672	10.456	10.289	10.158
6	13.745	10.925	9.7795	9.1483	8.7459	8.4661	8.2600	8.1017	7.9761
7	12.246	9.5466	8.4513	7.8466	7.4604	7.1914	6.9928	6.8400	6.7188
8	11.259	8.6491	7.5910	7.0061	6.6318	6.3707	6.1776	6.0289	5.9106
9	10.561	8.0215	6.9919	6.4221	6.0569	5.8018	5.6129	5.4671	5.3511
10	10.044	7.5594	6.5523	5.9943	5.6363	5.3858	5.2001	5.0567	4.9424
11	9.6460	7.2057	6.2167	5.6683	5.3160	5.0692	4.8861	4.7445	4.6315
12	9.3302	6.9266	5.9525	5.4120	5.0643	4.8206	4.6395	4.4994	4.3875
13	9.0738	6.7010	5.7394	5.2053	4.8616	4.6204	4.4410	4.3021	4.1911
14	8.8616	6.5149	5.5639	5.0354	4.6950	4.4558	4.2779	4.1399	4.0297
15	8.6831	6.3589	5.4170	4.8932	4.5556	4.3183	4.1415	4.0045	3.8948
16	8.5310	6.2262	5.2922	4.7726	4.4374	4.2016	4.0259	3.8896	3.7804
17	8.3997	6.1121	5.1850	4.6690	4.3359	4.1015	3.9267	3.7910	3.6822
18	8.2854	6.0129	5.0919	4.5790	4.2479	4.0146	3.8406	3.7054	3.5971
19	8.1849	5.9259	5.0103	4.5003	4.1708	3.9386	3.7653	3.6305	3.5225
20	8.0960	5.8489	4.9382	4.4307	4.1027	3.8714	3.6987	3.5644	3.4567
21	8.0166	5.7804	4.8740	4.3688	4.0421	3.8117	3.6396	3.5056	3.3981
22	7.9454	5.7190	4.8166	4.3134	3.9880	3.7583	3.5867	3.4530	3.3458
23	7.8811	5.6637	4.7649	4.2636	3.9392	3.7102	3.5390	3.4057	3.2986
24	7.8229	5.6136	4.7181	4.2184	3.8951	3.6667	3.4959	3.3629	3.2560
25	7.7698	5.5680	4.6755	4.1774	3.8550	3.6272	3.4568	3.3239	3.2172
26	7.7213	5.5263	4.6366	4.1400	3.8183	3.5911	3.4210	3.2884	3.1818
27	7.6767	5.4881	4.6009	4.1056	3.7848	3.5580	3.3882	3.2558	3.1494
28	7.6356	5.4529	4.5681	4.0740	3.7539	3.5276	3.3581	3.2259	3.1195
29	7.5977	5.4204	4.5378	4.0449	3.7254	3.4995	3.3303	3.1982	3.0920
30	7.5625	5.3903	4.5097	4.0179	3.6990	3.4735	3.3045	3.1726	3.0665
40	7.3141	5.1785	4.3126	3.8283	3.5138	3.2910	3.1238	2.9930	2.8876
60	7.0771	4.9774	4.1259	3.6490	3.3389	3.1187	2.9530	2.8233	2.7185
120	6.8509	4.7865	3.9491	3.4795	3.1735	2.9559	2.7918	2.6629	2.5586
∞	6.6349	4.6052	3.7816	3.3192	3.0173	2.8020	2.6393	2.5113	2.4073

Source: Table 5, of Pearson, E. S., and Hartley, H. O., editors; *Biometrika Tables for Statisticians.* Cambridge: Cambridge Univ. Press, 1966, with the kind permission of the trustees and publisher.

F Distribution (continued)

10	12	15	20	24	30	40	60	120	∞
6055.8	6106.3	6157.3	6208.7	6234.6	6260.6	6286.8	6313.0	6339.4	6365.9
99.399	99.416	99.433	99.449	99.458	99.466	99.474	99.482	99.491	99.499
27.229	27.052	26.872	26.690	26.598	26.505	26.411	26.316	26.221	26.125
14.546	14.374	14.198	14.020	13.929	13.838	13.745	13.652	13.558	13.463
10.051	9.8883	9.7222	9.5526	9.4665	9.3793	9.2912	9.2020	9.1118	9.0204
7.8741	7.7183	7.5590	7.3958	7.3127	7.2285	7.1432	7.0567	6.9690	6.8800
6.6201	6.4691	6.3143	6.1554	6.0743	5.9920	5.9084	5.8236	5.7373	5.6495
5.8143	5.6667	5.5151	5.3591	5.2793	5.1981	5.1156	5.0316	4.9461	4.8588
5.2565	5.1114	4.9621	4.8080	4.7290	4.6486	4.5666	4.4831	4.3978	4.3105
4.8491	4.7059	4.5581	4.4054	4.3269	4.2469	4.1653	4.0819	3.9965	3.9090
4.5393	4.3974	4.2509	4.0990	4.0209	3.9411	3.8596	3.7761	3.6904	3.6024
4.2961	4.1553	4.0096	3.8584	3.7805	3.7008	3.6192	3.5355	3.4494	3.3608
4.1003	3.9603	3.8154	3.6646	3.5868	3.5070	3.4253	3.3413	3.2548	3.1654
3.9394	3.8001	3.6557	3.5052	3.4274	3.3476	3.2656	3.1813	3.0942	3.0040
3.8049	3.6662	3.5222	3.3719	3.2940	3.2141	3.1319	3.0471	2.9595	2.8684
3.6909	3.5527	3.4089	3.2587	3.1808	3.1007	3.0182	2.9330	2.8447	2.7528
3.5931	3.4552	3.3117	3.1615	3.0835	3.0032	2.9205	2.8348	2.7459	2.6530
3.5082	3.3706	3.2273	3.0771	2.9990	2.9185	2.8354	2.7493	2.6597	2.5660
3.4338	3.2965	3.1533	3.0031	2.9249	2.8442	2.7608	2.6742	2.5839	2.4893
3.3682	3.2311	3.0880	2.9377	2.8594	2.7785	2.6947	2.6077	2.5168	2.4212
3.3098	3.1730	3.0300	2.8796	2.8010	2.7200	2.6359	2.5484	2.4568	2.3603
3.2576	3.1209	2.9779	2.8274	2.7488	2.6675	2.5831	2.4951	2.4029	2.3055
3.2106	3.0740	2.9311	2.7805	2.7017	2.6202	2.5355	2.4471	2.3542	2.2558
3.1681	3.0316	2.8887	2.7380	2.6591	2.5773	2.4923	2.4035	2.3100	2.2107
3.1294	2.9931	2.8502	2.6993	2.6203	2.5383	2.4530	2.3637	2.2696	2.1694
3.0941	2.9578	2.8150	2.6640	2.5848	2.5026	2.4170	2.3273	2.2325	2.1315
3.0618	2.9256	2.7827	2.6316	2.5522	2.4699	2.3840	2.2938	2.1985	2.0965
3.0320	2.8959	2.7530	2.6017	2.5223	2.4397	2.3535	2.2629	2.1670	2.0642
3.0045	2.8685	2.7256	2.5742	2.4946	2.4118	2.3253	2.2344	2.1379	2.0342
2.9791	2.8431	2.7002	2.5487	2.4689	2.3860	2.2992	2.2079	2.1108	2.0062
2.8005	2.6648	2.5216	2.3689	2.2880	2.2034	2.1142	2.0194	1.9172	1.8047
2.6318	2.4961	2.3523	2.1978	2.1154	2.0285	1.9360	1.8363	1.7263	1.6006
2.4721	2.3363	2.1915	2.0346	1.9500	1.8600	1.7628	1.6557	1.5330	1.3805
2.3209	2.1847	2.0385	1.8783	1.7908	1.6964	1.5923	1.4730	1.3246	1.0000

APPENDIX 7

Tukey Test

Percentage points (q) of the Studentized range

$(p = 0.05)$
n = the total number of means being compared
v = degrees of freedom of denominator of F test

v \ n	2	3	4	5	6	7	8	9	10
1	17.97	26.98	32.82	37.08	40.41	43.12	45.40	47.36	49.07
2	6.08	8.33	9.80	10.88	11.74	12.44	13.03	13.54	13.99
3	4.50	5.91	6.82	7.50	8.04	8.48	8.85	9.18	9.46
4	3.93	5.04	5.76	6.29	6.71	7.05	7.35	7.60	7.83
5	3.64	4.60	5.22	5.67	6.03	6.33	6.58	6.80	6.99
6	3.46	4.34	4.90	5.30	5.63	5.90	6.12	6.32	6.49
7	3.34	4.16	4.68	5.06	5.36	5.61	5.82	6.00	6.16
8	3.26	4.04	4.53	4.89	5.17	5.40	5.60	5.77	5.92
9	3.20	3.95	4.41	4.76	5.02	5.24	5.43	5.59	5.74
10	3.15	3.88	4.33	4.65	4.91	5.12	5.30	5.46	5.60
11	3.11	3.82	4.26	4.57	4.82	5.03	5.20	5.35	5.49
12	3.08	3.77	4.20	4.51	4.75	4.95	5.12	5.27	5.39
13	3.06	3.73	4.15	4.45	4.69	4.88	5.05	5.19	5.32
14	3.03	3.70	4.11	4.41	4.64	4.83	4.99	5.13	5.25
15	3.01	3.67	4.08	4.37	4.59	4.78	4.94	5.08	5.20
16	3.00	3.65	4.05	4.33	4.56	4.74	4.90	5.03	5.15
17	2.98	3.63	4.02	4.30	4.52	4.70	4.86	4.99	5.11
18	2.97	3.61	4.00	4.28	4.49	4.67	4.82	4.96	5.07
19	2.96	3.59	3.98	4.25	4.47	4.65	4.79	4.92	5.04
20	2.95	3.58	3.96	4.23	4.45	4.62	4.77	4.90	5.01
24	2.92	3.53	3.90	4.17	4.37	4.54	4.68	4.81	4.92
30	2.89	3.49	3.85	4.10	4.30	4.46	4.60	4.72	4.82
40	2.86	3.44	3.79	4.04	4.23	4.39	4.52	4.63	4.73
60	2.83	3.40	3.74	3.98	4.16	4.31	4.44	4.55	4.65
120	2.80	3.36	3.68	3.92	4.10	4.24	4.36	4.47	4.56
∞	2.77	3.31	3.63	3.86	4.03	4.17	4.29	4.39	4.47

Source: From Pearson, E. S., and Hartley, H. O., editors;
Biometrika Tables for Statisticians, ed. 3, vol. I. Cambridge:
Cambridge Univ. Press, 1966, with the kind permission
of the trustees and publishers.

Tukey Test
Percentage points (q) of the Studentized range (continued)

$(p = 0.05)$
  n = the total number of means being compared
  v = degrees of freedom of denominator of F test

v \ n	11	12	13	14	15	16	17	18	19	20
1	50.59	51.96	53.20	54.33	55.36	56.32	57.22	58.04	58.83	59.56
2	14.39	14.75	15.08	15.38	15.65	15.91	16.14	16.37	16.57	16.77
3	9.72	9.95	10.15	10.35	10.52	10.69	10.84	10.98	11.11	11.24
4	8.03	8.21	8.37	8.52	8.66	8.79	8.91	9.03	9.13	9.23
5	7.17	7.32	7.47	7.60	7.72	7.83	7.93	8.03	8.12	8.21
6	6.65	6.79	6.92	7.03	7.14	7.24	7.34	7.43	7.51	7.59
7	6.30	6.43	6.55	6.66	6.76	6.85	6.94	7.02	7.10	7.17
8	6.05	6.18	6.29	6.39	6.48	6.57	6.65	6.73	6.80	6.87
9	5.87	5.98	6.09	6.19	6.28	6.36	6.44	6.51	6.58	6.64
10	5.72	5.83	5.93	6.03	6.11	6.19	6.27	6.34	6.40	6.47
11	5.61	5.71	5.81	5.90	5.98	6.06	6.13	6.20	6.27	6.33
12	5.51	5.61	5.71	5.80	5.88	5.95	6.02	6.09	6.15	6.21
13	5.43	5.53	5.63	5.71	5.79	5.86	5.93	5.99	6.05	6.11
14	5.36	5.46	5.55	5.64	5.71	5.79	5.85	5.91	5.97	6.03
15	5.31	5.40	5.49	5.57	5.65	5.72	5.78	5.85	5.90	5.96
16	5.26	5.35	5.44	5.52	5.59	5.66	5.73	5.79	5.84	5.90
17	5.21	5.31	5.39	5.47	5.54	5.61	5.67	5.73	5.79	5.84
18	5.17	5.27	5.35	5.43	5.50	5.57	5.63	5.69	5.74	5.79
19	5.14	5.23	5.31	5.39	5.46	5.53	5.59	5.65	5.70	5.75
20	5.11	5.20	5.28	5.36	5.43	5.49	5.55	5.61	5.66	5.71
24	5.01	5.10	5.18	5.25	5.32	5.38	5.44	5.49	5.55	5.59
30	4.92	5.00	5.08	5.15	5.21	5.27	5.33	5.38	5.43	5.47
40	4.82	4.90	4.98	5.04	5.11	5.16	5.22	5.27	5.31	5.36
60	4.73	4.81	4.88	4.94	5.00	5.06	5.11	5.15	5.20	5.24
120	4.64	4.71	4.78	4.84	4.90	4.95	5.00	5.04	5.09	5.13
∞	4.55	4.62	4.68	4.74	4.80	4.85	4.89	4.93	4.97	5.01

Tukey Test (continued)

$(p = 0.01)$
n = the total number of means being compared
v = degrees of freedom of denominator of F test

v \ n	2	3	4	5	6	7	8	9	10
1	90.03	135.0	164.3	185.6	202.2	215.8	227.2	237.0	245.6
2	14.04	19.02	22.29	24.72	26.63	28.20	29.53	30.68	31.69
3	8.26	10.62	12.17	13.33	14.24	15.00	15.64	16.20	16.69
4	6.51	8.12	9.17	9.96	10.58	11.10	11.55	11.93	12.27
5	5.70	6.98	7.80	8.42	8.91	9.32	9.67	9.97	10.24
6	5.24	6.33	7.03	7.56	7.97	8.32	8.61	8.87	9.10
7	4.95	5.92	6.54	7.01	7.37	7.68	7.94	8.17	8.37
8	4.75	5.64	6.20	6.62	6.96	7.24	7.47	7.68	7.86
9	4.60	5.43	5.96	6.35	6.66	6.91	7.13	7.33	7.49
10	4.48	5.27	5.77	6.14	6.43	6.67	6.87	7.05	7.21
11	4.39	5.15	5.62	5.97	6.25	6.48	6.67	6.84	6.99
12	4.32	5.05	5.50	5.84	6.10	6.32	6.51	6.67	6.81
13	4.26	4.96	5.40	5.73	5.98	6.19	6.37	6.53	6.67
14	4.21	4.89	5.32	5.63	5.88	6.08	6.26	6.41	6.54
15	4.17	4.84	5.25	5.56	5.80	5.99	6.16	6.31	6.44
16	4.13	4.79	5.19	5.49	5.72	5.92	6.08	6.22	6.35
17	4.10	4.74	5.14	5.43	5.66	5.85	6.01	6.15	6.27
18	4.07	4.70	5.09	5.38	5.60	5.79	5.94	6.08	6.20
19	4.05	4.67	5.05	5.33	5.55	5.73	5.89	6.02	6.14
20	4.02	4.64	5.02	5.29	5.51	5.69	5.84	5.97	6.09
24	3.96	4.55	4.91	5.17	5.37	5.54	5.69	5.81	5.92
30	3.89	4.45	4.89	5.05	5.24	5.40	5.54	5.65	5.76
40	3.82	4.37	4.70	4.93	5.11	5.26	5.39	5.50	5.60
60	3.76	4.28	4.59	4.82	4.99	5.13	5.25	5.36	5.45
120	3.70	4.20	4.50	4.71	4.87	5.01	5.12	5.21	5.30
∞	3.64	4.12	4.40	4.60	4.76	4.88	4.99	5.08	5.16

Source: From Pearson, E. S., and Hartley, H. O., editors;
Biometrika Tables for Statisticians, ed. 3, vol. I. Cambridge:
Cambridge Univ. Press, 1966, with the kind permission
of the trustees and publishers.

Tukey Test (continued)

$(p = 0.01)$
n = the total number of means being compared
v = degrees of freedom of denominator of F test

v \ n	11	12	13	14	15	16	17	18	19	20
1	253.2	260.0	266.2	271.8	277.0	281.8	286.3	290.4	294.3	298.0
2	32.59	33.40	34.13	34.81	35.43	36.00	36.53	37.03	37.50	37.95
3	17.13	17.53	17.89	18.22	18.52	18.81	19.07	19.32	19.55	19.77
4	12.57	12.84	13.09	13.32	13.53	13.73	13.91	14.08	14.24	14.40
5	10.48	10.70	10.89	11.08	11.24	11.40	11.55	11.68	11.81	11.93
6	9.30	9.48	9.65	9.81	9.95	10.08	10.21	10.32	10.43	10.54
7	8.55	8.71	8.86	9.00	9.12	9.24	9.35	9.46	9.55	9.65
8	8.03	8.18	8.31	8.44	8.55	8.66	8.76	8.85	8.94	9.03
9	7.65	7.78	7.91	8.03	8.13	8.23	8.33	8.41	8.49	8.57
10	7.36	7.49	7.60	7.71	7.81	7.91	7.99	8.08	8.15	8.23
11	7.13	7.25	7.36	7.46	7.56	7.65	7.73	7.81	7.88	7.95
12	6.94	7.06	7.17	7.26	7.36	7.44	7.52	7.59	7.66	7.73
13	6.79	6.90	7.01	7.10	7.19	7.27	7.35	7.42	7.48	7.55
14	6.66	6.77	6.87	6.96	7.05	7.13	7.20	7.27	7.33	7.39
15	6.55	6.66	6.76	6.84	6.93	7.00	7.07	7.14	7.20	7.26
16	6.46	6.56	6.66	6.74	6.82	6.90	6.97	7.03	7.09	7.15
17	6.38	6.48	6.57	6.66	6.73	6.81	6.87	6.94	7.00	7.05
18	6.31	6.41	6.50	6.58	6.65	6.73	6.79	6.85	6.91	6.97
19	6.25	6.34	6.43	6.51	6.58	6.65	6.72	6.78	6.84	6.89
20	6.19	6.28	6.37	6.45	6.52	6.59	6.65	6.71	6.77	6.82
24	6.02	6.11	6.19	6.26	6.33	6.39	6.45	6.51	6.56	6.61
30	5.85	5.93	6.01	6.08	6.14	6.20	6.26	6.31	6.36	6.41
40	5.69	5.76	5.83	5.90	5.96	6.02	6.07	6.12	6.16	6.21
60	5.53	5.60	5.67	5.73	5.78	5.84	5.89	5.93	5.97	6.01
120	5.37	5.44	5.50	5.56	5.61	5.66	5.71	5.75	5.79	5.83
∞	5.23	5.29	5.35	5.40	5.45	5.49	5.54	5.57	5.61	5.65

APPENDIX 8

**Pearson product moment correlation values at the 0.05 and
0.01 levels of significnce**

d.f.	0.05	0.01
1	0.997	0.9999
2	0.950	0.990
3	0.878	0.959
4	0.811	0.917
5	0.754	0.874
6	0.707	0.834
7	0.666	0.798
8	0.632	0.765
9	0.602	0.735
10	0.576	0.708
11	0.553	0.684
12	0.532	0.661
13	0.514	0.641
14	0.497	0.623
15	0.482	0.606
16	0.468	0.590
17	0.456	0.575
18	0.444	0.561
19	0.433	0.549
20	0.423	0.537
21	0.413	0.526
22	0.404	0.515
23	0.396	0.505
24	0.388	0.496
25	0.381	0.487
26	0.374	0.479
27	0.367	0.471
28	0.361	0.463
29	0.355	0.456
30	0.349	0.449
32	0.339	0.436
34	0.329	0.424

d.f.	0.05	0.01
35	0.325	0.418
36	0.320	0.413
38	0.312	0.403
40	0.304	0.393
42	0.297	0.384
44	0.291	0.376
45	0.288	0.372
46	0.284	0.368
48	0.279	0.361
50	0.273	0.354
55	0.261	0.338
60	0.250	0.325
65	0.241	0.313
70	0.232	0.302
75	0.224	0.292
80	0.217	0.283
85	0.211	0.275
90	0.205	0.267
95	0.200	0.260
100	0.195	0.254
125	0.174	0.228
150	0.159	0.208
175	0.148	0.193
200	0.138	0.181
300	0.113	0.148
400	0.098	0.128
500	0.088	0.115
1,000	0.062	0.081

APPENDIX 9

Spearman rank correlation coefficient values

	Significance level (one-tailed test)	
N	0.05	0.01
4	1.000	
5	0.900	1.000
6	0.829	0.943
7	0.714	0.893
8	0.643	0.833
9	0.600	0.783
10	0.564	0.746
12	0.506	0.712
14	0.456	0.645
16	0.425	0.601
18	0.399	0.564
20	0.377	0.534
22	0.359	0.508
24	0.343	0.485
26	0.329	0.465
28	0.317	0.448
30	0.306	0.432

APPENDIX 10

Critical values of S for the Kendall rank correlation coefficient

S	Values of n				S	Values of n		
	4	**5**	**8**	**9**		**6**	**7**	**10**
0	0.625	0.592	0.540	0.540	1	0.500	0.500	0.500
2	0.375	0.408	0.452	0.460	3	0.360	0.386	0.431
4	0.167	0.242	0.360	0.381	5	0.235	0.281	0.364
6	0.042	0.117	0.274	0.306	7	0.136	0.191	0.300
8		0.042	0.199	0.238	9	0.068	0.119	0.242
10		$0.0^2 83$	0.138	0.179	11	0.028	0.068	0.190
12			0.089	0.130	13	$0.0^2 83$	0.035	0.146
14			0.054	0.090	15	$0.0^2 14$	0.015	0.108
16			0.031	0.060	17		$0.0^2 54$	0.078
18			0.016	0.038	19		$0.0^2 14$	0.054
20			$0.0^2 71$	0.022	21		$0.0^3 20$	0.036
22			$0.0^2 28$	0.012	23			0.023
24			$0.0^3 87$	$0.0^2 63$	25			0.014
26			$0.0^3 19$	$0.0^2 29$	27			$0.0^2 83$
28			$0.0^4 25$	$0.0^2 12$	29			$0.0^2 46$
30				$0.0^3 43$	31			$0.0^2 23$
32				$0.0^3 12$	33			$0.0^2 11$
34				$0.0^4 25$	35			$0.0^3 47$
36				$0.0^5 28$	37			$0.0^3 18$
					39			$0.0^4 58$
					41			$0.0^4 15$
					43			$0.0^5 28$
					45			$0.0^6 28$

Note.—Repeated zeros are indicated by powers, *e.g.* $0.0^3 47$ stands for 0.00047.

Source: Appendix Table 1, of Kendall, M. G. *Rank Correlation Methods*. London: Charles Griffin and Co. Ltd, 1970.

APPENDIX 11

**Critical values of the
multiple correlation coefficient**

critical region: $R \geq$ tabulated value

	5% significance level							1% significance level					
k / *n*	2	3	4	5	6	7	*k* / *n*	2	3	4	5	6	7
4	0.999						4	1.000					
5	0.975	0.999					5	0.995	1.000				
6	0.930	0.983	0.999				6	0.977	0.997	1.000			
7	0.881	0.950	0.987	1.000			7	0.949	0.983	0.997	1.000		
8	0.836	0.912	0.961	0.990	1.000		8	0.917	0.962	0.987	0.998	1.000	
9	0.795	0.874	0.930	0.968	0.991	1.000	9	0.886	0.937	0.970	0.990	0.998	1.000
10	0.758	0.839	0.898	0.942	0.973	0.993	10	0.855	0.911	0.949	0.975	0.991	0.999
11	0.726	0.807	0.867	0.914	0.950	0.977	11	0.827	0.885	0.927	0.957	0.979	0.992
12	0.697	0.777	0.838	0.886	0.925	0.956	12	0.800	0.860	0.904	0.938	0.963	0.981
13	0.671	0.750	0.811	0.860	0.900	0.934	13	0.776	0.837	0.882	0.918	0.946	0.967
14	0.648	0.726	0.786	0.835	0.876	0.911	14	0.753	0.814	0.861	0.898	0.928	0.952
15	0.627	0.703	0.763	0.812	0.854	0.889	15	0.732	0.793	0.840	0.878	0.909	0.935
16	0.608	0.683	0.741	0.790	0.832	0.868	16	0.712	0.773	0.821	0.859	0.891	0.919
17	0.590	0.664	0.722	0.770	0.812	0.848	17	0.694	0.755	0.802	0.841	0.874	0.902
18	0.574	0.646	0.703	0.751	0.792	0.829	18	0.677	0.737	0.785	0.824	0.857	0.886
19	0.559	0.630	0.686	0.733	0.774	0.811	19	0.662	0.721	0.768	0.807	0.841	0.870
20	0.545	0.615	0.670	0.717	0.757	0.793	20	0.647	0.706	0.752	0.791	0.825	0.855
21	0.532	0.601	0.655	0.701	0.741	0.777	21	0.633	0.691	0.738	0.776	0.810	0.840
22	0.520	0.587	0.641	0.687	0.726	0.762	22	0.620	0.678	0.724	0.762	0.796	0.825
23	0.509	0.575	0.628	0.673	0.712	0.747	23	0.607	0.665	0.710	0.749	0.782	0.812
24	0.498	0.563	0.615	0.660	0.698	0.733	24	0.596	0.652	0.697	0.736	0.769	0.799
25	0.488	0.552	0.604	0.647	0.686	0.720	25	0.585	0.641	0.685	0.723	0.757	0.786
26	0.479	0.542	0.593	0.636	0.673	0.707	26	0.574	0.630	0.674	0.712	0.745	0.774
27	0.470	0.532	0.582	0.624	0.662	0.696	27	0.565	0.619	0.663	0.700	0.733	0.762
28	0.462	0.523	0.572	0.614	0.651	0.684	28	0.555	0.609	0.653	0.690	0.722	0.751
29	0.454	0.514	0.562	0.604	0.640	0.673	29	0.546	0.600	0.643	0.679	0.711	0.740
30	0.446	0.506	0.553	0.594	0.630	0.663	30	0.538	0.590	0.633	0.669	0.701	0.730

Critical values of the
multiple correlation coefficient (continued)

critical region: $R \geq$ tabulated value

	5% significance level							1% significance level					
n \ k	2	3	4	5	6	7	n \ k	2	3	4	5	6	7
32	0.432	0.490	0.536	0.576	0.612	0.643	32	0.522	0.573	0.615	0.651	0.682	0.711
34	0.419	0.476	0.521	0.560	0.594	0.626	34	0.507	0.558	0.598	0.633	0.664	0.692
36	0.407	0.462	0.507	0.545	0.579	0.609	36	0.493	0.543	0.583	0.618	0.648	0.676
38	0.397	0.450	0.494	0.531	0.564	0.594	38	0.481	0.530	0.569	0.603	0.633	0.660
40	0.387	0.439	0.482	0.518	0.550	0.580	40	0.469	0.517	0.556	0.589	0.618	0.645
42	0.377	0.429	0.470	0.506	0.538	0.566	42	0.459	0.505	0.543	0.576	0.605	0.631
44	0.369	0.419	0.460	0.495	0.526	0.554	44	0.449	0.494	0.532	0.564	0.592	0.618
46	0.361	0.410	0.450	0.484	0.515	0.542	46	0.439	0.484	0.521	0.552	0.581	0.606
48	0.353	0.401	0.441	0.474	0.504	0.531	48	0.430	0.474	0.511	0.542	0.569	0.595
50	0.346	0.393	0.432	0.465	0.494	0.521	50	0.422	0.465	0.501	0.532	0.559	0.584
55	0.330	0.375	0.412	0.444	0.472	0.498	55	0.403	0.445	0.479	0.508	0.535	0.559
60	0.316	0.359	0.395	0.425	0.453	0.477	60	0.386	0.427	0.460	0.488	0.513	0.537
65	0.304	0.345	0.380	0.409	0.435	0.459	65	0.372	0.410	0.442	0.470	0.495	0.517
70	0.292	0.333	0.366	0.394	0.420	0.443	70	0.358	0.396	0.427	0.454	0.478	0.499
75	0.283	0.322	0.354	0.381	0.406	0.428	75	0.347	0.383	0.413	0.439	0.462	0.483
80	0.274	0.312	0.343	0.369	0.393	0.415	80	0.336	0.371	0.400	0.426	0.448	0.469
85	0.265	0.302	0.332	0.359	0.382	0.403	85	0.326	0.360	0.389	0.413	0.435	0.456
90	0.258	0.294	0.323	0.349	0.371	0.392	90	0.317	0.351	0.378	0.402	0.424	0.443
95	0.251	0.286	0.315	0.339	0.361	0.381	95	0.309	0.341	0.368	0.392	0.413	0.432
100	0.245	0.279	0.307	0.331	0.352	0.372	100	0.301	0.333	0.359	0.382	0.403	0.422
120	0.223	0.255	0.280	0.302	0.322	0.340	120	0.275	0.305	0.329	0.350	0.369	0.386
140	0.207	0.236	0.260	0.280	0.298	0.315	140	0.255	0.282	0.305	0.325	0.342	0.358
160	0.194	0.221	0.243	0.262	0.279	0.295	160	0.239	0.264	0.286	0.304	0.321	0.336
180	0.182	0.208	0.229	0.247	0.263	0.278	180	0.225	0.250	0.270	0.287	0.303	0.317
200	0.173	0.197	0.217	0.235	0.250	0.264	200	0.214	0.237	0.256	0.273	0.287	0.301

Source: Table 6.6 of Neave, H. R. *Statistics Tables.* London: George Allen and Unwin, 1978, p. 61.

APPENDIX 12

**Estimates of r_t for various
values of ad/bc**

r_{tet}	ad/bc	r_{tet}	ad/bc	r_{tet}	ad/bc
.00	0–1.00	.35	2.49–2.55	.70	8.50–8.90
.01	1.01–1.03	.36	2.56–2.63	.71	8.91–9.35
.02	1.04–1.06	.37	2.64–2.71	.72	9.36–9.82
.03	1.07–1.08	.38	2.72–2.79	.73	9.83–10.33
.04	1.09–1.11	.39	2.80–2.87	.74	10.34–10.90
.05	1.12–1.14	.40	2.88–2.96	.75	10.91–11.51
.06	1.15–1.17	.41	2.97–3.05	.76	11.52–12.16
.07	1.18–1.20	.42	3.06–3.14	.77	12.17–12.89
.08	1.21–1.23	.43	3.15–3.24	.78	12.90–13.70
.09	1.24–1.27	.44	3.25–3.34	.79	13.71–14.58
.10	1.28–1.30	.45	3.35–3.45	.80	14.59–15.57
.11	1.31–1.33	.46	3.46–3.56	.81	15.58–16.65
.12	1.34–1.37	.47	3.57–3.68	.82	16.66–17.88
.13	1.38–1.40	.48	3.69–3.80	.83	17.89–19.28
.14	1.41–1.44	.49	3.81–3.92	.84	19.29–20.85
.15	1.45–1.48	.50	3.93–4.06	.85	20.86–22.68
.16	1.49–1.52	.51	4.07–4.20	.86	22.69–24.76
.17	1.53–1.56	.52	4.21–4.34	.87	24.77–27.22
.18	1.57–1.60	.53	4.35–4.49	.88	27.23–30.09
.19	1.61–1.64	.54	4.50–4.66	.89	30.10–33.60
.20	1.65–1.69	.55	4.67–4.82	.90	33.61–37.79
.21	1.70–1.73	.56	4.83–4.99	.91	37.80–43.06
.22	1.74–1.78	.57	5.00–5.18	.92	43.07–49.83
.23	1.79–1.83	.58	5.19–5.38	.93	49.84–58.79
.24	1.84–1.88	.59	5.39–5.59	.94	58.80–70.95
.25	1.89–1.93	.60	5.60–5.80	.95	70.96–89.01
.26	1.94–1.98	.61	5.81–6.03	.96	89.02–117.54
.27	1.99–2.04	.62	6.04–6.28	.97	117.55–169.67
.28	2.05–2.10	.63	6.29–6.54	.98	169.68–293.12
.29	2.11–2.15	.64	6.55–6.81	.99	293.13–923.97
.30	2.16–2.22	.65	6.82–7.10	1.00	923.98 . . .
.31	2.23–2.28	.66	7.11–7.42		
.32	2.29–2.34	.67	7.43–7.75		
.33	2.35–2.41	.68	7.76–8.11		
.34	2.42–2.48	.69	8.12–8.49		

Source: M. D. Davidoff and H. W. Goheen. "A table for the rapid determination of the tetrachoric correlation coefficient." *Psychometrika*, 1953, *18*, 115–121. Reprinted with the permission of the authors and publisher.

"A note on a table for the rapid determination of the tetrachoric correlation coefficient." M. Davidoff, *Psychometrika*, Vol 19, no. 2 June 1954.

APPENDIX 13(a)

Binomial Test

N \ x	0	1	2	3	4	5	6	7	8	9	10	11	12	13	14	15
5	031	188	500	812	969	†										
6	018	109	344	656	891	984	†									
7	008	062	227	500	773	938	992	†								
8	004	035	145	363	637	855	965	996	†							
9	002	020	090	254	500	746	910	980	998	†						
10	001	011	055	172	377	623	828	945	989	999	†					
11		006	033	113	274	500	726	887	967	994	†	†				
12		003	019	073	194	387	613	806	927	981	997	†	†			
13		002	011	046	133	291	500	709	867	954	989	998	†	†		
14		001	006	029	090	212	395	605	788	910	971	994	999	†	†	
15			004	018	059	151	304	500	696	849	941	982	996	†	†	†
16			002	011	038	105	227	402	598	773	895	962	989	998	†	†
17			001	006	025	072	166	315	500	685	834	928	975	994	999	†
18			001	004	015	048	119	240	407	593	760	881	952	985	996	999
19				002	010	032	084	180	324	500	676	820	916	968	990	998
20				001	006	021	058	132	252	412	588	748	868	942	979	994
21				001	004	013	039	095	192	332	500	668	808	905	961	987
22					002	008	026	067	143	262	416	584	738	857	933	974
23					001	005	017	047	105	202	339	500	661	798	895	953
24					001	003	011	032	076	154	271	419	581	729	846	924
25						002	007	022	054	115	212	345	500	655	788	885

APPENDIX 13(b)

Table of Binomial Coefficients

N	$\binom{N}{0}$	$\binom{N}{1}$	$\binom{N}{2}$	$\binom{N}{3}$	$\binom{N}{4}$	$\binom{N}{5}$	$\binom{N}{6}$	$\binom{N}{7}$	$\binom{N}{8}$	$\binom{N}{9}$	$\binom{N}{10}$
0	1										
1	1	1									
2	1	2	1								
3	1	3	3	1							
4	1	4	6	4	1						
5	1	5	10	10	5	1					
6	1	6	15	20	15	6	1				
7	1	7	21	35	35	21	7	1			
8	1	8	28	56	70	56	28	8	1		
9	1	9	36	84	126	126	84	36	9	1	
10	1	10	45	120	210	252	210	120	45	10	1
11	1	11	55	165	330	462	462	330	165	55	11
12	1	12	66	220	495	792	924	792	495	220	66
13	1	13	78	286	715	1287	1716	1716	1287	715	286
14	1	14	91	364	1001	2002	3003	3432	3003	2002	1001
15	1	15	105	455	1365	3003	5005	6435	6435	5005	3003
16	1	16	120	560	1820	4368	8008	11440	12870	11440	8008
17	1	17	136	680	2380	6188	12376	19448	24310	24310	19448
18	1	18	153	816	3060	8568	18564	31824	43758	48620	43758
19	1	19	171	969	3876	11628	27132	50388	75582	92378	92378
20	1	20	190	1140	4845	15504	38760	77520	125970	167960	184756

Source: S. Siegel (1956) *Nonparametric Statistics For The Behavioural Sciences*, McGraw-Hill Book Co.

APPENDIX 14

Critical values of D in the Kolmogorov–Smirnov One Sample Test

Sample size N	Significance level	
	0.05	0.02
1	0.975	0.99
2	0.84	0.90
3	0.71	0.78
4	0.62	0.69
5	0.56	0.63
6	0.52	0.58
7	0.48	0.54
8	0.45	0.51
9	0.43	0.48
10	0.41	0.46
11	0.39	0.44
12	0.38	0.42
13	0.36	0.40
14	0.35	0.39
15	0.34	0.38
16	0.33	0.37
17	0.32	0.36
18	0.31	0.35
19	0.30	0.34
20	0.29	0.33
21	0.29	0.32
22	0.28	0.31
23	0.27	0.31
24	0.27	0.30
25	0.26	0.30
26	0.26	0.29
27	0.25	0.28
28	0.25	0.28
29	0.24	0.27
30	0.24	0.27

Source: Adapted from Table 4.4, Ray Meddis, *Statistical Handbook for Non-Statisticians*, London: McGraw-Hill, 1975, p. 63, with the kind permission of the author and publisher.

APPENDIX 15

Critical values of K

One-tailed test	0.05	0.025	0.01	0.001
Two-tailed test	0.1	0.05	0.02	0.002
K	1.22	1.36	1.51	1.86

Source: R. Meddis, *Statistical Handbook for Non-Statisticians*, London: McGraw-Hill, 1975, p. 62, with the kind permission of the author and publisher.

APPENDIX 16

Critical values of R in the runs test

n_1/n_2	2	3	4	5	6	7	8	9	10	11	12	13	14	15	16	17	18	19	20
2	—	—	—	—	—	—	—	—	—	—	2	2	2	2	2	2	2	2	2
	—	—	—	—	—	—	—	—	—	—	—	—	—	—	—	—	—	—	—
3	—	—	—	—	2	2	2	2	2	2	2	2	2	3	3	3	3	3	3
	—	—	—	—	—	—	—	—	—	—	—	—	—	—	—	—	—	—	—
4	—	—	—	2	2	2	3	3	3	3	3	3	3	3	4	4	4	4	4
	—	—	—	9	9	—	—	—	—	—	—	—	—	—	—	—	—	—	—
5	—	—	2	2	3	3	3	3	3	4	4	4	4	4	4	4	5	5	5
	—	—	9	10	10	11	11	—	—	—	—	—	—	—	—	—	—	—	—
6	—	2	2	3	3	3	3	4	4	4	4	5	5	5	5	5	5	6	6
	—	9	10	11	12	12	13	13	13	13	—	—	—	—	—	—	—	—	—
7	—	2	2	3	3	3	4	4	5	5	5	5	5	6	6	6	6	6	6
	—	—	11	12	13	13	14	14	14	14	15	15	15	—	—	—	—	—	—
8	—	2	3	3	3	4	4	5	5	5	6	6	6	6	6	7	7	7	7
	—	—	11	12	13	14	14	15	15	16	16	16	16	17	17	17	17	17	—
9	—	2	3	3	4	4	5	5	5	6	6	6	7	7	7	7	8	8	8
	—	—	—	13	14	14	15	16	16	16	17	17	18	18	18	18	18	18	18
10	—	2	3	4	4	5	5	6	6	7	7	7	8	8	8	9	9	9	9
	—	—	—	13	14	15	16	16	17	17	18	18	18	19	19	20	20	20	20
11	—	2	3	4	4	5	5	6	6	7	7	7	8	8	8	9	9	9	9
	—	—	—	13	14	15	16	17	17	18	19	19	19	20	20	21	21	21	22
12	2	2	3	4	4	5	6	6	7	7	7	8	8	8	9	9	10	10	10
	—	—	—	13	14	16	16	17	18	19	19	20	20	21	21	21	22	22	22
13	2	2	3	4	5	5	6	6	7	7	7	8	8	9	9	9	10	10	10
	—	—	—	—	15	16	17	18	19	19	20	20	21	22	22	23	23	23	24
14	2	2	3	4	5	5	6	7	7	8	8	9	9	9	10	10	10	11	11
	—	—	—	—	15	16	17	18	19	20	20	21	22	22	23	23	24	24	25
15	2	3	3	4	5	6	6	7	7	8	8	9	9	10	10	11	11	11	12
	—	—	—	—	15	16	18	18	19	20	21	22	22	23	23	24	24	25	25
16	2	3	4	4	5	6	6	7	8	8	9	9	10	10	11	11	11	12	12
	—	—	—	—	—	17	18	19	20	20	21	21	22	23	23	24	25	25	25
17	2	3	4	4	5	6	7	7	8	9	9	10	10	11	11	11	12	12	13
	—	—	—	—	—	17	18	19	20	21	21	22	23	23	24	25	25	26	26
18	2	3	4	5	5	6	7	8	8	9	9	10	10	11	11	12	12	13	13
	—	—	—	—	—	17	18	19	20	21	22	23	23	24	25	25	26	26	27
19	2	3	4	5	6	6	7	8	8	9	10	10	11	11	12	12	13	13	13
	—	—	—	—	—	17	18	20	21	22	23	23	24	25	25	26	26	27	27
20	2	3	4	5	6	6	7	8	9	9	10	10	11	12	12	13	13	13	14
	—	—	—	—	—	17	18	20	21	22	23	24	25	25	26	27	27	28	28

Each cell shows two values for n_1 and n_2. If R is equal to or less than the upper value or equal to or greater than the lower value in each cell, the sequence is nonrandom at $\alpha = 0.05$ (two-tailed test) or 0.025 (one-tailed test).

Dashes indicate that no decision is possible for the indicated values of n_1 and n_2.

Source: Table F of Runyon, R. P., *Nonparametric Statistics: A Contemporary Approach*, Reading, Mass.: Addison-Wesley Publishing Co., 1977, p. 161.

APPENDIX 17

Critical values of *T* in the Wilcoxon Test for Two Correlated Samples

Sample size	Levels of significance			
	One-tailed test			
	0.05	0.025	0.01	0.001
	Two-tailed test			
	0.1	0.05	0.02	0.002
N = 5	T ≤ 0			
6	2	0		
7	3	2	0	
8	5	3	1	
9	8	5	3	
10	10	8	5	0
11	13	10	7	1
12	17	13	9	2
13	21	17	12	4
14	25	21	15	6
15	30	25	19	8
16	35	29	23	11
17	41	34	27	14
18	47	40	32	18
19	53	46	37	21
20	60	52	43	26
21	67	58	49	30
22	75	65	55	35
23	83	73	62	40
24	91	81	69	45
25	100	89	76	51
26	110	98	84	58
27	119	107	92	64
28	130	116	101	71
30	151	137	120	86
31	163	147	130	94
32	175	159	140	103
33	187	170	151	112

Source: Adapted from Table 6.5, Ray Meddis, *Statistical Handbook for Non-Statisticians*, London: McGraw-Hill, 1975, p. 113, with the kind permission of the author and publisher.

APPENDIX 18

Critical values of χ_R^2 in the Friedman Two-Way Analysis of Variance

$k = 3$			$k = 3$			$k = 3$			$k = 4$			$k = 5$		
n	0.05	0.01	n	0.05	0.01	n	0.05	0.01	n	0.05	0.01	n	0.05	0.01
2	--	—	21	6.095	9.238	40	6.050	9.150	4	7.800	9.600	2	7.600	8.000
3	6.000	.	22	6.091	9.091	41	6.195	9.366	5	7.800	9.960	3	8.533	10.23
4	6.500	8.000	23	6.348	9.391	42	6.143	9.190	6	7.600	10.20	4	8.800	11.20
5	6.400	8.400	24	6.250	9.250	43	6.186	9.256	7	7.800	10.54	5	8.960	11.68
6	7.000	9.000	25	6.080	8.960	44	6.318	9.136	8	7.650	10.50	6	9.067	11.87
7	7.143	8.857	26	6.077	9.308	45	6.178	9.244	9	7.667	10.73	7	9.143	12.11
8	6.250	9.000	27	6.000	9.407	46	6.043	9.435	10	7.680	10.68	8	9.200	12.30
9	6.222	9.556	28	6.500	9.214	47	6.128	9.319	11	7.691	10.75	9	9.244	12.44
10	6.200	9.600	29	6.276	9.172	48	6.167	9.125	12	7.700	10.80			
11	6.545	9.455	30	6.200	9.267	49	6.041	9.184	13	7.800	10.85			
12	6.500	9.500	31	6.000	9.290	50	6.040	9.160	14	7.714	10.89	$k = 6$		
13	6.615	9.385	32	6.063	9.250				15	7.720	10.92			
14	6.143	9.143	33	6.061	9.152				16	7.800	10.95	n	0.05	0.01
15	6.400	8.933	34	6.059	9.176	$k = 4$			17	7.800	11.05			
16	6.500	9.375	35	6.171	9.314				18	7.733	10.93	2	9.357	9.929
17	6.118	9.294	36	6.167	9.389	n	0.05	0.01	19	7.863	11.02	3	9.857	11.76
18	6.333	9.000	37	6.054	9.243				20	7.800	11.10	4	10.39	12.82
19	6.421	9.579	38	6.158	9.053	2	6.000	--	21	7.800	11.06			
20	6.300	9.300	39	6.000	9.282	3	7.400	9.000	22	7.800	11.07			

Source: Table 4.3. Neave, H. R.: *Statistics Tables*, London: George Allen & Unwin, 1978, p. 49, with the kind permission of the author and publisher.

APPENDIX 19

Page's L trend test critical values of L at $p < .001$, $p < .01$ and $p < .05$

Subjects or sets of matched subjects	C Number of conditions				
	3	4	5	6	$p<$
2			109	178	.001
		60	106	173	.01
	28	58	103	166	.05
3		89	160	260	.001
	42	87	155	252	.01
	41	84	150	244	.05
4	56	117	210	341	.001
	55	114	204	331	.01
	54	111	197	321	.05
5	70	145	259	420	.001
	68	141	251	409	.01
	66	137	244	397	.05
6	83	172	307	499	.001
	81	167	299	486	.01
	79	163	291	474	.05
7	96	198	355	577	.001
	93	193	346	563	.01
	91	189	338	550	.05
8	109	225	403	655	.001
	106	220	393	640	.01
	104	214	384	625	.05
9	121	252	451	733	.001
	119	246	441	717	.01
	116	240	431	701	.05
10	134	278	499	811	.001
	131	272	487	793	.01
	128	266	477	777	.05
11	147	305	546	888	.001
	144	298	534	869	.01
	141	292	523	852	.05
12	160	331	593	965	.001
	156	324	581	946	.01
	153	317	570	928	.05

Source: *Journal of the American Statistical Association*, Vol. 58. 1963

APPENDIX 20

Mann–Whitney U Test values (two-tailed test)

n_L = larger sample size
n_S = smaller sample size

Values in the upper-right of each row (below the blank diagonal) correspond to the 0.05 table (top and right labels, $n_L \backslash n_S$); values in the lower-left correspond to the 0.01 table (bottom and left labels).

$n_S \backslash n_L$	2	3	4	5	6	7	8	9	10	11	12	13	14	15	16	17	18	19	20	21	22	23	24	25
2	–	–	–	–	–	–	0	0	0	0	1	1	1	1	1	2	2	2	2	2	3	3	3	3
3	–	–	0	1	1	2	2	3	3	4	4	5	5	6	6	7	7	8	8	9	9	10	10	11
4	–	–		1	2	3	4	4	5	6	7	8	9	10	11	11	12	13	14	15	16	17	17	18
5	–	–	–		3	5	6	7	8	9	11	12	13	14	15	17	18	19	20	22	23	24	25	27
6	–	–	0	1		6	8	10	11	13	14	16	17	19	21	22	24	25	27	29	30	32	33	35
7	–	–	0	1	3		10	12	14	16	18	20	22	24	26	28	30	32	34	36	38	40	42	44
8	–	–	1	2	4	6		15	17	19	22	24	26	29	31	34	36	38	41	43	45	48	50	53
9	–	0	1	3	5	7	9		20	23	26	28	31	34	37	39	42	45	48	50	53	56	59	62
10	–	0	2	4	6	9	11	13		26	29	33	36	39	42	45	48	52	55	58	61	64	67	71
11	–	0	2	5	7	10	13	16	18		33	37	40	44	47	51	55	58	62	65	69	73	76	80
12	–	1	3	6	9	12	15	18	21	24		41	45	49	53	57	61	65	69	73	77	81	85	89
13	–	1	3	7	10	13	17	20	24	27	31		50	54	59	63	67	72	76	80	85	89	94	98
14	–	1	4	7	11	15	18	22	26	30	34	38		59	64	69	74	78	83	88	93	98	102	107
15	–	2	5	8	12	16	20	24	29	33	37	42	46		70	75	80	85	90	96	101	106	111	117
16	–	2	5	9	13	18	22	27	31	36	41	45	50	55		81	86	92	98	103	109	115	120	126
17	–	2	6	10	15	19	24	29	34	39	44	49	54	60	65		93	99	105	111	117	123	129	135
18	–	2	6	11	16	21	26	31	37	42	47	53	58	64	70	75		106	112	119	125	132	138	145
19	0	3	7	12	17	22	28	33	39	45	51	57	63	69	74	81	87		119	126	133	140	147	154
20	0	3	8	13	18	24	30	36	42	48	54	60	67	73	79	86	92	99		134	141	149	156	163
21	0	3	8	14	19	25	32	38	44	51	58	64	71	78	84	91	98	105	112		150	157	165	173
22	0	4	9	14	21	27	34	40	47	54	61	68	75	82	89	96	104	111	118	125		166	174	182
23	0	4	9	15	22	29	35	43	50	57	64	72	79	87	94	102	109	117	125	132	140		183	192
24	0	4	10	16	23	30	37	45	52	60	68	75	83	91	99	107	115	123	131	139	147	155		201
25	0	5	10	17	24	32	39	47	55	63	71	79	87	96	104	112	121	129	138	146	155	163	172	

0.01 $\quad n_S \backslash n_L \quad$ 2 3 4 5 6 7 8 9 10 11 12 13 14 15 16 17 18 19 20 21 22 23 24 25

Mann–Whitney _U_ Test values (two-tailed test)

equal sample sizes

n	1	2	3	4	5	6	7	8	9	10	11	12	13	14	15	16	17	18	19	20	21	22	23	24	25
0.05	-	-	-	0	2	5	8	13	17	23	30	37	45	55	64	75	87	99	113	127	142	158	175	192	211
0.01	-	-	-	-	0	2	4	7	11	16	21	27	34	42	51	60	70	81	93	105	118	133	148	164	180

n	26	27	28	29	30	31	32	33	34	35	36	37	38	39	40	41	42	43	44	45	46	47	48	49	50
0.05	230	250	272	294	317	341	365	391	418	445	473	503	533	564	596	628	662	697	732	769	806	845	884	924	965
0.01	198	216	235	255	276	298	321	344	369	394	420	447	475	504	533	564	595	627	660	694	729	765	802	839	877

Source: Table 5.3, of Neave, H. R., _Statistics Tables_. London: George Allen & Unwin, 1978, p. 53, with the kind permission of the author and publisher.

APPENDIX 21

Table of critical values for the Walsh test

N	Significance level of tests		Tests	
			Two-tailed: accept $\mu_1 \neq 0$ if either	
	One-tailed	Two-tailed	One-tailed: accept $\mu_1 < 0$ if	One-tailed: accept $\mu_1 > 0$ if
4	.062	.125	$d_4 < 0$	$d_1 > 0$
5	.062	.125	$\frac{1}{2}(d_4 + d_5) < 0$	$\frac{1}{2}(d_1 + d_2) > 0$
	.031	.062	$d_5 < 0$	$d_1 > 0$
6	.047	.094	$\max [d_5, \frac{1}{2}(d_4 + d_6)] < 0$	$\min [d_2, \frac{1}{2}(d_1 + d_3)] > 0$
	.031	.062	$\frac{1}{2}(d_5 + d_6) < 0$	$\frac{1}{2}(d_1 + d_2) > 0$
	.016	.031	$d_6 < 0$	$d_1 > 0$
7	.055	.109	$\max [d_5, \frac{1}{2}(d_4 + d_7)] < 0$	$\min [d_3, \frac{1}{2}(d_1 + d_4)] > 0$
	.023	.047	$\max [d_6, \frac{1}{2}(d_5 + d_7)] < 0$	$\min [d_2, \frac{1}{2}(d_1 + d_3)] > 0$
	.016	.031	$\frac{1}{2}(d_6 + d_7) < 0$	$\frac{1}{2}(d_1 + d_2) > 0$
	.008	.016	$d_7 < 0$	$d_1 > 0$
8	.043	.086	$\max [d_6, \frac{1}{2}(d_4 + d_8)] < 0$	$\min [d_3, \frac{1}{2}(d_1 + d_5)] > 0$
	.027	.055	$\max [d_6, \frac{1}{2}(d_5 + d_8)] < 0$	$\min [d_3, \frac{1}{2}(d_1 + d_4)] > 0$
	.012	.023	$\max [d_7, \frac{1}{2}(d_6 + d_8)] < 0$	$\min [d_2, \frac{1}{2}(d_1 + d_3)] > 0$
	.008	.016	$\frac{1}{2}(d_7 + d_8) < 0$	$\frac{1}{2}(d_1 + d_2) > 0$
	.004	.008	$d_8 < 0$	$d_1 > 0$
9	.051	.102	$\max [d_6, \frac{1}{2}(d_4 + d_9)] < 0$	$\min [d_4, \frac{1}{2}(d_1 + d_6)] > 0$
	.022	.043	$\max [d_7, \frac{1}{2}(d_5 + d_9)] < 0$	$\min [d_3, \frac{1}{2}(d_1 + d_5)] > 0$
	.010	.020	$\max [d_8, \frac{1}{2}(d_5 + d_9)] < 0$	$\min [d_2, \frac{1}{2}(d_1 + d_5)] > 0$
	.006	.012	$\max [d_8, \frac{1}{2}(d_7 + d_9)] < 0$	$\min [d_2, \frac{1}{2}(d_1 + d_3)] > 0$
	.004	.008	$\frac{1}{2}(d_8 + d_9) < 0$	$\frac{1}{2}(d_1 + d_2) > 0$
10	.056	.111	$\max [d_6, \frac{1}{2}(d_4 + d_{10})] < 0$	$\min [d_5, \frac{1}{2}(d_1 + d_7)] > 0$
	.025	.051	$\max [d_7, \frac{1}{2}(d_5 + d_{10})] < 0$	$\min [d_4, \frac{1}{2}(d_1 + d_6)] > 0$
	.011	.021	$\max [d_8, \frac{1}{2}(d_6 + d_{10})] < 0$	$\min [d_3, \frac{1}{2}(d_1 + d_5)] > 0$
	.005	.010	$\max [d_9, \frac{1}{2}(d_6 + d_{10})] < 0$	$\min [d_2, \frac{1}{2}(d_1 + d_5)] > 0$

Table of critical values for the Walsh test (continued)

N	Significance level of tests		Tests	
			Two-tailed: accept $\mu_1 \neq 0$ if either	
	One-tailed	Two-tailed	One-tailed: accept $\mu_1 < 0$ if	One-tailed: accept $\mu_1 > 0$ if
11	.048	.097	$\max [d_7, \frac{1}{2}(d_4 + d_{11})] < 0$	$\min [d_5, \frac{1}{2}(d_1 + d_8)] > 0$
	.028	.056	$\max [d_7, \frac{1}{2}(d_5 + d_{11})] < 0$	$\min [d_5, \frac{1}{2}(d_1 + d_7)] > 0$
	.011	.021	$\max [\frac{1}{2}(d_6 + d_{11}), \frac{1}{2}(d_8 + d_9)] < 0$	$\min [\frac{1}{2}(d_1 + d_6), \frac{1}{2}(d_3 + d_4)] > 0$
	.005	.011	$\max [d_9, \frac{1}{2}(d_7 + d_{11})] < 0$	$\min [d_3, \frac{1}{2}(d_1 + d_5)] > 0$
12	.047	.094	$\max [\frac{1}{2}(d_4 + d_{12}), \frac{1}{2}(d_5 + d_{11})] < 0$	$\min [\frac{1}{2}(d_1 + d_9), \frac{1}{2}(d_2 + d_8)] > 0$
	.024	.048	$\max [d_8, \frac{1}{2}(d_5 + d_{12})] < 0$	$\min [d_5, \frac{1}{2}(d_1 + d_8)] > 0$
	.010	.020	$\max [d_9, \frac{1}{2}(d_6 + d_{12})] < 0$	$\min [d_4, \frac{1}{2}(d_1 + d_7)] > 0$
	.005	.011	$\max [\frac{1}{2}(d_7 + d_{12}), \frac{1}{2}(d_9 + d_{10})] < 0$	$\min [\frac{1}{2}(d_1 + d_6), \frac{1}{2}(d_3 + d_4)] > 0$
13	.047	.094	$\max [\frac{1}{2}(d_4 + d_{13}), \frac{1}{2}(d_5 + d_{12})] < 0$	$\min [\frac{1}{2}(d_1 + d_{10}), \frac{1}{2}(d_2 + d_9)] > 0$
	.023	.047	$\max [\frac{1}{2}(d_5 + d_{13}), \frac{1}{2}(d_6 + d_{12})] < 0$	$\min [\frac{1}{2}(d_1 + d_9), \frac{1}{2}(d_2 + d_8)] > 0$
	.010	.020	$\max [\frac{1}{2}(d_6 + d_{13}), \frac{1}{2}(d_9 + d_{10})] < 0$	$\min [\frac{1}{2}(d_1 + d_8), \frac{1}{2}(d_4 + d_5)] > 0$
	.005	.010	$\max [d_{10}, \frac{1}{2}(d_7 + d_{13})] < 0$	$\min [d_4, \frac{1}{2}(d_1 + d_7)] > 0$
14	.047	.094	$\max [\frac{1}{2}(d_4 + d_{14}), \frac{1}{2}(d_5 + d_{13})] < 0$	$\min [\frac{1}{2}(d_1 + d_{11}), \frac{1}{2}(d_2 + d_{10})] > 0$
	.023	.047	$\max [\frac{1}{2}(d_5 + d_{14}), \frac{1}{2}(d_6 + d_{13})] < 0$	$\min [\frac{1}{2}(d_1 + d_{10}), \frac{1}{2}(d_2 + d_9)] > 0$
	.010	.020	$\max [d_{10}, \frac{1}{2}(d_6 + d_{14})] < 0$	$\min [d_5, \frac{1}{2}(d_1 + d_9)] > 0$
	.005	.010	$\max [\frac{1}{2}(d_7 + d_{14}), \frac{1}{2}(d_{10} + d_{11})] < 0$	$\min [\frac{1}{2}(d_1 + d_8), \frac{1}{2}(d_4 + d_5)] > 0$
15	.047	.094	$\max [\frac{1}{2}(d_4 + d_{15}), \frac{1}{2}(d_5 + d_{14})] < 0$	$\min [\frac{1}{2}(d_1 + d_{12}), \frac{1}{2}(d_2 + d_{11})] > 0$
	.023	.047	$\max [\frac{1}{2}(d_5 + d_{15}), \frac{1}{2}(d_6 + d_{14})] < 0$	$\min [\frac{1}{2}(d_1 + d_{11}), \frac{1}{2}(d_2 + d_{10})] > 0$
	.010	.020	$\max [\frac{1}{2}(d_6 + d_{15}), \frac{1}{2}(d_{10} + d_{11})] < 0$	$\min [\frac{1}{2}(d_1 + d_{10}), \frac{1}{2}(d_5 + d_6)] > 0$
	.005	.010	$\max [d_{11}, \frac{1}{2}(d_7 + d_{15})] < 0$	$\min [d_5, \frac{1}{2}(d_1 + d_9)] > 0$

Source: Walsh J. E. 1949. *Journal of the American Statistical Association*, 44, p. 343

APPENDIX 22

Wald–Wolfowitz number-of-runs test

critical region: $R \leq$ tabulated value

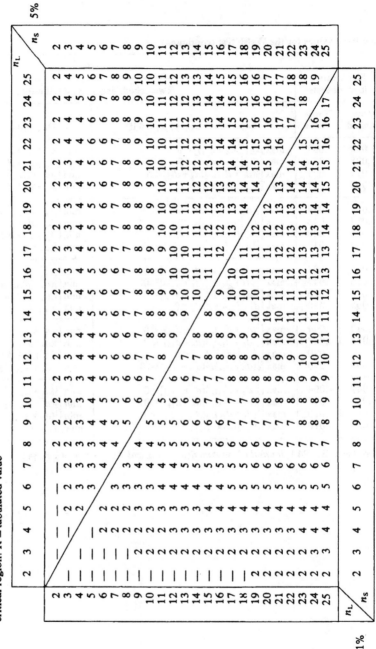

5%

1%

equal sample sizes

n	3	4	5	6	7	8	9	10–12	13–15	16	17–21	22	23–27	28–31	32
5%	—	2	3	3	4	5	6	6–8	9–11	11	12–16	17	17–21	22–25	25
1%	—	—	2	2	3	4	4	5–7	7–9	10	10–14	14	15–19	19–22	23

n	33	34–41	42	43–49	50–53	54–58	59–67	68–78	79–81	82–89	90–98	99	100
5%	26	27–34	35	35–41	42–45	45–49	50–58	58–68	69–71	71–78	79–87	87	88
1%	24	24–31	31	32–38	38–41	42–46	46–54	54–64	64–66	67–74	74–82	83	84

Source: Table 5.5 of Neave, H. R. *Statistics Tables*. London: George Allen and Unwin, 1978, p. 55.

APPENDIX 23

Table of critical values of D (or C) in the Fisher test

Totals in right margin		B (or A)†	Level of significance			
			.05	.025	.01	.005
$A + B = 3$	$C + D = 3$	3	0	—	—	—
$A + B = 4$	$C + D = 4$	4	0	0	—	—
	$C + D = 3$	4	0	—	—	—
$A + B = 5$	$C + D = 5$	5	1	1	0	0
		4	0	0	—	—
	$C + D = 4$	5	1	0	0	—
		4	0	—	—	—
	$C + D = 3$	5	0	0	—	—
	$C + D = 2$	5	0	—	—	—
$A + B = 6$	$C + D = 6$	6	2	1	1	0
		5	1	0	0	—
		4	0	—	—	—
	$C + D = 5$	6	1	0	0	0
		5	0	0	—	—
		4	0	—	—	—
	$C + D = 4$	6	1	0	0	0
		5	0	0	—	—
	$C + D = 3$	6	0	0	—	—
		5	0	—	—	—
	$C + D = 2$	6	0	—	—	—
$A + B = 7$	$C + D = 7$	7	3	2	1	1
		6	1	1	0	0
		5	0	0	—	—
		4	0	—	—	—
	$C + D = 6$	7	2	2	1	1
		6	1	0	0	0
		5	0	0	—	—
		4	0	—	—	—
	$C + D = 5$	7	2	1	0	0
		6	1	0	0	—
		5	0	—	—	—
	$C + D = 4$	7	1	1	0	0
		6	0	0	—	—
		5	0	—	—	—
	$C + D = 3$	7	0	0	0	—
		6	0	—	—	—
	$C + D = 2$	7	0	—	—	—

† 'When B is entered in the middle column, the significance levels are for D. When A is used in place of B, the significance levels are for C.'

Table of critical values of _D_ (or _C_) in the Fisher test (continued)

Totals in right margin		_B_ (or _A_)†	Level of significance			
			.05	.025	.01	.005
_A_ + _B_ = 8	_C_ + _D_ = 8	8	4	3	2	2
		7	2	2	1	0
		6	1	1	0	0
		5	0	0	—	—
		4	0	—	—	—
	_C_ + _D_ = 7	8	3	2	2	1
		7	2	1	1	0
		6	1	0	0	—
		5	0	0	—	—
	_C_ + _D_ = 6	8	2	2	1	1
		7	1	1	0	0
		6	0	0	0	—
		5	0	—	—	—
	_C_ + _D_ = 5	8	2	1	1	0
		7	1	0	0	0
		6	0	0	—	—
		5	0	—	—	—
	_C_ + _D_ = 4	8	1	1	0	0
		7	0	0	—	—
		6	0	—	—	—
	_C_ + _D_ = 3	8	0	0	0	—
		7	0	0	—	—
	_C_ + _D_ = 2	8	0	0	—	—
_A_ + _B_ = 9	_C_ + _D_ = 9	9	5	4	3	3
		8	3	3	2	1
		7	2	1	1	0
		6	1	1	0	0
		5	0	0	—	—
		4	0	—	—	—
	_C_ + _D_ = 8	9	4	3	3	2
		8	3	2	1	1
		7	2	1	0	0
		6	1	0	0	—
		5	0	0	—	—
	_C_ + _D_ = 7	9	3	3	2	2
		8	2	2	1	0
		7	1	1	0	0
		6	0	0	—	—
		5	0	—	—	—

Table of critical values of D (or C) in the Fisher test (continued)

Totals in right margin		B (or A)†	Level of significance			
			.05	.025	.01	.005
$A+B=9$	$C+D=6$	9	3	2	1	1
		8	2	1	0	0
		7	1	0	0	—
		6	0	0	—	—
		5	0	—	—	—
	$C+D=5$	9	2	1	1	1
		8	1	1	0	0
		7	0	0	—	—
		6	0	—	—	—
	$C+D=4$	9	1	1	0	0
		8	0	0	0	—
		7	0	0	—	—
		6	0	—	—	—
	$C+D=3$	9	1	0	0	0
		8	0	0	—	—
		7	0	—	—	—
	$C+D=2$	9	0	0	—	—
$A+B=10$	$C+D=10$	10	6	5	4	3
		9	4	3	3	2
		8	3	2	1	1
		7	2	1	1	0
		6	1	0	0	—
		5	0	0	—	—
		4	0	—	—	—
	$C+D=9$	10	5	4	3	3
		9	4	3	2	2
		8	2	2	1	1
		7	1	1	0	0
		6	1	0	0	—
		5	0	0	—	—
	$C+D=8$	10	4	4	3	2
		9	3	2	2	1
		8	2	1	1	0
		7	1	1	0	0
		6	0	0	—	—
		5	0	—	—	—
	$C+D=7$	10	3	3	2	2
		9	2	2	1	1
		8	1	1	0	0
		7	1	0	0	—
		6	0	0	—	—
		5	0	—	—	—

Table of critical values of D (or C) in the Fisher test (continued)

Totals in right margin		B (or A)†	Level of significance			
			.05	.025	.01	.005
$A+B=10$	$C+D=6$	10	3	2	2	1
		9	2	1	1	0
		8	1	1	0	0
		7	0	0	—	—
		6	0	—	—	—
	$C+D=5$	10	2	2	1	1
		9	1	1	0	0
		8	1	0	0	—
		7	0	0	—	—
		6	0	—	—	—
	$C+D=4$	10	1	1	0	0
		9	1	0	0	0
		8	0	0	—	—
		7	0	—	—	—
	$C+D=3$	10	1	0	0	0
		9	0	0	—	—
		8	0	—	—	—
	$C+D=2$	10	0	0	—	—
		9	0	—	—	—
$A+B=11$	$C+D=11$	11	7	6	5	4
		10	5	4	3	3
		9	4	3	2	2
		8	3	2	1	1
		7	2	1	0	0
		6	1	0	0	—
		5	0	0	—	—
		4	0	—	—	—
	$C+D=10$	11	6	5	4	4
		10	4	4	3	2
		9	3	3	2	1
		8	2	2	1	0
		7	1	1	0	0
		6	1	0	0	—
		5	0	—	—	—
	$C+D=9$	11	5	4	4	3
		10	4	3	2	2
		9	3	2	1	1
		8	2	1	1	0
		7	1	1	0	0
		6	0	0	—	—
		5	0	—	—	—

Table of critical values of D (or C) in the Fisher test (continued)

Totals in right margin		B (or A)†	Level of significance			
			.05	.025	.01	.005
$A + B = 11$	$C + D = 8$	11	4	4	3	3
		10	3	3	2	1
		9	2	2	1	1
		8	1	1	0	0
		7	1	0	0	—
		6	0	0	—	—
		5	0	—	—	—
	$C + D = 7$	11	4	3	2	2
		10	3	2	1	1
		9	2	1	1	0
		8	1	1	0	0
		7	0	0	—	—
		6	0	0	—	—
	$C + D = 6$	11	3	2	2	1
		10	2	1	1	0
		9	1	1	0	0
		8	1	0	0	—
		7	0	0	—	—
		6	0	—	—	—
	$C + D = 5$	11	2	2	1	1
		10	1	1	0	0
		9	1	0	0	0
		8	0	0	—	—
		7	0	—	—	—
	$C + D = 4$	11	1	1	1	0
		10	1	0	0	0
		9	0	0	—	—
		8	0	—	—	—
	$C + D = 3$	11	1	0	0	0
		10	0	0	—	—
		9	0	—	—	—
	$C + D = 2$	11	0	0	—	—
		10	0	—	—	—
$A + B = 12$	$C + D = 12$	12	8	7	6	5
		11	6	5	4	4
		10	5	4	3	2
		9	4	3	2	1
		8	3	2	1	1
		7	2	1	0	0
		6	1	0	0	—
		5	0	0	—	—
		4	0	—	—	—

Table of critical values of *D* (or *C*) in the Fisher test (continued)

Totals in right margin		*B* (or *A*)†	Level of significance			
			.05	.025	.01	.005
A + *B* = 12	*C* + *D* = 11	12	7	6	5	5
		11	5	5	4	3
		10	4	3	2	2
		9	3	2	2	1
		8	2	1	1	0
		7	1	1	0	0
		6	1	0	0	—
		5	0	0	—	—
	C + *D* = 10	12	6	5	5	4
		11	5	4	3	3
		10	4	3	2	2
		9	3	2	1	1
		8	2	1	0	0
		7	1	0	0	0
		6	0	0	—	—
		5	0	—	—	—
	C + *D* = 9	12	5	5	4	3
		11	4	3	3	2
		10	3	2	2	1
		9	2	2	1	0
		8	1	1	0	0
		7	1	0	0	—
		6	0	0	—	—
		5	0	—	—	—
	C + *D* = 8	12	5	4	3	3
		11	3	3	2	2
		10	2	2	1	1
		9	2	1	1	0
		8	1	1	0	0
		7	0	0	—	—
		6	0	0	—	—
	C + *D* = 7	12	4	3	3	2
		11	3	2	2	1
		10	2	1	1	0
		9	1	1	0	0
		8	1	0	0	—
		7	0	0	—	—
		6	0	—	—	—

Table of critical values of D (or C) in the Fisher test (continued)

Totals in right margin		B (or A)†	Level of significance			
			.05	.025	.01	.005
$A+B=12$	$C+D=6$	12	3	3	2	2
		11	2	2	1	1
		10	1	1	0	0
		9	1	0	0	0
		8	0	0	—	—
		7	0	0	—	—
		6	0	—	—	—
	$C+D=5$	12	2	2	1	1
		11	1	1	1	0
		10	1	0	0	0
		9	0	0	0	—
		8	0	0	—	—
		7	0	—	—	—
	$C+D=4$	12	2	1	1	0
		11	1	0	0	0
		10	0	0	0	—
		9	0	0	—	—
		8	0	—	—	—
	$C+D=3$	12	1	0	0	0
		11	0	0	0	—
		10	0	0	—	—
		9	0	—	—	—
	$C+D=2$	12	0	0	—	—
		11	0	—	—	—
$A+B=13$	$C+D=13$	13	9	8	7	6
		12	7	6	5	4
		11	6	5	4	3
		10	4	4	3	2
		9	3	3	2	1
		8	2	2	1	0
		7	2	1	0	0
		6	1	0	0	—
		5	0	0	—	—
		4	0	—	—	—
	$C+D=12$	13	8	7	6	5
		12	6	5	5	4
		11	5	4	3	3
		10	4	3	2	2
		9	3	2	1	1
		8	2	1	1	0
		7	1	1	0	0
		6	1	0	0	—
		5	0	0	—	—

Table of critical values of *D* (or *C*) in the Fisher test (continued)

Totals in right margin		B (or A)†	Level of significance			
			.05	.025	.01	.005
A + B = 13	C + D = 11	13	7	6	5	5
		12	6	5	4	3
		11	4	4	3	2
		10	3	3	2	1
		9	3	2	1	1
		8	2	1	0	0
		7	1	0	0	0
		6	0	0	—	—
		5	0	—	—	—
	C + D = 10	13	6	6	5	4
		12	5	4	3	3
		11	4	3	2	2
		10	3	2	1	1
		9	2	1	1	0
		8	1	1	0	0
		7	1	0	0	—
		6	0	0	—	—
		5	0	—	—	—
	C + D = 9	13	5	5	4	4
		12	4	4	3	2
		11	3	3	2	1
		10	2	2	1	1
		9	2	1	0	0
		8	1	1	0	0
		7	0	0	—	—
		6	0	0	—	—
		5	0	—	—	—
	C + D = 8	13	5	4	3	3
		12	4	3	2	2
		11	3	2	1	1
		10	2	1	1	0
		9	1	1	0	0
		8	1	0	0	—
		7	0	0	—	—
		6	0	—	—	—
	C + D = 7	13	4	3	3	2
		12	3	2	2	1
		11	2	2	1	1
		10	1	1	0	0
		9	1	0	0	0
		8	0	0	—	—
		7	0	0	—	—
		6	0	—	—	—

Table of critical values of D (or C) in the Fisher test (continued)

Totals in right margin		B (or A)†	.05	.025	.01	.005
				Level of significance		
$A+B=13$	$C+D=6$	13	3	3	2	2
		12	2	2	1	1
		11	2	1	1	0
		10	1	1	0	0
		9	1	0	0	—
		8	0	0	—	—
		7	0	—	—	—
	$C+D=5$	13	2	2	1	1
		12	2	1	1	0
		11	1	1	0	0
		10	1	0	0	—
		9	0	0	—	—
		8	0	—	—	—
	$C+D=4$	13	2	1	1	0
		12	1	1	0	0
		11	0	0	0	—
		10	0	0	—	—
		9	0	—	—	—
	$C+D=3$	13	1	1	0	0
		12	0	0	0	—
		11	0	0	—	—
		10	0	—	—	—
	$C+D=2$	13	0	0	0	—
		12	0	—	—	—
$A+B=14$	$C+D=14$	14	10	9	8	7
		13	8	7	6	5
		12	6	6	5	4
		11	5	4	3	3
		10	4	3	2	2
		9	3	2	2	1
		8	2	2	1	0
		7	1	1	0	0
		6	1	0	0	—
		5	0	0	—	—
		4	0	—	—	—

Table of critical values of D (or C) in the Fisher test (continued)

Totals in right margin		B (or A)†	Level of significance			
			.05	.025	.01	.005
$A + B = 14$	$C + D = 13$	14	9	8	7	6
		13	7	6	5	5
		12	6	5	4	3
		11	5	4	3	2
		10	4	3	2	2
		9	3	2	1	1
		8	2	1	1	0
		7	1	1	0	0
		6	1	0	—	—
		5	0	0	—	—
	$C + D = 12$	14	8	7	6	6
		13	6	6	5	4
		12	5	4	4	3
		11	4	3	3	2
		10	3	3	2	1
		9	2	2	1	1
		8	2	1	0	0
		7	1	0	0	—
		6	0	0	—	—
		5	0	—	—	—
	$C + D = 11$	14	7	6	6	5
		13	6	5	4	4
		12	5	4	3	3
		11	4	3	2	2
		10	3	2	1	1
		9	2	1	1	0
		8	1	1	0	0
		7	1	0	0	—
		6	0	0	—	—
		5	0	—	—	—
	$C + D = 10$	14	6	6	5	4
		13	5	4	4	3
		12	4	3	3	2
		11	3	3	2	1
		10	2	2	1	1
		9	2	1	0	0
		8	1	1	0	0
		7	0	0	0	—
		6	0	0	—	—
		5	0	—	—	—

Table of critical values of D (or C) in the Fisher test (continued)

Totals in right margin		B (or A)†	.05	.025	.01	.005
				Level of significance		
$A = B = 14$	$C + D = 9$	14	6	5	4	4
		13	4	4	3	3
		12	3	3	2	2
		11	3	2	1	1
		10	2	1	1	0
		9	1	1	0	0
		8	1	0	0	—
		7	0	0	—	—
		6	0	—	—	—
	$C + D = 8$	14	5	4	4	3
		13	4	3	2	2
		12	3	2	2	1
		11	2	2	1	1
		10	2	1	0	0
		9	1	0	0	0
		8	0	0	0	—
		7	0	0	—	—
		6	0	—	—	—
	$C + D = 7$	14	4	3	3	2
		13	3	2	2	1
		12	2	2	1	1
		11	2	1	1	0
		10	1	1	0	0
		9	1	0	0	—
		8	0	0	—	—
		7	0	—	—	—
	$C + D = 6$	14	3	3	2	2
		13	2	2	1	1
		12	2	1	1	0
		11	1	1	0	0
		10	1	0	0	—
		9	0	0	—	—
		8	0	0	—	—
		7	0	—	—	—
	$C + D = 5$	14	2	2	1	1
		13	2	1	1	0
		12	1	1	0	0
		11	1	0	0	0
		10	0	0	—	—
		9	0	0	—	—
		8	0	—	—	—

Table of critical values of D (or C) in the Fisher test (continued)

Totals in right margin		B (or A)†	Level of significance			
			.05	.025	.01	.005
$A+B=14$	$C+D=4$	14	2	1	1	1
		13	1	1	0	0
		12	1	0	0	0
		11	0	0	—	—
		10	0	0	—	—
		9	0	—	—	—
	$C+D=3$	14	1	1	0	0
		13	0	0	0	—
		12	0	0	—	—
		11	0	—	—	—
	$C+D=2$	14	0	0	0	—
		13	0	0	—	—
		12	0	—	—	—
$A+B=15$	$C+D=15$	15	11	10	9	8
		14	9	8	7	6
		13	7	6	5	5
		12	6	5	4	4
		11	5	4	3	3
		10	4	3	2	2
		9	3	2	1	1
		8	2	1	1	0
		7	1	1	0	0
		6	1	0	0	—
		5	0	0	—	—
		4	0	—	—	—
	$C+D=14$	15	10	9	8	7
		14	8	7	6	6
		13	7	6	5	4
		12	6	5	4	3
		11	5	4	3	2
		10	4	3	2	1
		9	3	2	1	1
		8	2	1	1	0
		7	1	1	0	0
		6	1	0	—	—
		5	0	—	—	—

347

Table of critical values of D (or C) in the Fisher test (continued)

Totals in right margin		B (or A)†	.05	.025	.01	.005
				Level of significance		
$A+B=15$	$C+D=13$	15	9	8	7	7
		14	7	7	6	5
		13	6	5	4	4
		12	5	4	3	3
		11	4	3	2	2
		10	3	2	2	1
		9	2	2	1	0
		8	2	1	0	0
		7	1	0	0	—
		6	0	0	—	—
		5	0	—	—	—
	$C+D=12$	15	8	7	7	6
		14	7	6	5	4
		13	6	5	4	3
		12	5	4	3	2
		11	4	3	2	2
		10	3	2	1	1
		9	2	1	1	0
		8	1	1	0	0
		7	1	0	0	—
		6	0	0	—	—
		5	0	—	—	—
	$C+D=11$	15	7	7	6	5
		14	6	5	4	4
		13	5	4	3	3
		12	4	3	2	2
		11	3	2	2	1
		10	2	2	1	1
		9	2	1	0	0
		8	1	1	0	0
		7	1	0	0	—
		6	0	0	—	—
		5	0	—	—	—
	$C+D=10$	15	6	6	5	5
		14	5	5	4	3
		13	4	4	3	2
		12	3	3	2	2
		11	3	2	1	1
		10	2	1	1	0
		9	1	1	0	0
		8	1	0	0	—
		7	0	0	—	—
		6	0	—	—	—

Table of critical values of D (or C) in the Fisher test (continued)

Totals in right margin		B (or A)‡	Level of significance			
			.05	.025	.01	.005
$A + B = 15$	$C + D = 9$	15	6	5	4	4
		14	5	4	3	3
		13	4	3	2	2
		12	3	2	2	1
		11	2	2	1	1
		10	2	1	0	0
		9	1	1	0	0
		8	1	0	0	—
		7	0	0	—	—
		6	0	—	—	—
	$C + D = 8$	15	5	4	4	3
		14	4	3	3	2
		13	3	2	2	1
		12	2	2	1	1
		11	2	1	1	0
		10	1	1	0	0
		9	1	0	0	—
		8	0	0	—	—
		7	0	—	—	—
		6	0	—	—	—
	$C + D = 7$	15	4	4	3	3
		14	3	3	2	2
		13	2	2	1	1
		12	2	1	1	0
		11	1	1	0	0
		10	1	0	0	0
		9	0	0	—	—
		8	0	0	—	—
		7	0	—	—	—
	$C + D = 6$	15	3	3	2	2
		14	2	2	1	1
		13	2	1	1	0
		12	1	1	0	0
		11	1	0	0	0
		10	0	0	0	—
		9	0	0	—	—
		8	0	—	—	—
	$C + D = 5$	15	2	2	2	1
		14	2	1	1	1
		13	1	1	0	0
		12	1	0	0	0
		11	0	0	0	—
		10	0	0	—	—
		9	0	—	—	—

Table of critical values of D (or C) in the Fisher test (continued)

Totals in right margin		B (or A)†	Level of significance			
			.05	.025	.01	.005
$A+B=15$	$C+D=4$	15	2	1	1	1
		14	1	1	0	0
		13	1	0	0	0
		12	0	0	0	—
		11	0	0	—	—
		10	0	—	—	—
	$C+D=3$	15	1	1	0	0
		14	0	0	0	0
		13	0	0	—	—
		12	0	0	—	—
		11	0	—	—	—
	$C+D=2$	15	0	0	0	—
		14	0	0	—	—
		13	0	—	—	—

Source: Finney, D. J. 1948 *Biometrika* 35, pp. 149–54

APPENDIX 24

Probabilities associated with values as large as observed values of *H* in the Kruskal–Wallis One-Way Analysis of Variance by Ranks

Sample sizes	0.05	0.01	Sample sizes	0.05	0.01	Sample sizes	0.05	0.01	Sample sizes	0.05	0.01
k = 3			**k = 3**			**k = 4**			**k = 4**		
2 2 2	—	—	5 5 3	5.705	7.578	2 2 1 1	—	—	4 4 4 1	6.725	8.588
3 2 1	—	—	5 5 4	5.666	7.823	2 2 2 1	5.679	—	4 4 4 2	6.957	8.871
3 2 2	4.714	—	5 5 5	5.780	8.000	2 2 2 2	6.167	6.667	4 4 4 3	7.142	9.075
3 3 1	5.143	—	6 1 1	—	—	3 1 1 1	—	—	4 4 4 4	7.235	9.287
3 3 2	5.361	—	6 2 1	4.822	—	3 2 1 1	—	—			
3 3 3	5.600	7.200	6 2 2	5.345	6.655	3 2 2 1	5.833	—	**k = 5**		
4 2 1	—	—	6 3 1	4.855	6.873	3 2 2 2	6.333	7.133			
4 2 2	5.333	—	6 3 2	5.348	6.970	3 3 1 1	6.333	—	Sample sizes	0.05	0.01
4 3 1	5.208	—	6 3 3	5.615	7.410	3 3 2 1	6.244	7.200			
4 3 2	5.444	6.444	6 4 1	4.947	7.106	3 3 2 2	6.527	7.636	2 2 1 1 1	—	—
4 3 3	5.791	6.745	6 4 2	5.340	7.340	3 3 3 1	6.600	7.400	2 2 2 1 1	6.750	—
4 4 1	4.967	6.667	6 4 3	5.610	7.500	3 3 3 2	6.727	8.015	2 2 2 2 1	7.133	7.533
4 4 2	5.455	7.036	6 4 4	5.681	7.795	3 3 3 3	7.000	8.538	2 2 2 2 2	7.418	8.291
4 4 3	5.598	7.144	6 5 1	4.990	7.182	4 1 1 1	—	—	3 1 1 1 1	—	—
4 4 4	5.692	7.654	6 5 2	5.338	7.376	4 2 1 1	5.833	—	3 2 1 1 1	6.583	—
5 2 1	5.000	—	6 5 3	5.602	7.590	4 2 2 1	6.133	7.000	3 2 2 1 1	6.800	7.600
5 2 2	5.160	6.533	6 5 4	5.661	7.936	4 2 2 2	6.545	7.391	3 2 2 2 1	7.309	8.127
5 3 1	4.960	—	6 5 5	5.729	8.028	4 3 1 1	6.178	7.067	3 2 2 2 2	7.682	8.682
5 3 2	5.251	6.909	6 6 1	4.945	7.121	4 3 2 1	6.309	7.455	3 3 1 1 1	7.111	—
5 3 3	5.648	7.079	6 6 2	5.410	7.467	4 3 2 2	6.621	7.871	3 3 2 1 1	7.200	8.073
5 4 1	4.985	6.955	6 6 3	5.625	7.725	4 3 3 1	6.545	7.758	3 3 2 2 1	7.591	8.576
5 4 2	5.273	7.205	6 6 4	5.724	8.000	4 3 3 2	6.795	8.333	3 3 2 2 2	7.910	9.115
5 4 3	5.656	7.445	6 6 5	5.765	8.124	4 3 3 3	6.984	8.659	3 3 3 1 1	7.576	8.424
5 4 4	5.657	7.760	6 6 6	5.801	8.222	4 4 1 1	5.945	7.909	3 3 3 2 1	7.769	9.051
5 5 1	5.127	7.309	7 7 7	5.819	8.378	4 4 2 1	6.386	7.909	3 3 3 2 2	8.044	9.505
5 5 2	5.338	7.338	8 8 8	5.805	8.465	4 4 2 2	6.731	8.346	3 3 3 3 1	8.000	9.451
						4 4 3 1	6.635	8.231	3 3 3 3 2	8.200	9.876
						4 4 3 2	6.874	8.621	3 3 3 3 3	8.333	10.20
						4 4 3 3	7.038	8.876			

Source: Table 4.2, Neave, H. R.; *Statistics Tables*, London: George Allen & Unwin, 1978, p. 49, with the kind permission of the author and publisher.

APPENDIX 25

Table of Critical values of s in the Kendall Coefficient of Concordance

k	N 3	4	5	6	7	Additional values for N = 3 k	s
			Values at the .05 level of significance				
3			64.4	103.9	157.3	9	54.0
4		49.5	88.4	143.3	217.0	12	71.9
5		62.6	112.3	182.4	276.2	14	83.8
6		75.7	136.1	221.4	335.2	16	95.8
8	48.1	101.7	183.7	299.0	453.1	18	107.7
10	60.0	127.8	231.2	376.7	571.0		
15	89.8	192.9	349.8	570.5	864.9		
20	119.7	258.0	468.5	764.4	1,158.7		
			Values at the .01 level of significance				
3			75.6	122.8	185.6	9	75.9
4		61.4	109.3	176.2	265.0	12	103.5
5		80.5	142.8	229.4	343.8	14	121.9
6		99.5	176.1	282.4	422.6	16	140.2
8	66.8	137.4	242.7	388.3	579.9	18	158.6
10	85.1	175.3	309.1	494.0	737.0		
15	131.0	269.8	475.2	758.2	1,129.5		
20	177.0	364.2	641.2	1,022.2	1,521.9		

Source: Adapted from Friedman, M. A. Comparison of Alternative Tests of Significance for the Problem of m Rankings, *Annals of Mathematical Statistics* 11, p. 91.

APPENDIX 26

Jonckheere Trend Test: minimum values of S (one-tailed)*

Significance level $p < .05$									
$C \backslash n$	2	3	4	5	6	7	8	9	10
3	10	17	24	33	42	53	64	76	88
4	14	26	38	51	66	82	100	118	138
5	20	34	51	71	92	115	140	166	194
6	26	44	67	93	121	151	184	219	256
Significance level $p < .01$									
$C \backslash n$	2	3	4	5	6	7	8	9	10
3	—	23	32	45	59	74	90	106	124
4	20	34	50	71	92	115	140	167	195
5	26	48	72	99	129	162	197	234	274
6	34	62	94	130	170	213	260	309	361

* For a two-tailed test (where direction of trend is not predicted) significance levels should be doubled.
Source: *Biometrika* 41 (1954). Jonckheere Trend Test.

APPENDIX 27

Critical values of A

For any given value of $n-1$, the table shows the values of A corresponding to various levels of probability. A is significant at a given level if it is equal to or *less than* the value shown in the table.

	Level of significance for one-tailed test					
	.05	.025	.01	.005	.0005	
	Level of significance for two-tailed test					
n−1*	.10	.05	.02	.01	.001	n−1*
1	0.5125	0.5031	0.50049	0.50012	0.5000012	1
2	0.412	0.369	0.347	0.340	0.334	2
3	0.385	0.324	0.286	0.272	0.254	3
4	0.376	0.304	0.257	0.238	0.211	4
5	0.372	0.293	0.240	0.218	0.184	5
6	0.370	0.286	0.230	0.205	0.167	6
7	0.369	0.281	0.222	0.196	0.155	7
8	0.368	0.278	0.217	0.190	0.146	8
9	0.368	0.276	0.213	0.185	0.139	9
10	0.368	0.274	0.210	0.181	0.134	10
11	0.368	0.273	0.207	0.178	0.130	11
12	0.368	0.271	0.205	0.176	0.126	12
13	0.368	0.270	0.204	0.174	0.124	13
14	0.368	0.270	0.202	0.172	0.121	14
15	0.368	0.269	0.201	0.170	0.119	15
16	0.368	0.268	0.200	0.169	0.117	16
17	0.368	0.268	0.199	0.168	0.116	17
18	0.368	0.267	0.198	0.167	0.114	18
19	0.368	0.267	0.197	0.166	0.113	19
20	0.368	0.266	0.197	0.165	0.112	20
21	0.368	0.266	0.196	0.165	0.111	21
22	0.368	0.266	0.196	0.164	0.110	22
23	0.368	0.266	0.195	0.163	0.109	23
24	0.368	0.265	0.195	0.163	0.108	24
25	0.368	0.265	0.194	0.162	0.108	25
26	0.368	0.265	0.194	0.162	0.107	26
27	0.368	0.265	0.193	0.161	0.107	27
28	0.368	0.265	0.193	0.161	0.106	28
29	0.368	0.264	0.193	0.161	0.106	29
30	0.368	0.264	0.193	0.160	0.105	30
40	0.368	0.263	0.191	0.158	0.102	40
60	0.368	0.262	0.189	0.155	0.099	60
120	0.367	0.261	0.187	0.153	0.095	120
∞	0.367	0.260	0.185	0.151	0.092	∞

* n = number of pairs

Source: J. Sandler (1955), A test of the significance of the difference between the means of correlated measures based on a simplification of Student's t. *British Journal of Psychology* **46**, 225–226.

APPENDIX 28

Natural Cosines

Subtract

	0'	6'	12'	18'	24'	30'	36'	42'	48'	54'	1'	2'	3'	4'	5'
0°	1.0000	1.000	1.000	1.000	1.000	1.000	**9999**	**9999**	**9999**	**9999**					
1	.9998	9998	9998	9997	9997	9997	9996	9996	9995	9995					
2	.9994	9993	9993	9992	9991	9990	9990	9989	9988	9987	0	0	0	0	1
3	.9986	9985	9984	9983	9982	9981	9980	9979	9978	9977	0	0	1	1	1
4	.9976	9974	9973	9972	9971	9969	9968	9966	9965	9963	0	0	1	1	1
5	.9962	9960	9959	9957	9956	9954	9952	9951	9949	9947	0	1	1	1	1
6	.9945	9943	9942	9940	9938	9936	9934	9932	9930	9928	0	1	1	1	2
7	.9925	9923	9921	9919	9917	9914	9912	9910	9907	9905	0	1	1	2	2
8	.9903	9900	9898	9895	9893	9890	9888	9885	9882	9880	0	1	1	2	2
9	.9877	9874	9871	9869	9866	9863	9860	9857	9854	9851	0	1	1	2	2
10	.9848	9845	9842	9839	9836	9833	9829	9826	9823	9820	1	1	2	2	3
11	.9816	9813	9810	9806	9803	9799	9796	9792	9789	9785	1	1	2	2	3
12	.9781	9778	9774	9770	9767	9763	9759	9755	9751	9748	1	1	2	3	3
13	.9744	9740	9736	9732	9728	9724	9720	9715	9711	9707	1	1	2	3	3
14	.9703	9699	9694	9690	9686	9681	9677	9673	9668	9664	1	1	2	3	4
15	.9659	9655	9650	9646	9641	9636	9632	9627	9622	9617	1	2	2	3	4
16	.9613	9608	9603	9598	9593	9588	9583	9578	9573	9568	1	2	2	3	4
17	.9563	9558	9553	9548	9542	9537	9532	9527	9521	9516	1	2	3	4	4
18	.9511	9505	9500	9494	9489	9483	9478	9472	9466	9461	1	2	3	4	5
19	.9455	9449	9444	9438	9432	9426	9421	9415	9409	9403	1	2	3	4	5
20	.9397	9391	9385	9379	9373	9367	9361	9354	9348	9342	1	2	3	4	5
21	.9336	9330	9323	9317	9311	9304	9298	9291	9285	9278	1	2	3	4	5
22	.9272	9265	9259	9252	9245	9239	9232	9225	9219	9212	1	2	3	4	6
23	.9205	9198	9191	9184	9178	9171	9164	9157	9150	9143	1	2	3	5	6
24	.9135	9128	9121	9114	9107	9100	9092	9085	9078	9070	1	2	4	5	6
25	.9063	9056	9048	9041	9033	9026	9018	9011	9003	8996	1	3	4	5	6
26	.8988	8980	8973	8965	8957	8949	8942	8934	8926	8918	1	3	4	5	6
27	.8910	8902	8894	8886	8878	8870	8862	8854	8846	8838	1	3	4	5	7
28	.8829	8821	8813	8805	8796	8788	8780	8771	8763	8755	1	3	4	6	7
29	.8746	8738	8729	8721	8712	8704	8695	8686	8678	8669	1	3	4	6	7
30	.8660	8652	8643	8634	8625	8616	8607	8599	8590	8581	1	3	4	6	7
31	.8572	8563	8554	8545	8536	8526	8517	8508	8499	8490	2	3	5	6	8
32	.8480	8471	8462	8453	8443	8434	8425	8415	8406	8396	2	3	5	6	8
33	.8387	8377	8368	8358	8348	8339	8329	8320	8310	8300	2	3	5	6	8
34	.8290	8281	8271	8261	8251	8241	8231	8221	8211	8202	2	3	5	7	8
35	.8192	8181	8171	8161	8151	8141	8131	8121	8111	8100	2	3	5	7	8
36	.8090	8080	8070	8059	8049	8039	8028	8018	8007	7997	2	3	5	7	9
37	.7986	7976	7965	7955	7944	7934	7923	7912	7902	7891	2	4	5	7	9
38	.7880	7869	7859	7848	7837	7826	7815	7804	7793	7782	2	4	5	7	9
39	.7771	7760	7749	7738	7727	7716	7705	7694	7683	7672	2	4	6	7	9
40	.7660	7649	7638	7627	7615	7604	7593	7581	7570	7559	2	4	6	8	9
41	.7547	7536	7524	7513	7501	7490	7478	7466	7455	7443	2	4	6	8	10
42	.7431	7420	7408	7396	7385	7373	7361	7349	7337	7325	2	4	6	8	10
43	.7314	7302	7290	7278	7266	7254	7242	7230	7218	7206	2	4	6	8	10
44	.7193	7181	7169	7157	7145	7133	7120	7108	7096	7083	2	4	6	8	10
	0'	6'	12'	18'	24'	30'	36'	42'	48'	54'	1'	2'	3'	4'	5'

The **bold** type indicates that the integer changes.

Natural Cosines

Subtract

	0'	6'	12'	18'	24'	30'	36'	42'	48'	54'	1'	2'	3'	4'	5'
45°	.7071	7059	7046	7034	7022	7009	6997	6984	6972	6959	2	4	6	8	10
46	.6947	6934	6921	6909	6896	6884	6871	6858	6845	6833	2	4	6	8	11
47	.6820	6807	6794	6782	6769	6756	6743	6730	6717	6704	2	4	6	9	11
48	.6691	6678	6665	6652	6639	6626	6613	6600	6587	6574	2	4	7	9	11
49	.6561	6547	6534	6521	6508	6494	6481	6468	6455	6441	2	4	7	9	11
50	.6428	6414	6401	6388	6374	6361	6347	6334	6320	6307	2	4	7	9	11
51	.6293	6280	6266	6252	6239	6225	6211	6198	6184	6170	2	5	7	9	11
52	.6157	6143	6129	6115	6101	6088	6074	6060	6046	6032	2	5	7	9	12
53	.6018	6004	5990	5976	5962	5948	5934	5920	5906	5892	2	5	7	9	12
54	.5878	5864	5850	5835	5821	5807	5793	5779	5764	5750	2	5	7	9	12
55	.5736	5721	5707	5693	5678	5664	5650	5635	5621	5606	2	5	7	10	12
56	.5592	5577	5563	5548	5534	5519	5505	5490	5476	5461	2	5	7	10	12
57	.5446	5432	5417	5402	5388	5373	5358	5344	5329	5314	2	5	7	10	12
58	.5299	5284	5270	5255	5240	5225	5210	5195	5180	5165	2	5	7	10	12
59	.5150	5135	5120	5105	5090	5075	5060	5045	5030	5015	3	5	8	10	13
60	.5000	4985	4970	4955	4939	4924	4909	4894	4879	4863	3	5	8	10	13
61	.4848	4833	4818	4802	4787	4772	4756	4741	4726	4710	3	5	8	10	13
62	.4695	4679	4664	4648	4633	4617	4602	4586	4571	4555	3	5	8	10	13
63	.4540	4524	4509	4493	4478	4462	4446	4431	4415	4399	3	5	8	10	13
64	.4384	4368	4352	4337	4321	4305	4289	4274	4258	4242	3	5	8	11	13
65	.4226	4210	4195	4179	4163	4147	4131	4115	4099	4083	3	5	8	11	13
66	.4067	4051	4035	4019	4003	3987	3971	3955	3939	3923	3	5	8	11	13
67	.3907	3891	3875	3859	3843	3827	3811	3795	3778	3762	3	5	8	11	13
68	.3746	3730	3714	3697	3681	3665	3649	3633	3616	3600	3	5	8	11	14
69	.3584	3567	3551	3535	3518	3502	3486	3469	3453	3437	3	5	8	11	14
70	.3420	3404	3387	3371	3355	3338	3322	3305	3289	3272	3	5	8	11	14
71	.3256	3239	3223	3206	3190	3173	3156	3140	3123	3107	3	6	8	11	14
72	.3090	3074	3057	3040	3024	3007	2990	2974	2957	2940	3	6	8	11	14
73	.2924	2907	2890	2874	2857	2840	2823	2807	2790	2773	3	6	8	11	14
74	.2756	2740	2723	2706	2689	2672	2656	2639	2622	2605	3	6	8	11	14
75	.2588	2571	2554	2538	2521	2504	2487	2470	2453	2436	3	6	8	11	14
76	.2419	2402	2385	2368	2351	2334	2317	2300	2284	2267	3	6	8	11	14
77	.2250	2233	2215	2198	2181	2164	2147	2130	2113	2096	3	6	9	11	14
78	.2079	2062	2045	2028	2011	1994	1977	1959	1942	1925	3	6	9	11	14
79	.1908	1891	1874	1857	1840	1822	1805	1788	1771	1754	3	6	9	11	14
80	.1736	1719	1702	1685	1668	1650	1633	1616	1599	1582	3	6	9	11	14
81	.1564	1547	1530	1513	1495	1478	1461	1444	1426	1409	3	6	9	12	14
82	.1392	1374	1357	1340	1323	1305	1288	1271	1253	1236	3	6	9	12	14
83	.1219	1201	1184	1167	1149	1132	1115	1097	1080	1063	3	6	9	12	14
84	.1045	1028	1011	0993	0976	0958	0941	0924	0906	0889	3	6	9	12	14
85	.0872	0854	0837	0819	0802	0785	0767	0750	0732	0715	3	6	9	12	14
86	.0698	0680	0663	0645	0628	0610	0593	0576	0558	0541	3	6	9	12	15
87	.0523	0506	0488	0471	0454	0436	0419	0401	0384	0366	3	6	9	12	15
88	.0349	0332	0314	0297	0279	0262	0244	0227	0209	0192	3	6	9	12	15
89	.0175	0157	0140	0122	0105	0087	0070	0052	0035	0017	3	6	9	12	15
	0'	6'	12'	18'	24'	30'	36'	42'	48'	54'	1'	2'	3'	4'	5'

The **bold** type indicates that the integer changes.

APPENDIX 29

Tables of critical values in the k-sample slippage test

Enter the part of the table for the current number of samples (k). Read across the row corresponding to the size (n) of each sample. The critical values for various levels of significance are found at the intersection of (k) and (n) under p = .80; .90; .95; .98; .99. Reject the null hypothesis when the obtained value *exceeds* the selected critical value.

	k = 2					k = 3					k = 4				
	p=.80	.90	.95	.98	.99	p=.80	.90	.95	.98	.99	p=.80	.90	.95	.98	.99
n = 3	2	2				2									
4	2	3	3			3	3				3	3			
5	2	3	3	4	4	3	3	4	4		3	4	4		
6	2	3	4	4	5	3	4	4	5	5	3	4	4	5	5
7	2	3	4	5	5	3	4	4	5	5	3	4	5	5	6
8	2	3	4	5	5	3	4	4	5	6	3	4	5	5	6
9	3	3	4	5	5	3	4	5	5	6	3	4	5	6	6
10	3	3	4	5	6	3	4	5	6	6	4	4	5	6	6
12	3	3	4	5	6	3	4	5	6	6	4	4	5	6	7
14	3	3	4	5	6	3	4	5	6	7	4	5	5	6	7
16	3	4	4	5	6	3	4	5	6	7	4	5	5	6	7
18	3	4	4	5	6	3	4	5	6	7	4	5	6	7	7
20	3	4	4	6	6	3	4	5	6	7	4	5	6	7	7
25	3	4	5	6	6	3	4	5	6	7	4	5	6	7	8
30	3	4	5	6	7	3	4	5	7	7	4	5	6	7	8
35	3	4	5	6	7	3	4	5	7	7	4	5	6	7	8
40	3	4	5	6	7	3	4	5	7	8	4	5	6	7	8
Approximation for n > 40	3	4	5	6	7	4	5	6	7	8	4	5	6	8	9

	k = 5					k = 6					k = 7				
	p=.80	.90	.95	.98	.99	p=.80	.90	.95	.98	.99	p=.80	.90	.95	.98	.99
n = 4	4	4				3	3				3				
5	3	4	4			3	4	4			3	4	4		
6	3	4	4	5	5	4	4	5	5	5	4	4	5	5	
7	3	4	5	5	6	4	4	5	5	6	4	4	5	6	6
8	4	4	5	6	6	4	4	5	6	6	4	5	5	6	6
9	4	4	5	6	6	4	5	5	6	7	4	5	5	6	7
10	4	5	5	6	7	4	5	5	6	7	4	5	6	6	7
12	4	5	5	6	7	4	5	6	7	7	4	5	6	7	7
14	4	5	6	7	7	4	5	6	7	7	4	5	6	7	8
16	4	5	6	7	7	4	5	6	7	8	4	5	6	7	8
18	4	5	6	7	8	4	5	6	7	8	4	5	6	7	8
20	4	5	6	7	8	4	5	6	7	8	4	5	6	7	8
25	4	5	6	7	8	4	5	6	7	8	5	6	6	8	8
30	4	5	6	7	8	4	5	6	8	8	5	6	7	8	9
35	4	5	6	7	8	4	5	6	8	9	5	6	7	8	9
40	4	5	6	7	8	4	6	7	8	9	5	6	7	8	9
Approximation for n > 40	4	6	7	8	9	5	6	7	8	9	5	6	7	9	10

	k = 8					k = 9					k = 10				
	p=.80	.90	.95	.98	.99	p=.80	.90	.95	.98	.99	p=.80	.90	.95	.98	.99
n = 4	3					3					3				
5	3	4	4			4	4	4			4	4	4		
6	4	4	5	5		4	4	5	5		4	4	5	5	
7	4	5	5	6	6	4	5	5	6	6	4	5	5	6	6
8	4	5	5	6	6	4	5	5	6	6	4	5	5	6	7
9	4	5	5	6	7	4	5	6	6	7	4	5	6	6	7
10	4	5	6	6	7	4	5	6	7	7	4	5	6	7	7
12	4	5	6	7	7	4	5	6	7	7	5	5	6	7	8
14	4	5	6	7	8	5	5	6	7	8	5	6	6	7	8
16	4	5	6	7	8	5	6	6	7	8	5	6	6	7	8
18	5	6	6	7	8	5	6	6	8	8	5	6	7	8	8
20	5	6	6	8	8	5	6	7	8	8	5	6	7	8	8
25	5	6	7	8	9	5	6	7	8	9	5	6	7	8	9
30	5	6	7	8	9	5	6	7	8	9	5	6	7	8	9
35	5	6	7	8	9	5	6	7	8	9	5	6	7	8	9
40	5	6	7	8	9	5	6	7	8	9	5	6	7	8	9
Approximation for n > 40	5	6	7	9	10	5	6	8	9	10	5	7	8	9	10

Source: W. J. Conover, *Practical Nonparametric Statistics*. New York: John Wiley and Sons, 1971, pages 411–13.

BIBLIOGRAPHY

H. M. Blalock, Jnr., *Social Statistics.* New York: McGraw-Hill, 1979

J. V. Bradley, *Distribution-Free Statistical Tests.* Englewood Cliffs, N.J.: Prentice Hall, 1968

A. Bryman and D. Cramer, *Quantitative Data Analysis for Social Scientists.* London: Routledge, 1990

W. G. Cochran and G. M. Cox, *Experimental Design.* New York: John Wiley and Sons, 1957

L. Cohen and M. Holliday, *Statistics for Education and Physical Education.* London: Paul Chapman Publishing, 1979

N. J. Conover, *Practical Nonparametric Statistics.* New York: John Wiley and Sons, 1971

M. Cooper, 'An Exact Probability Test for Use with Likert Scales.' *Educational and Psychological Measurement,* **36**(1976), 647–55

M. D. Davidoff and H. W. Goheen, 'A Table for the Rapid Determination of the Tetrachoric Correlation Coefficient.' *Psychometrika,* **18**(1953), 115-121

R. G. Davies, *Computer Programming in Quantitative Biology.* London: Academic Press, 1971

P. H. Dubois, An *Introduction to Psychological Statistics.* New York: Harper and Row, 1965

E. S. Edgington, *Statistical Inference—The Distribution-Free Approach.* New York: McGraw-Hill, 1969

A. L. Edwards, *Statistical Analysis.* New York: Holt, Rinehart and Winston, 1958

A. L. Edwards, *Experimental Design in Psychological Research.* New York: Holt, Rinehart and Winston, 1968

B. H. Erickson and T. A. Nosanchuk, *Understanding Data.* Milton Keynes: Open University Press, 1979

B. F. Everitt, *The Analysis of Contingency Tables.* London: Chapman and Hall, 1977

R. A. Fisher, *Statistical Methods for Research Workers.* Edinburgh: Oliver and Boyd, 1941

M. Friedman, 'The Use of Ranks to Avoid the Assumption of Normality Implicit in Analysis of Variance.' *Journal of the American Statistical Assn.,* **32**(1937), 675-701

H. E. Garrett, *Statistics in Psychology and Education.* London: Longman, 1966

E. S. Gellman, *Statistics for Teachers.* New York: Harper and Row, 1973

J. D. Gibbons, *Nonparametric Methods for Quantitative Analysis.* New York: Holt, Rinehart and Winston, 1976

L. A. Goodman and W. H. Kruskal, 'Measures of Association for Cross Classifications.' *Journal of the American Statistical Assn.,* **49**(1954), 732-764

R. A. Groeneveld, An *Introduction to Probability and Statistics Using Basic.* New York: Marcel Dekker, Inc., 1979

J. P. Guilford, *Psychometric Methods.* New York: McGraw-Hill, 1954

M. Hamburg, *Basic Statistics: A Modern Approach.* New York: Harcourt, Brace, Jovanovich, 1974

N. L. Hays, *Statistics for the Social Sciences.* New York: Holt, Rinehart and Winston, 1973

M. Hollander and D. A. Wolfe, *Nonparametric Statistical Methods.* New York: John Wiley and Sons, 1973

M. G. Kendall, 'A Comparison of Alternative Tests of Significance for the Problem of m Rankings.' *Annals of Mathematical Statistics,* **11**(1940), 86-92

M. G. Kendall, *Rank Correlation Methods.* New York: Hafner Publishing Co., 1970

F. N. Kerlinger, *Foundations of Behavioural Research.* London: Holt, Rinehart and Winston, 1973

R. Kirk, *Experimental Design in Psychological Research.* Belmont, Calif., Brooks/Cole, 1968

F. J. Kohout, *Statistics for Social Scientists*. New York: John Wiley and Sons, 1974

C. Kraft and C. Van Feden, A *Nonparametric Introduction to Statistics*. New York: The Macmillan Co., 1968

C. Leach, *Introduction to Statistics, A Nonparametric Approach for the Social Sciences*. Chichester: John Wiley and Sons, 1979

J. D. Lee and T. D. Lee, *Statistics and Computer Methods in Basic*. London: Van Nostrand Reinhold Co., 1982

D. G. Lewis, *Experimental Design in Education*, London: University of London Press, 1968

E. F. Lindquist, *Design and Analysis of Experiments in Psychology and Education*. Boston: Houghton Kiflin Co., 1953

W. A. Lindner, *Statistics for Students in the Behavioral Sciences*. London: Benjamin/Cummings Publishing Co. Inc., 1979

H. B. Mann and D. R. Whitney, 'On a Test Whether One of Two Random Variables is Stochastically Larger Than the Other.' *Annals of Mathematical Statistics*, **18**(1947), 50-60

A. E. Maxwell, *Analysing Qualitative Data*. London: Methuen 1961

R. B. McCall, *Fundamental Statistics for Psychology*. New York: Harcourt, Brace and World, 1970

D. McKay, N. Schofield and P. Whiteley, *Data Analysis and Social Sciences*. London: Frances Pinter, 1983

R. Meddis, *Statistical Handbook for Non-Statisticians*. London: McGraw-Hill, 1975

F. Mosteller, 'Association and Estimation in Contingency Tables.' *Journal of the American Statistical Association*, **63**(1968), 1–28

H. R. Neave, *Statistics Tables*. London: George Allen and Unwin, 1978

Open University, *Education Course E341*. Open University Ptess, Milton Keynes

A. Pierce, *Fundamentals of Nonparametric Statistics*. Belmont, Calif., Dickenson Publishing Co., 1970

W. R. Pirie and M. A. Hamden, 'Some Revised Continuity Corrections for Discrete Data.' *Biometrics*, **28**(1972), 693–701

W. J. Popham and K. Sirotnik, *Educational Statistics*. New York: Harper and Row, 1973

M. W. Riley, *Sociological Research I: A Case Approach*. New York: Harcourt, Brace and World, Inc. 1963

C. Robson, *Experiment, Design and Statistics in Psychology*. London: Penguin, 1973

R. P. Runyan, *Nonparametric Statistics: A Contemporary Approach*. Reading, Massachusetts: Addison-Wesley, 1977

S. Siegel, *Nonparametric Statistics for the Behavioural Sciences*. New York: McGraw-Hill, 1956

M. J. Schmidt, *Understanding and Using Statistics*. Basic Concepts, Lexington, D.C. Health and Company, 1979

F. S. Swed and C. Eiserhart, 'Tables for Testing Randomness of Grouping in a Sequence of Alternatives.' Annals *of Mathematical Statistics*. 18(1947), 52-54

M. Tate and R. C. Clennand, *Nonparametric and Short-Cut Statistics*. Danville, Ill.: Interstate Printers and Publishers, t957

A. Wald and J. Wolfowitz, 'On a Test Whether Two Samples are Drawn from the same Population.' *Annals of Mathematical Statistics*, 11(1940), 147-162

J. E. Walsh, 'Applications of Some Significance Tests for the Median are Valid Under Very General Conditions.' *Journal of the American Statistical Assn.*, 44(1949), 343

R. S. Weiss, *Statistics in Social Research*. New York: John Wiley and Sons, 1968

P. Whiteley, 'The Analysis of Contingency Tables in D. McKay, N. Schofield and P. Whiteley *Data Analysis and the Social Sciences*. London: Frances Pinter, 1983, pp. 72–119

B. J. Winer, *Statistical Principles in Experimental Design*. New York: McGraw-Hill, t971

R. A. Zelder and G. Carmines. *Statistical Analysis of Social Data*. Chicago: Rand McNally College Publishing Co., 1978

INDEX

366